# Galileo on the World Systems

# Galileo on the World Systems

## A New Abridged Translation and Guide

Maurice A. Finocchiaro

UNIVERSITY OF CALIFORNIA PRESS

*Berkeley / Los Angeles / London*

University of California Press
Berkeley and Los Angeles, California

University of California Press, Ltd.
London, England



Some of the material in this book has appeared in other forms or in earlier versions.
Selection 7 in: "A Landmark in Critical Thinking: Galileo's *Dialogue*," *Inquiry: Critical Thinking across the Disciplines* 12 (1993).
Parts of Appendix 1 in: *Galileo and the Art of Reasoning* (Dordrecht: Kluwer, 1980).
Parts of Appendix 3 in: "Varieties of Rhetoric in Science," *History of the Human Sciences* 3 (1990).

Library of Congress Cataloging-in-Publication Data

Galilei, Galileo, 1564–1642.
Dialogo dei massimi sistemi. English
Galileo on the world systems: a new abridged translation and guide / Maurice A. Finocchiaro.
p. cm.
Includes bibliographical references and index.
ISBN 0-520-20548-0 (alk. paper).—ISBN 0-520-20646-0 (pbk.: alk. paper)
1. Galilei, Galileo, 1564–1642. Dialogo dei massimi sistemi. 2. Astronomy—Early works to 1800. 3. Solar system—Early works to 1800. I. Finocchiaro, Maurice A., 1942– . II. Title.
QB41.G173 1997
520—dc20 96-4256
CIP

Printed in the United States of America
9 8 7 6 5 4 3 2 1

The paper used in this publication meets the minimum requirements of American National Standards for Information Sciences—Permanence of Paper for Printed Library Materials, ANSI Z39.48-1984.

# Contents

# Editorial Preface and Acknowledgments

This book contains a new abridged translation of Galileo Galilei's *Dialogue on the Two Chief World Systems, Ptolemaic and Copernican*, together with a considerable amount of commentary designed to provide an introduction to the reading, understanding, appreciation, and criticism of this classic work. My commentary emphasizes the themes of critical reasoning and methodological reflection, and so it is also meant to provide a concrete introduction to critical thinking.

The translation has been made from the Italian text provided in volume 7 of the critical National Edition of Galileo's complete works edited by Antonio Favaro (1890–1909). I have, however, also consulted the text of the original edition published in 1632; this has been especially valuable in regard to matters of punctuation, which was almost completely reworked by Favaro; the original punctuation provides additional clues to the syntactical and logical structure of the text. Selected passages are identified in my translated text by the insertion of page numbers from that National Edition; these are designated by the numerals in square brackets interspersed within my translated text. Omitted passages are indicated by ellipses, and the notes give information about what is being left out. Occasionally there is an ellipsis within Galileo's own Italian text, and I have noted the fact in the notes.

In making my translation, I have consulted and benefited from all previous English translations of Galileo's book, namely Webbe (1635), Salusbury (1661), Santillana (1953), and Drake (1967). However, I have tried to improve the accuracy of these without making my translation so literal as to lose readability; I have also aimed to facilitate comprehension rather than mere reading; and I have attempted to steer a middle course between an excessively free and easy-flowing translation and an excessively literal and hard-to-read one.

The selections have been made in accordance with the following principles. First, the focus is on the Copernican controversy; by this I do not mean exclusively or even primarily the issues pertaining to technical, mathematical, planetary astronomy, but rather those pertaining to cosmology, physics, methodology, and qualitative astronomy. Second, I include primarily passages which are explicit discussions of objections against or arguments in favor of the Copernican system. A third reason for inclusion stems from whether the passage is illustrative of an important philosophical or scientific idea, or informative about some significant historical incident. A fourth criterion is that each selection be a relatively self-contained unit which can be read and understood independently of the others. A fifth principle is that I want to maximize the variety of distinct scientific and philosophical issues discussed, and avoid excessive overlappings or near duplications. Finally, I have been sensitive to the requirement that the selections should not include too much of the original work, in order not to make the present book excessively and self-defeatingly long; the selections amount to about 38 percent of the original.

Galileo's text lacks any kind of subdivision other than the dedication, preface, and four chapters called Days. To help readers find their way, I have numbered my selections and given a title to each. These numbers and titles are placed in brackets in the text, in order to indicate that they are editorial additions. The titles are meant to be as descriptive and informative as possible while avoiding pedantry.

The Galilean text is preceded by an outline that subdivides each selection into a number of parts, in order to give a better overview of the content of each selection. These parts are listed in the order in which the respective topics arise in Galileo's discussion; but the correspondence is inexact, partly because the discussion in the dialogue does not proceed uniformly in a single direction, and partly because the various topics overlap to some extent. Nevertheless, the outline aims to provide a useful indication of the key points of any particular selection.

The notes are partly historical and partly philosophical; partly informative, partly interpretive, and partly critical. I have tended to avoid purely erudite commentary, but I have included scholarly points and bibliographical references which I felt were helpful. In regard to controversial issues, I have tried to indicate the nature of the issues rather than present just one side or my own resolution of the problem. My comments are meant to be relatively elementary and introductory, by which I mean that they involve simplifications and approximations, all of which are liable to further refinement and deeper analysis; however, I aim to distinguish simplifications from oversimplifications and approximations from distortions. In writing the notes, I have benefited from previous editions of Galileo's book and often adapted without acknowledgment the information they provide; therefore, here I should like to take this opportunity to explicitly acknowledge my debt to Strauss (1891), Santillana (1953), Pagnini (1964), Drake (1967), Sosio (1970), and Sexl and von Meyenn (1982). In those notes where my debt is more direct, I have explicitly acknowledged the fact, giving appropriate references and quotations from their works.

The introduction is meant to be a systematic and elementary exposition of the background historical information. It provides interpretive and narrative accounts of those developments in cultural, intellectual, and scientific history to which Galileo's book is directly connected, namely the Copernican Revolution, the Scientific Revolution, and the Galileo Affair. It also contains a preliminary sketch of those universal and perennial traits and activities of the human mind that are recorded in Galileo's book and make it a living document; these are the mental skills of critical reasoning, methodological reflection, and verbal rhetoric.

However, a systematic exposition of these concepts and their connection with critical thinking, together with a corresponding critical interpretation of the *Dialogue*, is to be found in the appendix rather than in the introduction. The appendix thus provides abstract articulations and textual illustrations of the guiding conceptual framework which I have both used in, and derived from, my critical reading of the book.

The commentary in the notes, introduction, and appendix explains most of the Galilean terms that are technical, semi-technical, archaic, personal names, or geographical names. It also explains special terms introduced by the present editor, for their explanations are usually given in the context where they are first used. However, there are a number of terms and concepts (both Galilean and editorial) which, for one reason or another, could not be efficiently or adequately explained in that commentary; they are explained in the glossary. The largest group of such terms are those Galilean words that are used in more than one selection; in these cases, if the explanation is not found in that commentary, the reader should look in the glossary. Another important group of glossary entries are those editorial terms that are especially ubiquitous in my commentary. In any case, the reader can consult the index for references to definitions and occurrences of all terms.

It should be noted that bibliographical references to Galileo's works in the notes are given by using his first name, by which he is generally referred to in English. This is done for the sake of uniformity since most such notes also use his first name in a nonbibliographical context. However, the bibliography lists his works in the standard manner under his last name.

Finally, in creating this work, I have benefited from many persons and institutions, and here I should like to express my acknowledgments to them.

I first conceived this book in 1980 at the time of the publication of my *Galileo and the Art of Reasoning* as a kind of sequel to that work; at that time I received the encouragement of Stillman Drake (Toronto). Later, after the publication of my *Galileo Affair* in 1989, I received the encouragement of Daniel Jones (National Endowment for the Humanities); he rekindled my interest, for I could then see myself doing with the text of the *Dialogue* something analogous to what my recent book had done with the documents of Galileo's trial. Then John L. Heilbron (Berkeley) responded favorably to the idea, and I am indebted to him for his constant support. Ron Naylor (Greenwich, England), William Shea (McGill), and Michael Segre (Munich) provided valuable encouragement and support at various stages of the project. Frederic L. Holmes (Yale) and Ernan McMullin (Notre Dame) made helpful comments on various progress reports. I also received valuable comments, suggestions, and constructive criticism from several scholars who read part or all of the book manuscript at various stages of its development: I. Bernard Cohen (Harvard), Allan Franklin (Colorado at Boulder), David Hill (Augustana College), Peter Machamer (Pittsburgh), Connie Missimer (independent scholar, Seattle), Albert Van Helden (Rice), and William A. Wallace (Maryland and Catholic University). My students Ioana Gal, John Ketchum, and Alan Rhoda provided invaluable assistance by being meticulous proofreaders and by playing the role of intelligent consumers and perceptive nonexpert readers.

Acknowledgments are also due to the National Endowment for the Humanities for a research grant (number RH-20980-91) from 1992 to 1995, without whose support the required work would not have been done; this is especially true in the light of the fact that this grant was awarded by the special program in "guided studies of historically significant scientific writings from antiquity to the twentieth century" in the Humanities, Science, and Technology Category of the Division of Interpretive Research. At my own institution, the University of Nevada, Las Vegas, my gratitude goes to the following: the Sabbatical Leave Committee, for a sabbatical leave in fall 1994, during which this book was completed; and the College of Liberal Arts and the Department of Philosophy for a reduction in teaching load and assignment to the Center for Advanced Research in the spring semester of 1994.

# Introduction

## 1. Significance of Galileo's *Dialogue*

1.1. From prehistoric times until the middle of the sixteenth century, almost all thinkers believed that the earth stood still at the center of the universe and that all heavenly bodies revolved around it. By the end of the seventeenth century, most thinkers had come to believe that the earth is the third planet circling the sun once a year and spinning around its own axis once a day. Nowadays, after three more centuries of accumulating knowledge, this modern view is known to be true beyond any reasonable doubt. But the earlier view had been a very plausible belief; for two millennia the earth's motion had been inconceivable or untenable, and then for a century and a half, the discussion of the relative merits of the two views was the subject of heated debate. In fact, the transition was a slow, difficult, and controversial process. We may fix its beginning with the publication in 1543 of Nicolaus Copernicus's book *On the Revolutions of the Heavenly Spheres* and its completion with the publication in 1687 of Isaac Newton's *Mathematical Principles of Natural Philosophy*.

The discovery of the motion and noncentral location of the earth involved not only a key astronomical fact, but was interwoven with the discovery of the most basic laws of nature, such as the laws of inertia, of force and acceleration, of action and reaction, and of universal gravitation. This discovery was also connected with the clarification of some key principles of scientific method. It represents, therefore, the most significant breakthrough in the history of science; thus, the series of developments starting with Copernicus in 1543 and ending with Newton in 1687 may be labeled the Scientific Revolution.[1]

More generally, it would perhaps be no exaggeration to say that this transition represents the most important intellectual transformation in human history.[2] One reason for this involves the worldwide repercussions of the Scientific Revolution itself; science seems to be the only cultural force which has managed to dominate human societies in all parts of the earth. Another reason stems from the interdisciplinary character of the transition from a geocentric to a geokinetic world view; the transformation involved not only many branches of science but also other disciplines and activities such as philosophy, theology, religion, art, literature, technology, industry, and commerce; indeed it changed mankind's self-image in general. We may thus also call this transition the Copernican Revolution, if we want a label which leaves open its broad ramifications outside science; this label also gives due credit to the one thinker whose contribution initiated the process.[3]

Galileo Galilei (1564–1642) was a key protagonist of these historical developments. In physics, he pioneered the experimental investigation of motion; he formulated, clarified, and systematized many of the basic concepts needed for the theoretical analysis of motion; and he discovered the laws of falling bodies. In astronomy, he introduced the telescope as an instrument for systematic observation; he made several crucial observational discoveries; and he understood the cosmological significance of these observational facts and gave essentially correct interpretations of many of them. Galileo was also an inventor, making significant contributions to the devising and improvement of such instruments as the telescope, microscope, thermometer, and pendulum clock. In regard to scientific method, he pioneered several important practices, such as the use of artificial instruments (like the telescope) to learn new facts about the world, and the active intervention into and exploratory manipulation of physical phenomena to gain access to aspects of nature which are not detectable without such experimentation; he also contributed to the establishment and extension of other more traditional methodological practices, such as the use of a quantitative approach in

1. For some valuable accounts centered around this theme, see H. F. Cohen (1994), Hall (1954), and Lindberg and Westman (1990).

2. This thesis is generally attributed to Butterfield (1949).

3. For a classic example of this type of general account, see Kuhn (1957).

the study of motion; and he contributed to the explicit formulation and clarification of important methodological principles, such as the disregard of biblical assertions and religious authority in scientific inquiry.[4]

Galileo's *Dialogue on the Two Chief World Systems, Ptolemaic and Copernican* (1632) is one of the most important texts of the Copernican or Scientific Revolution. It also constitutes his mature synthesis of astronomy, physics, and scientific methodology. From the viewpoint of the transition from geocentrism to Copernicanism, the book may be summarized by saying that it strengthened the geokinetic theory by means of theoretical considerations based on Galileo's new physics, observational evidence stemming from his telescopic discoveries, and concrete methodological analyses.

Galileo's *Dialogue* is also the book which triggered his trial by the Roman Catholic Inquisition in 1633, ending with his condemnation as a heretic and the banning of the book. The trial was the climax of a series of events that began in 1613 and included a related series of Inquisition proceedings in 1615–1616. This twenty-year sequence of developments is now known as the Galileo Affair, and so the book is also an important document in this tragic but instructive episode (of which more below).

Because the Galileo Affair involved a conflict between one of the founders of modern science and one of the world's great religious institutions, it has traditionally been taken as an example of the warfare between science and religion. Whether this is really so is one of the main issues in the controversy that has arisen about the interpretation and evaluation of the Galileo Affair. This issue cannot be resolved here, but two comments are in order. First, even a cursory reading of the relevant documents shows that many churchmen were on his side and many scientists were on the opposite side; thus, there was a split within both science and religion, along the lines of what may be called conservation and innovation; so the real conflict was between a conservative attitude and a progressive one. Second, many of the problems between Galileo and the Church stemmed from the fact that the Church was then not just a religious authority but also a political power and a social institution; thus, the episode illustrates the conflict or interaction between science and politics, between science and society, and between individual freedom and institutional authority.

4. For some good scholarly general accounts see Drake (1978; 1990) and Geymonat (1965); for some interesting popular general accounts see Reston (1994), Ronan (1974), and Seeger (1966).

These remarks highlight the historical significance of the *Dialogue* by way of its connection with the Scientific Revolution, Copernican Revolution, and Galileo Affair. These episodes have such a perennial interest and universal relevance that the book thereby acquires perennial and universal significance. So far this means only that it is worth reading by all educated persons and by every generation in order for them to acquire factual information, come to their own interpretation, formulate their own evaluation, and derive their own lessons. But its perennial and universal importance can be elaborated in more abstract ways that connect it to several general activities of the human mind.

1.2. First, Galileo wrote the book in vernacular Italian rather than in scholarly Latin. His main motive was that he wanted to appeal to a broad audience of educated nonexperts and to suggest that the issues raised were of general cultural significance. This intention does not mean that he was addressing the book only to such readers; rather he wrote it for a professional audience of scholars and specialists, as well as for a lay audience of educated, intelligent, studious, and curious persons. However, what deserves emphasis here is that the book has an explicitly universalist aim.

Moreover, Galileo wrote the book in the form of a dialogue among three speakers: Salviati, an expert who takes the Copernican side; Simplicio, a scholar who takes the geocentric viewpoint; and Sagredo, an intelligent, educated, and inquisitive layman who knows little about the topic but wants to listen to both sides and make up his mind as a result of the critical scrutiny of what they have to say. This feature is in part connected with the Galileo Affair and will be discussed again below. The point to stress here is that the dialogue form reinforces the book's universal appeal inasmuch as the speakers personify the abstract intellectual issues and make it easier for readers to relate to them. But the dialogue form is also directly connected with another important aspect of the book, to which we now turn.

The book's most striking feature is critical reasoning, taking this term to mean reasoning aimed at the analysis, evaluation, and/or self-reflective presentation of arguments.[5] In fact, Galileo was writing before the

5. This definition has been inspired by Michael Scriven, although it does not represent a mere adoption of his exact definition, but rather a formulation of a concept that I needed to make sense of Galileo's book; cf. Scriven and Fisher (forthcoming). For example, it should be noted that here I am talking of critical *reasoning*, whereas Scriven is talking about critical *thinking*, which I take to be a broader concept and to include also what I call methodological reflection; for more details see the appendix (1.6, 1.8, 2.1, and 2.7).

new Copernican view was conclusively established, when the situation was fluid and controversial; thus, to form an intelligent opinion on the topic required more than mere observation, experiment, calculation, or deduction; it required reasoning, judgment, analysis, evaluation, and argumentation. So it is not surprising that he felt the most fruitful thing to do was to undertake a critical examination of the arguments on both sides of the controversy. His overall conclusion was clearly that the Copernican side was preferable to the geocentric side. But this means only that Copernicanism was more probable or more likely to be true than geocentrism; that is, that the pro-Copernican arguments were stronger than the anti-Copernican ones; or again, that the reasons for believing the earth to be in motion were better than those for believing it to stand still; or finally, that the evidence or support favoring the geokinetic idea outweighed the evidence or support favoring the geostatic one. Galileo's conclusion was not that Copernicanism was clearly true or certainly true or absolutely true or demonstrably true;[6] nor was it that there were no reasons for believing the earth to stand still; nor was it that the geostatic arguments were worthless. The point is that the book's key thesis is one about the relative merits of the arguments on each side, that this thesis is substantiated and not merely asserted, and that the substantiation proceeds by the reasoned presentation, analysis, and evaluation of the arguments. In short, critical reasoning is a key part of both the book's content and the book's approach.

Despite the prevalence of critical reasoning, we should not be too one-sided about it; for the *Dialogue* is also full of methodological reflections. By methodological reflections I mean[7] discussions meant to formulate, clarify, evaluate, and use general principles about the nature of truth or knowledge and about the proper procedure to follow in the search for truth and the quest for knowledge; by calling them reflections I mean to stress that they arise in the context of a particular investigation

6. This interpretation will be elaborated and defended later in the introduction (5) and the appendix (1).

7. This definition of methodological reflection, together with the earlier definition of critical reasoning, makes clear that the two are different; thus I will usually treat them as distinct activities. But for certain purposes it is useful to associate critical reasoning and methodological reflection under the single heading of critical thinking; while the two remain distinct, this association interrelates them; this is discussed in the appendix (1.8, 2.1, and 2.7). On the other hand, I will usually treat "methodological reflection" and "epistemological reflection" as interchangeable terms; and the same holds for "methodology" and "epistemology"; I will do so even though these two things are not completely identical and in some contexts a distinction is drawn between them; this is discussed in the appendix (2.1).

about what is physically true or what we know of physical reality, and so they function to help us understand better what we are doing and decide what we should be doing. That is, issues and principles about truth, method, and knowledge are constantly discussed in the book, not because Galileo intends or pretends to write an abstract treatise about the nature of these concepts, but because the specific scientific questions (about whether or not the earth is standing still at the center of the universe) are so basic that they raise questions about how one is proceeding and about the proper way to proceed. For example, there are discussions about the nature and proper role of authority, observation versus intellectual theorizing, the limitations of human understanding, independent-mindedness and open-mindedness, simplicity, probability, experiments, mathematics, artificial instruments, the Bible, divine purpose and human interest, and causal explanation.

The *Dialogue* is also a gold mine of rhetoric, but here one must be especially careful. In this context, I take rhetoric to mean[8] the theory and practice of verbal communication, involving not only persuasive argumentation but also such forms and techniques as emotional expression, beautiful language, imaginative description, bare assertion, nuanced assertion, repetition, wit, satire, humor, and ridicule. The wealth and complexity of the book's rhetoric derive in part from its universalist aim, which implies that Galileo is addressing several audiences at once; they derive in part from its dialogue form, which means that there is a certain amount of drama unfolding before the reader; the book's rhetoric also stems from the controversial character of the scientific and methodological issues discussed, which means that we are witnessing a polemical discussion; it also stems from the context of Galileo's struggle with the Church, which means that in writing the book he was taking considerable risk and could not always say what he meant or mean what he said; the rhetoric originates to some extent from the fact that the practice of science at that time was socially and financially dependent for the most part on the patronage of princes, which means generally that Galileo's career was partly that of a courtier and specifically that his book represented an action in an intricate network of patronage involving the Tuscan Medici court in Florence and the Vatican court of Pope Urban VIII in Rome;[9] finally, the rhetoric originates to some extent from the fact that he was a gifted writer who poured his heart and soul into this work,

8. For more details see the appendix (3).
9. For a brilliant account of this aspect of Galileo's career, see Biagioli (1993).

so much so that many passages achieve a high degree of literary and aesthetic value. Notice that I am *not* equating rhetoric with the art of deception in general, and the skill of making the weaker argument appear stronger in particular;[10] so understood, rhetoric would be an inherently objectionable activity, whereas my definition allows both good and bad rhetoric. Nevertheless, rhetoric does not easily mix or coexist with scientific inquiry, critical reasoning, and methodological reflection; it considerably complicates the proper understanding and evaluation of the text. In this regard, the important thing to do is not to deny the existence of rhetoric in the book, nor to overstress it and neglect the book's other aspects, nor to conflate it with these other aspects, but to learn to detect, analyze, evaluate, and appreciate it. Fortunately, the rhetoric enhances the readability of the book and the enjoyment one can derive from the experience of reading it.

In short, Galileo's *Dialogue* can and should be read for what it tells us about the history of the Copernican Revolution, the Scientific Revolution, and the Galileo Affair, and for what it can teach us in general about critical reasoning, scientific methodology, and the art of rhetoric.

## 2. The Geostatic Worldview

2.1. The worldview accepted until the middle of the sixteenth century contained two main theses. One was that the earth is motionless, and so we may speak of the *geostatic* worldview, or more simply of geostaticism. The other asserted that the earth is located at the center of the universe, and so we may call it the *geocentric* theory, or more simply, geocentrism.

Although it is now known that the geocentric view is not true, it corresponds, even today, to everyday observation and common sense intuition; and, although it has this natural appeal, its technical elaboration was the result of arduous work by some of the greatest thinkers of antiquity. Two individuals made contributions which were so important

10. This seems to be the notion of rhetoric presupposed by Feyerabend (1975), Hill (1984), and Koestler (1959). They do not literally say this, but they tend to focus on situations where rhetoric is allegedly used for the purpose of deception; it is this tendency of theirs that creates the impression. Clearly, it is as wrong to claim that scientists never use deceptive rhetoric as to claim that they always do; my point is simply that scholars who usually study cases of (allegedly) deceptive rhetoric convey the impression that this is all there is to rhetoric.

that their names became synonymous with this view of the universe. Aristotle (384–322 B.C.) was a pupil of Plato who lived in Athens during the period of classical Greek civilization; he contributed primarily by elaborating the cosmology, the physics, the general philosophical principles, and the qualitative astronomical ideas of the geostatic worldview. Ptolemy lived in Alexandria in the second century A.D., at the end of the Hellenistic phase of Greek culture; he contributed primarily the mathematics and the quantitative details of the astronomical system, forging a synthesis of the observational, mathematical, and theoretical discoveries of the five intervening centuries. Thus, we may also label the old view the *Aristotelian* or the *Ptolemaic* theory of the universe. Furthermore, since the Aristotelians acquired the nickname of Peripatetics, geocentrism was also traditionally labeled the Peripatetic worldview.

These remarks suggest that the geostatic worldview was not just an astronomical theory, but contained parts belonging to philosophy, physics, and cosmology. The explanation of its details will make this interdisciplinary mixture more obvious.

Moreover, the geocentric view was not a monolithic entity, but rather a theory that underwent two thousand years of explicit historical development comprising five centuries before and fifteen centuries after the birth of Christ (not to speak of its prehistory). Thus, there are many versions of the theory; for example, Aristotle's and Ptolemy's versions differ not only in emphasis but also in substantive detail. The version expounded below is not a synopsis of any one work, but rather a reconstruction of the most widespread beliefs at the start of the sixteenth century, in a form useful for understanding the *Dialogue* and its role in the historical events it represents.[11]

2.2. Let us begin with the question of the earth's *shape*. The geostatic view held that the earth is a sphere, so that its surface is not flat but round; this is, of course, true. In fact, the arguments proving this fact were known to Aristotle and can be found in his writings. Although uneducated persons or primitive peoples at the time of Aristotle or Galileo may have believed that the earth was flat, scholars had settled the question a long time ago; thus, it should be clear that the Copernican controversy had nothing to do with the shape of the earth but rather was concerned with its behavior and location.

11. My account has been inspired by Galileo's own *Treatise on the Sphere, or Cosmography*, a short elementary textbook of traditional geostatic astronomy that he wrote and used in the early part of his teaching career but never published (cf. Favaro 2:205–55). My account also relies on I. B. Cohen (1960), Kuhn (1957), and Toulmin and Goodfield (1961).

Similarly, the maritime voyages and geographical discoveries of Columbus and others at the end of the fifteenth century and thereafter did provide additional confirmation of the earth's spherical shape; but this was only a more direct, experiential proof of the earth's roundness. Those voyages also provided new evidence about the earth's size, structure, and composition; and this evidence affected cosmological and astronomical thought.[12] But this means that the geographical discoveries may have been a factor in the Copernican Revolution, not that the issue was about the earth's shape or size.

The size and shape which did become part of the dispute were those of the whole universe. The old view held that the universe was a sphere much larger than the earth but of *finite* size, the size being slightly larger than the orbit of the outermost planet; that is, the distance from the outermost planet to the stars was about the same as the distance between one planet and another. The stars were all at the same distance from the center, attached to the surface of the *stellar sphere*. This sphere (also called *celestial sphere*) enclosed the whole universe, and outside of it, there was nothing physical. That is, the size and shape of the universe were defined in terms of the size and shape of a sphere to which were attached about six thousand fixed stars visible with the naked eye. This contrasts with the classical modern view that the universe is infinite, space goes on without end, stars are scattered everywhere in infinite space, and so it does not even make sense to speak of the shape, size, or center of the universe.

The finite spherical universe was based on the same set of observations that led to the belief that at the center of the stellar sphere was the motionless earth. This was the phenomenon of *apparent diurnal motion*: the earth feels to be at rest; the whole universe appears to move daily around the earth in a westward direction; thousands of stars visible with the naked eye at night appear to undergo no change in size or brightness, but rather seem to be at a fixed distance from us; they appear to

12. For example, Margolis (1987) suggests that learning about the existence of large land masses in the western hemisphere made it difficult to continue believing that the terrestrial globe consisted essentially of a series of concentric spherical layers of the elements earth, water, air, and fire; as explained below, the latter thesis was an important part of the geostatic view. The connection was that if one believed that land emerged out of the oceans only in a small part of the earth's surface or on one side of the globe, then one could regard this as a minor exception to the rule that the natural place of the element earth is below the element water; but if one knows that there is another continent in the western hemisphere, this implies that land emerges out of the oceans on opposite sides of the earth, and it becomes harder to believe that the normal arrangement is or should be to have the element earth below water.

move in unison, so that their relative positions remain fixed; they appear to move in circles that are larger for stars lying closer to the equator and smaller for those lying closer to the poles. In short, the stars appear to move as if they were attached to a sphere that rotates daily westward around a motionless earth at the center. Given the plausible principle that what appears to normal observation corresponds to reality, one had the fundamental argument in support of the basic tenets of the geostatic worldview.

In the spherical finite universe, position or location or place had an absolute meaning. The geometrical center of the stellar sphere was a definite and unique place, and so was its surface or circumference; and between the center and the circumference, various layers or spherical shells defined various intermediate positions. The part of the universe outside the earth was called *heaven* in general, and to distinguish one heavenly region from another, one spoke of different heavens (in the plural); for example, the stellar sphere was the highest heaven, which meant the most distant one from the earth and which was also called the firmament; whereas the closest heaven was the spherical layer to which the nearest heavenly body (the moon) was attached, and so the lunar sphere or sphere of the moon was the first heaven. Between the lunar and the stellar spheres, six other particular heavens or heavenly spheres were distinguished; one was for the sun, and there was one for each of the other five known planets (Mercury, Venus, Mars, Jupiter, and Saturn). Details about the motion of the planets will be discussed later.

Here it is important to distinguish a *heavenly sphere* from a *heavenly body*: a heavenly sphere was one of the eight nested spherical layers surrounding the central earth, each of which was the region occupied by a particular heavenly body or group of heavenly bodies, and to each of which these heavenly bodies were respectively attached; whereas a heavenly body was a term referring to either the sun, the moon, a planet, or one of the fixed stars. The two terms are confusing because heavenly bodies were considered to be spherical in shape, and so they were spheres in their own right; but the term heavenly sphere referred only to one of the spheres concentric with the center of the universe to which the (spherical) bodies of the sun, moon, planets, and fixed stars were attached.[13]

13. This clarification has been made by Rosen (1959; 1992), who stressed that the title of Copernicus's book refers to spheres concentric with the center of the universe and not to heavenly bodies.

The terrestrial region too had its own layered structure. This is related to a threefold meaning for the term *earth*. In saying earlier that the earth is a sphere, I was referring to the terrestrial globe consisting of land and oceans; this globe is a sphere, not in the sense of a perfect sphere, but only approximately because the land is above the water and is full of mountains and valleys; such an approximation is very good because the height of even the tallest mountain is insignificant compared to the earth's radius. But it was only natural to distinguish water from earth, taking the latter term to mean just land, rocks, sand, and minerals; when so understood, earth was obviously only a part of the whole globe. It was also natural to count the air or atmosphere surrounding the globe as part of the terrestrial region; and so by earth one could also mean the whole region of the universe near the terrestrial globe, up to but excluding the moon and the lunar sphere. In short, *earth* had three increasingly broad meanings: it could refer to just the solid part of the terrestrial world; or to the globe consisting of both land and oceans; or to land plus oceans plus atmosphere.

Terminology aside, the substantive point is that the earth (namely, the place where mankind lives) is not a body of uniform composition, but contains three main parts: a solid, a liquid, and a gaseous part. These three parts (earth, water, and air) were labeled *elements* to signify their fundamental importance. In regard to their arrangement, the element earth sinks in water, and so earth must extend to the central inner core of the world and must make up most of what exists below the surface; but most of the surface of the globe is covered with water, and the element water mostly surrounds the element earth. This was expressed theoretically by claiming that the *natural place* of the element earth was a sphere immediately surrounding the center of the universe, and that the natural place of the element water was a spherical layer surrounding the innermost sphere. As for air, simple observation tells us that it surrounds the spheres of the first two elements, and so its natural place was a third sphere surrounding the first two.

There was a fourth terrestrial element, which was called fire; but it required a more roundabout explanation. Just as we see earth sink in water and water fall through air, we see flames shoot upwards through air when something is burning, currents of heat move upwards through air during hot summer days, and smoke generally rise; we also see trapped fire escape upwards in volcanic eruptions. Such observations were taken as evidence that the natural place of fire was a fourth spherical layer above the atmosphere and just below the lunar sphere.

The existence of the element fire was also derived from considerations about basic physical qualities.[14] There were two fundamental pairs of physical opposites: hot and cold, and humid and dry. The element earth was a combination of cold and dry; the element water a combination of cold and humid; the element air a combination of hot and humid; so there had to be a combination of hot and dry, and that was what constituted the element fire.

In summary, from the point of view of location in the geostatic finite universe, there were twelve natural places, each consisting of a sphere or spherical layer with a common center. The four terrestrial spheres were the natural places of the four terrestrial elements (earth, water, air, and fire). The eight heavenly spheres were the natural places of the heavenly bodies; they ranged from the lunar sphere to the stellar sphere, with six intermediate spheres for the sun and the five planets. The stellar sphere enclosed everything, and the earth was at the focus of it all.

As with position, direction had an absolute meaning in the finite universe. There were three basic directions: toward the center of the universe, called *downward*; away from the center of the universe, called *upward*; and around the center of the universe. Thus, one important way of classifying motions was in these cosmological terms: bodies could and did move toward, away from, and around the center of the universe.

Geometrically, motion could be simple or mixed. Simple motion was motion along a simple line. A simple line was defined as a line every part of which is congruent with any other part. Thus, there were only two such lines—circles and straight lines; and there were two types of simple motion—straight and circular. Mixed motion was motion which is neither straight nor circular.

Another way to classify motions was in terms of the motions characteristic of the elements, namely, the motions that the elements underwent spontaneously. This categorization was meant to correspond with the two other classifications as follows. For example, earth and water characteristically moved straight downward, while air and fire characteristically moved straight upward. Now, since heavenly spheres and heavenly bodies moved characteristically with circular motion around the center, this meant that they were composed of a fifth element; the term *aether* or *quintessence* was used to refer to this heavenly element.

Finally, another important classification was in terms of the opposition between natural and violent motions. *Violent motion* was motion

14. Galileo, *Cosmography*, in Favaro (2:213).

caused by some external action; *natural motion* was motion that a body underwent because of its nature, so that the cause was internal. For example, the downward motion of earth and water, the upward motion of air and fire, and the circular motion of heavenly spheres and heavenly bodies were all cases of natural motion; whereas rocks thrown upward, rain blown sideways by the wind, a cart pulled by a horse, and a ship sailing over the sea were all cases of violent motion.

More fundamentally, motion was the opposite of rest. Rest was the natural state of bodies, and so all motion presupposed a force in some way. Natural motion was essentially the motion of a body toward or within its proper place; only when displaced from its proper place by some force would a terrestrial body engage in natural motion up or down; and only if started by some mover would a heavenly sphere rotate around the center of the universe, thus carrying its planet or stars in circular motion. On the other hand, violent motion was motion that was not toward the body's proper place, and such motion could happen only by the constant operation of a force.

From what has been said, it is apparent that earth and heaven were very different; indeed this radical difference was enshrined in an idea that needs to be made explicit and that deserves a special label. The key term is the *earth-heaven dichotomy*; but one can equivalently speak of the dichotomy between the earthly or terrestrial or sublunary or elemental region of the universe on the one hand, and the heavenly or celestial or superlunary or aethereal region on the other. We have already seen that one difference between the two regions was location, which was absolute in the finite spherical universe: terrestrial bodies occupied the central region of the universe below the moon, whereas heavenly bodies occupied the outer region from the lunar to the stellar sphere. Similarly, there was another difference in regard to natural motions: earthly bodies moved naturally straight toward or away from the center of the universe, whereas celestial bodies moved circularly around the same center.

We have also seen that the two regions differed in regard to the elements of which bodies were composed. Sublunary bodies were made of earth, water, air, and fire. On the other hand, in the superlunary region things were made of aether, or various concentrations thereof; that is, aether in low concentration made up the heavenly spheres, which were invisible; whereas, aether in a highly concentrated state generated the moon, sun, planets, and stars, which were the visible heavenly bodies.

Just as the natural places and the natural motions of the two regions obviously corresponded to each other, the elements in the two regions also corresponded to the natural places and motions. That is, the natural places and the natural motions of terrestrial bodies could be conceived as defining the essential properties of the terrestrial elements, while the natural places and motions of celestial bodies could be conceived as defining the essential properties of aether. Other differences between earth and heaven could be defined in terms of additional properties of the different elements. For example, whereas superlunary substances had no weight, sublunary bodies obviously did; or to be more exact, whereas aether was weightless, earth and water had weight (and so they were called *heavy bodies*), and air and fire had *levity* (namely, the tendency to go up) and so they were called *light bodies*. Moreover, aether was intrinsically luminous, namely, capable of giving off its own light; but, earthly elements were dark, namely, incapable of emitting their own light; even fire did not emit an inherent light of its own but only temporarily produced light when it was in the process of escaping from lower regions to move to its natural place just below the lunar sphere.

Of the many differences between earth and heaven, two deserve special attention: natural motion and susceptibility to qualitative change. Natural motion has always been regarded as an essential or defining characteristic of a physical body. This seems to have remained unchanged even by the Copernican Revolution; from this viewpoint, what changed was the type of natural motion attributed to bodies. Since the geocentric view attributed different natural motions to terrestrial and celestial bodies, it is no surprise that it included the earth-heaven dichotomy.

To elaborate, one must first understand why the geostatic universe was *not* a trichotomy, given that there were three visible kinds of natural motions (namely, downward for earth and water, upward for air and fire, and around the center of the universe for aether). The answer implied by the discussion above is that downward and upward natural motions were both straight, and so were conceived as two minor subspecies of the same fundamental kind, namely, rectilinear motion. Geometrically, there were only two lines with the property that all parts are congruent with any other part—the circle and the straight line; thus, what was common to both upward and downward natural motions (straightness) was more important than what distinguished them (toward and away from the universal center).

However, this geometrical reason was not the only justification for making the essential distinction to be the twofold one between straight and circular natural motions rather than the threefold one among upward, downward, and around. There was also the cosmological reason that, unlike circular natural motion, straight natural motion could not be perpetual. For once a rock had reached the center of the universe, its nature would make it remain there rather than continue moving past the center, which would constitute upward and thus unnatural motion for the rock; similarly, once a fiery body had reached the region above the terrestrial atmosphere just below the lunar sphere, it had reached its natural place; it had no place to go because to continue moving would bring it into the first heavenly sphere reserved for the aethereal moon where the element fire could not subsist.

Finally, there was a theoretical reason why upward and downward natural motions could belong to the same fundamental region of the universe but were essentially different from natural circular motion. The theory in question was the theory of change as contrariety, according to which all change derives from contrariety, and no change can exist where there is no contrariety; by contrariety was meant such oppositions as hot and cold, dry and humid. Now, up and down, together with the related pair of light and heavy, was another fundamental contrariety. Thus, a region full of bodies some of which moved naturally downward and some upward was bound to be full of all sorts of qualitative changes; and indeed observation obviously revealed that the terrestrial world is full of birth, growth, decay, generation, destruction, weather and climactic change, and so on. On the other hand, the circular natural motion of the heavenly bodies was thought to have no contrary; consequently, the heavenly region lacked an essential condition for the existence of change. Add to this the belief that the opposition between hot and cold and between dry and humid belonged only within the four terrestrial elements, and one could claim that the region of aether lacked any of the proper conditions for change. And observation confirmed that claim, too, because no physical or organic or chemical changes were easily detected in the heavens, and none were said to have ever been seen; the only essential phenomenon in the heavens was motion, but all heavenly motion was regular and involved the rotation of concentric spheres, which thus remained in place, so that there was not even change of place; what changed was only the relative position of the various bodies attached to these celestial spheres.

This analysis clarifies how natural motion and qualitative change provided the basis for the earth-heaven dichotomy. There were many differences between earth and heaven, but two interrelated ones were crucial: in the terrestrial world bodies moved naturally with straight motion and underwent qualitative change, whereas in the celestial region things moved naturally with circular motion and were not subject to qualitative change.

To summarize the discussion so far, the Aristotelians and Ptolemaics believed that the earth was spherical, motionless, and located at the center of the universe; that the universe was finite, was bounded at the outer limit by the stellar sphere, and was structured into a series of a dozen nested spheres, all inside the stellar sphere and surrounding the central sphere of the solid earth; that there was a fundamental division in the universe between the earthly and the heavenly regions; and that these regions consisted of bodies with very different properties and behavior, such as different natural places, natural motions, elemental composition, and possibilities for qualitative change. Two things must now be added to this general cosmological picture: the details of the physics of the motion of terrestrial bodies and the astronomical details of the heavenly bodies. Let us begin with the former.

2.3. In the terrestrial region, the natural *state* of bodies was rest. To be more exact, it was rest at the proper place, depending on the elemental composition of the body: at the innermost core for the element earth; just above that for water; above water for air; and above air for fire. This meant that, whereas no cause was sought to explain why a body rested at its proper place, when a body was in motion or at rest outside its proper element, an explanation was required.

The explanation for why a body was in motion could be that it was going to rest at its proper place; this was the case of natural motion like rocks and rain falling or smoke rising though air. Or the explanation could be that the body was being made to move by an external agent; this was the case of violent motion, for example, a cart pulled by a horse, a boat sailing over the water, rain blown by the wind, or weights being lifted from the ground to the top of a building. But both natural and violent motions required a force; the only difference was that in natural motion the force was internal to the body, whereas in violent motion the force was external. For example, falling bodies fell because of their inherent tendency to go to their natural place if they were not already there; this internal force was termed *gravity* and was measured by the weight of an object. On the other hand, for a sailboat, the wind was obviously the external force, and for a cart, the horse.

Sometimes "violent motion" was equated with "forced motion," but in such cases it was understood that by "forced motion" one meant motion caused by an *external* as distinct from an *internal* force. Since all motion was forced, the term "forced motion" was sometimes regarded as redundant if taken to mean caused motion, and it was found informative only if taken to mean externally caused motion. That is, the term *force* was ambiguous and could mean either any cause of motion or an external cause of motion; this may generate some confusion, but the context usually clarifies the meaning.

All motion, then, whether natural or violent, was caused by a force, whether internal or external. But another condition was required by all motion, namely, resistance. That is, motion was the overcoming of resistance. This was so partly because all space was filled and there was no vacuum or void, so that whenever a body moved it could move only through some medium, be it air, water, oil, molasses, sand, or soil. Even the heavenly region, interplanetary and interstellar space, was not devoid of matter; it was filled with aether.

Moreover, it was argued that if there were no resistance to overcome, then a force (however small) would make a body move instantaneously, namely, with infinite speed; and this was an absurdity since it meant that the body would occupy different places at the same time, indeed, many different places at the same time. This argument depended on the idea that speed was inversely proportional to resistance, for this idea would provide the justification of why motion without resistance would be instantaneous; that is, not only was resistance required for motion to occur, but motion was correspondingly slower with greater resistance and faster with lesser resistance.

This quantitative relationship between speed and resistance was taken seriously for the extreme case of zero resistance and used as just indicated in the above argument. But the relationship was not taken equally seriously for the other end of the spectrum, namely, for very strong resistance. That is, when the resistance was very strong, rather than saying that a given force would cause some motion, perhaps at very slow speed, it was held that there was a threshold for motion to occur at all; the force had to be sufficient to overcome the resistance in the first place, and if that was the case, then the speed was inversely proportional to the resistance. Here, the typical example was that of a single man trying to pull a ship into dry dock by himself; it is clear that he would not be able to move the ship at all, not even one hundred times slower than a team of the one hundred men required to accomplish the task.

The relationship between force and speed (when the resistance is constant) could also be expressed quantitatively. The formula was that at constant resistance, the speed is directly proportional to the force. Here the paradigm example was the fall of heavy objects through a fluid like water: heavier objects sink faster than lighter ones, and do so more or less in proportion to their weight; and weight in this case is the (internal) force.

Combining the two relationships, one obtained the formula that, given that the force can overcome the resistance, the body moved at a speed which was directly proportional to the force and inversely proportional to the resistance: *speed* = *constant* × (*force*/*resistance*).[15]

These ideas were plausible and largely in accordance with observation, except for situations like free fall through air and violent projectile motion. For free fall, the Aristotelian theory implied that a lead ball fell much faster than a rock, so that when both were dropped from the same height, the lead would reach the ground earlier than the rock; and for a given object, its speed of fall should not increase indefinitely with time because its maximum value depended only on its fixed weight and the fixed resistance of the air. The problem of projectiles involved the motion of such things as arrows shot from bows, and the question was what was the force making them move after the projectiles left the ejector. The Aristotelians were aware of these problems and tried to solve them, but their solutions were found to be increasingly inadequate. The discussion of these problems provided one line of development in the rejection of the old physics and the construction of the new one.[16]

2.4. Let us now go on to the main astronomical details of the geostatic worldview. I have already mentioned that the whole universe outside the earth moved around it daily in a westward direction. This phenomenon was called the *diurnal motion* and was regarded as directly observable. The observation is that all heavenly bodies appear to revolve westward around us; this is most obvious for the case of the sun, whose rising in the east and setting in the west causes the cycle of night and day; the moon is also easily seen to do the same; and at night each fixed star appears to follow the same westward trajectory as the previous night.

Since the universe was spherical, this diurnal motion was conceived as the daily rotation of a sphere around a line, called the *axis* of diurnal ro-

15. This formula is a modern way of expressing the combination of the two relationships and is not one that was available to the ancients or medievals; their concepts of quantity and proportion were such that they thought only magnitudes of the same kind could be compared.

16. For more details on the physics of the geostatic world view, see, for example, I. B. Cohen (1960, 22–35), Franklin (forthcoming), Lindberg (1992, 58–62 and 290–307), and Toulmin and Goodfield (1961, 93–103).

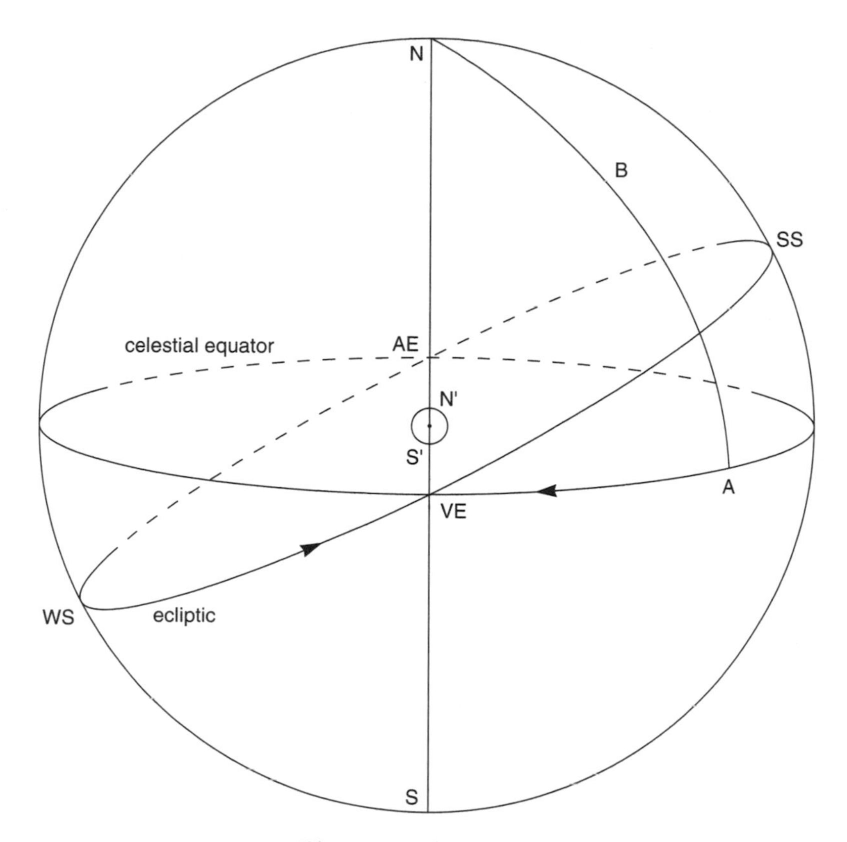


Figure 1. Celestial sphere

tation, which went through the north and the south celestial *poles* (N and S, in fig. 1);[17] this line also intersected the earth's center and two points on its surface, the north and the south poles of the earth (N′ and S′). From an observational viewpoint, the celestial poles were the two points in the heavens which appeared to be motionless; the north celestial pole seemed motionless to observers in the earth's northern hemisphere, and the south celestial pole seemed motionless to observers in the southern hemisphere; and the circular paths of the fixed stars appeared to be centered at the respective poles. On the surface of the celestial sphere, midway between the poles was a great circle of special importance, called the

17. This diagram is adapted from Kuhn (1957, 31–36) and from the *New Columbia Encyclopedia* (Harris and Levey 1975, 883).

*celestial equator*; it too had a terrestrial counterpart (the earth's equator), which could be defined as the intersection of the plane of the celestial equator with the earth's surface, or as the great circle on the earth's surface halfway between the north and south terrestrial poles.

One reason for the importance of the celestial poles and equator was that they yielded a fixed frame of reference to define the position of the heavenly bodies, and correspondingly the position of points on the earth's surface. One could measure the angular position of a star north or south of the celestial equator, which was called *declination* (AB, in fig. 1); similarly, the angular distance from the terrestrial equator of a point on the earth's surface was called *latitude*. For each declination or latitude one could conceive a plane parallel to the equator whose intersection with the surface of the two spheres generated circles (called *parallels*) which became smaller as one moved toward a pole. The east-west position of a star (B) required first the drawing of a *meridian*, namely, a great circle (partially shown as NBA) through the star and the poles; then one would measure the *ascension*, namely, the angular distance from this meridian to some particular meridian (for example, A to VE); it was analogous for positions on the earth's surface, except that this east-west angular distance was called *longitude*.

There were two kinds of heavenly bodies, called *fixed stars* and *wandering stars*. A fixed star was a heavenly body that moved daily around the earth in such a way that its position relative to other heavenly bodies did not change; for example, its declination remained constant, and so did its angular distance from any other fixed star. A wandering star was a heavenly body that not only moved daily around the earth but also changed its position relative to other heavenly bodies; that is, the wandering stars were those heavenly bodies which, besides undergoing the diurnal motion, appeared to move in other ways (to be discussed presently). There were only seven wandering stars, which were also called planets; indeed the word *planet* originally meant literally "wandering star." Because wandering stars were often called simply planets, fixed stars were often called simply stars.[18]

Thousands of fixed stars were visible at night with the naked eye; they were catalogued both in terms of apparent brightness (called *magnitude*)

18. That is, though one broad meaning of the word *star* was synonymous with the term *heavenly body*, one narrow meaning of *star* was identical to the term *fixed star*; the point is that the term *fixed* was often dropped when the context made it clear that one was indeed referring to fixed stars.

and in terms of shapes or patterns formed by groups of stars close to each other (called *constellations*). The naked eye could be trained to distinguish six magnitudes; stars of the first magnitude were the brightest, and those of the sixth magnitude were the faintest. The brightest star was named Sirius or the Dog Star; it was located near the equator and was part of the constellation of Canis Major. A star of the second magnitude was especially important because it was so close to the north celestial pole that, for practical purposes (such as navigation), it could be regarded to be the pole; it was called Polaris or the North Star and was part of the constellation of the Little Dipper.

Both the sun and moon were planets because they moved ("wandered") in relation to the fixed stars. Because of their brilliance and their relatively large size, they were called the two luminaries. The other known planets were named Mercury, Venus, Mars, Jupiter, and Saturn. We now know that there are three other planets (Neptune, Uranus, and Pluto) circling the sun in orbits beyond Saturn, but they were unknown not only to the ancients but even to Copernicus and Galileo; so they played no role in the Copernican Revolution.

The most basic point about the planets was that, out of the thousands of heavenly bodies, there were seven which circled the earth westward once a day like all others, but did not do so in unison with them; these seven bodies also revolved slowly eastward, so that from day to day their position shifted. That is, whereas a fixed star revolved around the earth in such a way that after twenty-four hours it returned to the same exact position it had before, after twenty-four hours a planet did not quite return to the earlier position but had fallen behind somewhat, being located slightly eastward. This can be seen most easily for the case of the moon by observing its position on succeeding nights at midnight; relative to the fixed stars, it appears to move eastward. The planets seemed to behave as if their motion were a combination of two circular motions in opposite directions: they circled the motionless earth westward with the universal diurnal motion, and in addition they simultaneously moved slowly eastward.

The planets moved eastward at different rates. For example, the moon took about a month to return to the same position relative to the fixed stars; the sun took one year; Mars about two years; and Saturn about twenty-nine years. Thus, the planets moved not only relative to the earth and the fixed stars, but also relative to each other; each planet had its own distinctive motion, besides the universal diurnal motion. Since the westward diurnal motion was common to all, when one spoke of planetary motions one usually referred to the distinctive individual motions of the planets. Note that, while all the individual planetary motions were eastward, this direction was opposite to that of the diurnal rotation, which was westward.

The planetary motion of the moon, which took about a month, was the most readily observable one since it was connected with the cycle of its phases; a full moon is easily seen and the period from one full moon to the next is an obvious unit of time that can be used as the basis for a calendar. The planetary motion of the sun was also easy to observe since it is related to the cycle of the seasons of the year; hence, it was called the *annual motion*. Because of its crucial importance, I shall discuss it in some detail.

Everyone can easily observe that in the course of a year the sun slowly moves in a north-south direction. In the northern hemisphere, the key observations are as follows. Sometimes it rises near due east and sets near due west, which is to say that it is seen on the celestial equator; this happens around March 21, which is the time of the *vernal equinox*; it also occurs around September 23, the time of the *autumnal equinox*. Sometimes it rises and sets about 23.5 degrees north of due east and due west (namely, north of the celestial equator); this happens around June 22, the time of the *summer solstice*. Sometimes it rises and sets about 23.5 degrees south of due east and due west (south of the celestial equator); this occurs around December 22, the time of the *winter solstice*. One can also observe from a given location on the earth's surface the elevation above the horizon of the sun at noon; in the course of a year this elevation changes daily and ranges about 47 degrees, being highest around June 22 and lowest around December 22.

This annual northward and southward motion of the sun indicates that its position relative to the fixed stars changes along a north and south direction since, as stated earlier, the fixed stars remain at a constant distance from the celestial equator. In short, the declination of the sun changes by about 47 degrees during a year, while the declination of a fixed star never changes; so this north-south motion of the sun is part of its wandering among the fixed stars.

Though this apparent solar motion was the one most easily observed, it was not exactly identical to its planetary motion mentioned earlier; for the latter was eastward, whereas the former was northward and southward. The two were related as follows. The sun's eastward revolution in its planetary orbit did not take place in the plane of the celestial equator but in a plane inclined to it by 23.5 degrees. The point was that the sun's motion among the fixed stars was not *exactly* eastward, but *mostly* eastward; its trajectory was slanted north and south. The sun moved eastward and southward for six months, and eastward and northward for the other six months. This can be made clear by means of a diagram, but before explaining it, let us mention a simple kind of observation to detect the sun's eastward motion.

The difficulty in observing the sun's eastward motion among the fixed stars stems from the fact that they cannot be seen when it is visible. What one can do is to observe some star located near the celestial equator and rising in the east just after the sun sets in the west; this means that the sun and star are diametrically opposed, or about 180 degrees apart. Observe the position of the same star just after sunset about a month later; it will be seen to be not just rising, but high in the sky and about 30 degrees west of its previous position; that means that the sun is now only about 150 degrees away, which is to say that sun has moved eastward about 30 degrees closer to the star. About six months after the first observation, the star will appear and immediately set in the west just after sunset. Twelve months later, the star will again rise in the east when the sun sets in the west.[19]

The planetary motion of the sun may be pictured as in figure 1. Imagine a large sphere surrounding a small one at its center, and let the small sphere represent the earth and the large one the stellar sphere. On the stellar sphere, picture a great circle lying in an horizontal plane to represent the celestial equator, and also a vertical line perpendicular to the equatorial plane and going through the center to represent the axis of diurnal rotation; this axis intersects the stellar sphere at two points, the north celestial pole (N) and the south celestial pole (S). Now imagine looking at the large sphere from the north celestial pole, and picture the large sphere rotating clockwise around the motionless small central sphere to represent the westward diurnal rotation of the stellar sphere around the earth. Next, imagine a great circle on the stellar sphere in a plane cutting the equatorial one at an angle of 23.5 degrees, to represent the sun's geocentric orbit projected onto the stellar sphere; in accordance with standard terminology, let us use the term *ecliptic* to refer to this actual orbit, or the corresponding great circle on the stellar sphere, or the plane on which they both lie. The intersection of the ecliptic and the equator on the stellar sphere defines two special points, called the vernal equinox (VE) and the autumnal equinox (AE); and halfway

19. This example is adapted from Galileo, *Cosmography*, in Favaro (2:214).

around the ecliptic between the equinoxes are two other special points, the summer solstice (SS) at the northern end, and the winter solstice (WS) at the southern end; these four points thus divide the ecliptic circle into four equal quadrants. Now, imagine the sun moving counterclockwise around the ecliptic at a rate that makes it traverse the whole circumference in one year; then the sun will be at VE around March 21, at SS around June 22, at AE around September 23, and at WS around December 22.

Let us now combine the clockwise rotation of the whole stellar sphere with the counterclockwise revolution of the sun along the ecliptic. The result is that the sun in reality moved in a spiral path which in one year looped clockwise around the earth about 365 times (days of the year), but which in any one day corresponded almost but not quite to one of the parallels on the stellar sphere. I say "almost but not quite" first because the parallel circle was not completely traversed by the sun, but fell short by about one degree (1/360 of a circle, which approximately equals 1/365 of a year); and second because the end of the daily path rises northward or drops southward relative to the beginning of the same daily path by one-quarter of a degree on the average (23.5 degrees every 3 months, or 23.5 degrees every 90 days).

The ecliptic was important not only because it represented the yearly eastward path of the sun among the stars, but also because it was used to define a frame of reference, distinct from the equatorial one mentioned earlier. For example, one could draw a line perpendicular to the center of the ecliptic (called the axis of the ecliptic); one could then speak of the poles of the ecliptic as the points where its axis intersected the celestial sphere; one could define the position of a star in terms of its angular distance from the ecliptic toward one of the poles; and one could also plot the position of a body in terms of east-west position along the ecliptic.

This ecliptic frame of reference was especially important for the other six planets because they are never seen to wander much away from the ecliptic; that is, planets are always observed to be somewhere inside a narrow belt extending 8 degrees above and below the ecliptic. This was the result of the fact that the individual circular paths of the planets took place in planes which, while not identical with the ecliptic, intersected it at small angles no larger than 8 degrees. This narrow belt on the stellar sphere along which the planets revolved was called the *zodiac*. It was subdivided into 12 equal parts of 30 degrees each, and each part happened to be the location of a group of stars which seemed to be arranged into a distinct pattern. These twelve patterns were the constellations of the zodiac and were named: Aquarius, Pisces, Aries, Taurus, Gemini, Cancer, Leo, Virgo, Libra, Scorpio, Sagittarius, and Capricorn. The sun, moon, and other planets were at all times found somewhere in one of these constellations, and they moved from one constellation to the next in the order just listed. This order corresponded to what we have called an eastward direction (from the viewpoint of terrestrial observation) or counterclockwise (in connection with the pictorial diagram just described); the key point, however, was that the order of the signs of the zodiac was a direction of motion opposite to that of the diurnal rotation.

When projected onto the stellar sphere, the eastward motion of the planets could be described in terms of great circles on the surface of that sphere, all of which were within the zodiac and intersected one another at small angles. But the planets were not believed to be attached to the stellar sphere like the fixed stars; unlike the fixed stars, the planets were not regarded to be equidistant from the earth. The fact that the planets appeared to move relative to the fixed stars and that this motion took place at different rates implied that each planet was attached to its own sphere that rotated eastward at its own rate while being carried westward daily by the diurnal rotation of the stellar sphere. There was no direct way to measure the sizes of the various planetary spheres or orbits (namely, the distances of the various planets from the earth), but the relative determination was done on the basis of the length of time required for a given planet to complete one circular journey among the stars. The principle used was that the bigger ones of these nested planetary spheres rotated at slower rates, and the smaller ones at faster rates; that is, the bigger the orbit, the slower the period of revolution. This principle was combined with the observation that the periods of revolution ranged from one month for the moon to one year for the sun and twenty-nine years for Saturn. The result was that in order of increasing distance from the earth, the planets were most commonly arranged as follows: moon, Mercury, Venus, sun, Mars, Jupiter, and Saturn. Thus, as mentioned earlier, between the stellar sphere and the earth, there were seven other nested spheres, whose rotation carried the corresponding planets in their own individual eastward orbits, while they were all being carried in a westward daily whirl by the diurnal rotation of the stellar sphere.

One last topic about the planets must now be discussed to complete but also to complicate this picture of the geocentric universe. Careful observation revealed that no planet moved at a uniform rate in its orbit, but that its speed appeared to vary. Moreover, though the sun and moon always moved eastward in their planetary revolutions, periodically the other five planets were seen to slow down, stop, and reverse course and briefly move westward relative to the fixed stars; this reversed movement was called *retrogression* or *retrograde planetary motion*. Finally, during retrogression, the planets appeared brighter, as if they were nearer the earth than at other times. These observations meant that a planet could not be simply attached to a rotating heavenly sphere, for in that case neither the distance nor the direction of revolution nor the speed should change. The device most commonly used to explain retrograde motion and variation in brightness and speed was a mechanism consisting of *deferents* and *epicycles*. To understand this, it is best at first to disregard the nested spheres.

A deferent was defined as a geocentric circle whose circumference (ABCD, in fig. 2) rotated around the earth (E). An epicycle was defined as a circle (FGHI) whose center (A) lay on the circumference of the deferent, and whose circumference rotated in the same direction as the deferent. The planet was located on the circumference of the epicycle. Thus, when the rotation of the epicycle carried the planet on the far side (F) of the epicycle from the earth, its distance was the sum of the radii of the deferent and the epicycle; whereas when the epicyclic rotation carried the planet on the near side (H) of the epicycle from the earth, its distance was the difference between the two radii. Thus, in its geocentric revolution, the distance of the planet from the earth changed by an amount equal to the diameter of the epicycle. This difference accounted for the variation in brightness.

Moreover, the planet's motion was the result of its motion along the epicycle and the motion of the center of the epicycle along the deferent. Thus, when the planet was on the far side (F), its speed was the sum of the deferent speed and the epicycle speed; then its speed was faster than its average speed. But when the planet was on the near side (H) of the deferent, its speed was the difference between the two; in this case, if the epicycle speed was greater than that of the deferent, the planet appeared to move backwards (clockwise or westward). Retrograde planetary motion then resulted.

For each planet, the relative sizes of deferent and epicycle and their relative rates of rotation could be adjusted so that their combination yielded mathematically the observed details about retrogression and changes in brightness and speed. For example, if the planet was observed to retrogress twice while revolving through its complete orbit once, then the epicycle was assumed to rotate twice as fast as the deferent; this yielded a path that in reality was looped (fig. 2); but from the earth (E)

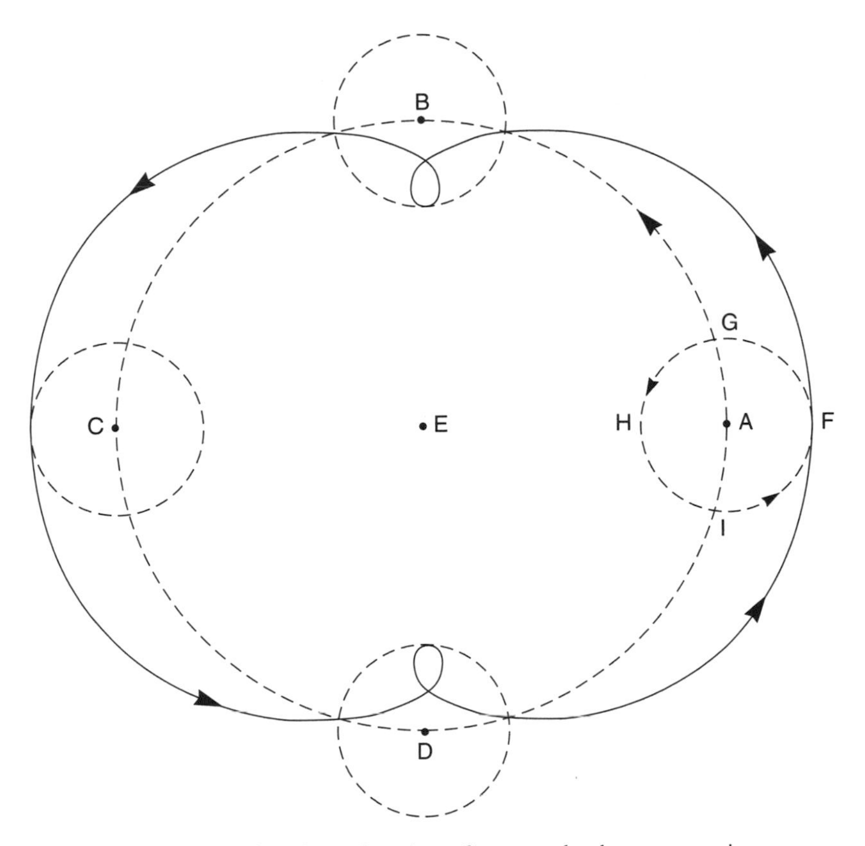


Figure 2. Ptolemaic explanation of retrograde planetary motion

the loop was not seen, and instead the planet would appear brighter and retrogressing near B and D.

This framework of deferents and epicycles could also be combined with the framework of nested spheres mentioned earlier. One could conceive a planetary sphere not as a mere spherical surface on which the planet was attached, but as a spherical shell of considerable thickness, having an inner and an outer surface; the thickness of the shell could then be adjusted so that the inner surface of the sphere corresponded to the minimum distance between the planet and the earth, and the outer surface to the maximum distance; and within the shell, the planet could perform its epicyclic motion while the whole shell rotated around the earth.

The framework of deferents and epicycles was a very powerful instrument for the analysis of planetary motion. There was much more that an astronomer could do besides adjusting the relative sizes and speeds of a deferent and its epicycle. For example, one could add a second epicycle on the first epicycle; one could make the center of the deferent different from the center of the earth, in which case the deferent was called an *eccentric*; and one could even make the center of the deferent move in some way, perhaps in a small circle around the earth's center. For many centuries before Copernicus, such calculations, adjustments, and refinements involving deferents, epicycles, and eccentrics constituted the primary theoretical and mathematical task of planetary astronomy. This enhanced the power of the theory, but it also rendered the whole system increasingly more complicated. Moreover, the relationship between the framework of deferents, epicycles, and eccentrics and the system of nested spheres became increasingly unclear. However, the geostatic-geocentric system of Aristotle, as elaborated by the Ptolemaic system of deferents, epicycles, and eccentrics yielded plausible explanations and useful predictions; in short, it worked. For about two thousand years no one was able to devise anything better. All this changed with Copernicus.

## 3. The Copernican Controversy

3.1. In 1543 Copernicus published his epoch-making book, *On the Revolutions of the Heavenly Spheres*. In it he elaborated the details of a theory that may be sketched as follows.

The earth was still spherical and the universe was still finite and spherical; the fixed stars were still attached to the stellar sphere and equidistant from the center. But the stellar sphere was motionless and did not revolve around the earth with westward diurnal rotation; instead, the diurnal rotation belonged to the earth, though its direction was eastward in order to result in the observational appearance of the whole universe rotating westward. To stress this feature of the earth's rotation, the Copernican worldview may be labeled *geokinetic*.

The earth was given a second motion, an orbital revolution around the sun with a period of one year, and also in an eastward direction. That is, the annual motion was shifted from the sun to the earth (with the direction remaining unchanged), thus making the earth a planet rather than the sun. This terrestrial orbital revolution meant that the earth was located off center, the center being instead the sun; to stress this feature, the Copernican worldview may be labeled *heliocentric*.

The moon remained a body which circles the earth eastward once a month. The other five planets continued to be planets, but their orbits were centered on the sun rather than the earth. Around the sun there thus revolved six planets in the order: Mercury, Venus, earth, Mars, Jupiter, and Saturn.

What Copernicus did was to update an idea which had been advanced in various forms by the Pythagoreans, by Aristarchus, and by other astronomers in ancient Greece, but had been almost universally rejected; that is, the idea that the earth moves by rotating on its own axis daily, and by revolving around the sun once a year. In a sense, Copernicus's accomplishment was to give a *new* argument in support of this *old* idea which had been considered and rejected earlier. His theory was not primarily based on new observational evidence, but was essentially a novel and detailed reinterpretation of available data. He demonstrated that the *known* details about the motions of the heavenly bodies (especially the planets) could be explained *more simply* and *more coherently* if the sun rather than the earth is assumed to be at the center, and the earth is taken to be the third planet circling the sun yearly and spinning daily on its own axis.

For example, there are thousands fewer moving parts in the geokinetic system since the apparent daily motion of all heavenly bodies around the earth is explained by the earth's axial rotation, and thus there is only one thing moving daily (the earth), rather than thousands of stars. Thus, insofar as simplicity depends on the number of moving bodies, the geokinetic system is simpler than the geostatic.

A similar point can be made with regard to the number of directions of motion. Fewer are needed in the Copernican than in the Ptolemaic system: in the geostatic system there are *two* opposite directions, but in the geokinetic system all bodies rotate or revolve in the same direction. In the geostatic system, while all the heavenly bodies revolved around the earth with the diurnal motion from *east to west*, the seven planets (moon, Mercury, Venus, sun, Mars, Jupiter, and Saturn) also simultaneously revolved around it from *west to east*, each in a different period of time. But in the geokinetic system there is only one direction of motion since, for example, if the apparent diurnal motion from east to west is explained by attributing to the earth an axial rotation, then the direction of the latter has to be reversed (west to east); whereas, if the apparent annual motion of the sun from west to east is explained by attributing to the earth an orbital revolution around the sun, then the same direction has to be retained.[20]

20. For more details, see Kuhn (1957, 160–65) and the notes to selection 6, p. [143] and selection 12, p. [384].

A third reason for the greater simplicity of the geokinetic system was that it had a single uniform pattern in the relationship between size of orbit and period of revolution, namely, the bigger the orbit, the slower the period of revolution. But in the geostatic system this pattern was only partially valid because, although the planetary motions did occur in accordance with it, the diurnal rotation of the stellar sphere broke the uniform pattern insofar as it was the largest sphere and yet revolved at the fastest rate.

With regard to explanatory coherence, this concept means the ability to explain many phenomena in detail by means of one's basic principles without having to add artificial and ad hoc assumptions. For Copernicus, the basic principles referred to the earth's motion, while the explained phenomena were primarily the various known facts about the motions and the orbits of the planets. But in the geostatic system, the thesis of a motionless central earth had to be combined with a whole series of unrelated assumptions in order to explain what is observed to happen.[21]

The best example of explanatory coherence is the Copernican explanation of retrograde planetary motion and of planetary variation in brightness. These phenomena are explained without the Ptolemaic ad hoc postulation and construction of epicycles, as needed to fit the observations. Instead, when the earth and another planet reach points in their orbits which are on the same side of the sun and are thus at the minimum distance from one another, their different speeds make the other planet seem to move backward (westward). If the other planet is a superior one (namely, one with an orbit larger than the earth's), this happens when the planet's apparent position on the celestial sphere is opposite to that of the sun; the earth's faster eastward motion leaves behind the other planet, which thus appears to move westward relative to the fixed stars. For example, in figure 3, while the earth moves through points B, C, D, E, F, G, H, I, K, L, and M along the smaller orbit, the superior planet moves along its bigger orbit through a corresponding set of points comprising a shorter distance due to its slower speed; but the apparent position of the planet against the background of the fixed stars changes in the order P, Q, R, S, T, U, V, W, X, Y, and Z.[22] If the other

21. For more details on explanatory coherence, see Lakatos and Zahar (1975, 368–81), Millman (1976), and Thomason (1992), but note that their terminology is different.

22. This diagram is adapted from a passage in the *Dialogue* omitted from my selections; cf. Favaro (7:371) and Galileo (1967, 343).

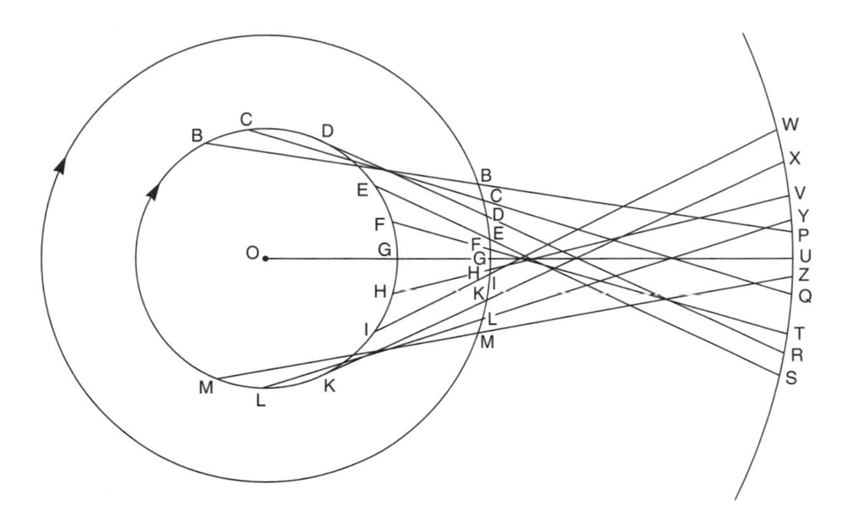


Figure 3. Copernican explanation of retrograde planetary motion

planet is an inferior one (namely, one with a smaller orbit), the phenomenon occurs when its apparent position on the celestial sphere is near that of the sun; the earth's slower motion enables the other planet to overtake the earth relative to the fixed stars, thus generating the appearance that the planet is moving in a direction opposite to that of the sun; since the latter always appears to move eastward relative to the fixed stars, the planet appears to move westward.

Despite these advantages of the geokinetic theory from the points of view of simplicity and explanatory coherence, as a *proof* of the earth's motion, Copernicus's argument was far from conclusive. Notice first that his argument is a hypothetical one. That is, it is based on the claim that *if* the earth were in motion *then* the observed phenomena would result; but from this it does not follow with logical necessity that the earth is in motion; all we would be entitled to infer is that the earth's motion offers an explanation of observed facts. Given the greater simplicity and coherence just mentioned, we could add that the earth's motion offered a simpler and more coherent explanation of heavenly phenomena. This does provide *two reasons* for preferring the geokinetic idea, but they are not decisive reasons. They would be decisive only in the absence of reasons for rejecting the idea. In short, one has to look at counterarguments, and there were plenty of them.

3.2. Almost all these objections are mentioned, stated, analyzed, criticized, or otherwise discussed in the *Dialogue*, many in our selections; there is no point repeating them here, except for some preliminary highlights. These arguments can be classified into various groups, depending on the branch of learning or type of principle from which they stemmed.

Let us begin with the viewpoint of physics, mechanics, or the science of how bodies move. It was argued that the earth's axial rotation and orbital revolution were physically impossible because they contradicted the principle that the natural motion of terrestrial bodies (consisting of the elements earth and water) is straight toward the center of the universe (selection 2). It was also argued that if the earth rotated, bodies could not fall vertically, as they are seen to do (selection 8). Another objection claimed that axial rotation would scatter terrestrial bodies off the earth's surface toward the heavens, which obviously is not observed (selection 9).

From the standpoint of cosmology, the main objection was that if the earth were the third planet revolving around the sun, then it would share with the other planets many other physical properties; but this contradicted the earth-heaven dichotomy (see selection 3).

Astronomically speaking, it could be objected (selection 11) that if Copernicanism were correct, then several consequences would follow in regard to the planets: the apparent diameter of Mars would undergo periodic changes by a factor of eight; the apparent diameter of Venus would change periodically by a factor of six; Venus would also exhibit the full cycle of phases similar to those of the moon; but none of these were observed (until the telescope). It would also follow that the moon, which clearly revolves in a monthly geocentric orbit, would be left behind as the earth moved in its annual motion. Moreover, the earth's annual motion would imply annual changes in the apparent brightness, diameter, and position of the fixed stars; and these were not observed, even with Galileo's telescope (selections 13 and 14).

The earth's motion was also a problem from the viewpoint of epistemology, the branch of philosophy that studies the nature of knowledge. To understand this problem, note that Copernicus did not claim that he could either feel, see, or otherwise perceive the earth's motion by means of the senses. Like everyone else's, Copernicus's senses told him that the earth is at rest. Thus, if his theory were true, then the human senses would be lying to us. But it was regarded as absurd that the senses should deceive us about such a basic phenomenon as the state of rest or motion of the terrestrial globe on which we live. That is, the geokinetic theory seemed to be in flat contradiction with direct sense experience, and to violate the fundamental methodological principle claiming that under normal conditions the senses provide us with an access to reality (see selection 10).

Finally, there were theological and religious objections.[23] One of these, labeled the *biblical objection*, appealed to the authority of the Bible. It claimed that the idea of the earth moving is heretical because it contradicts many biblical passages stating or implying that the earth stands still. For example, Psalm 104:5 says that the Lord "laid the foundations of the earth, that it should not be removed for ever"; and this seems to say explicitly that the earth is motionless. Other passages were less explicit, but they seemed to attribute motion to the sun and so to presuppose the geostatic system. For example, Ecclesiastes 1:5 states that "the sun also riseth, and the sun goeth down, and hasteth to the place where he ariseth."

The biblical objection had greater appeal to those (like Protestants) who took a literal interpretation of the Bible seriously. But for those (like Catholics) less inclined in this direction, the same conclusion could be reinforced by appeal to the *consensus of Church Fathers*; these were the theologians and churchmen who had played an influential role in the establishment of Christianity. The argument claimed that Church Fathers were unanimous in interpreting relevant biblical passages in accordance with the geostatic view; thus, the geostatic system is binding on all believers, and to claim otherwise (as Copernicans did) is heretical.

A third theological objection was based crucially on the idea that God is all-powerful, and it may be labeled the *divine omnipotence argument*. One of its most famous proponents was Pope Urban VIII, during whose reign Galileo was tried and condemned in 1633. A version of the argument seemed to claim that since God is all-powerful, he could have created any one of a number of worlds, for example, one in which the earth is motionless; therefore, regardless of how much evidence there is supporting the earth's motion, one can never assert that this must be so, for that would be to want to limit God's power to do otherwise. Another version seemed to argue that divine omnipotence implies that God

23. Cf. selections 1, 12, and 16. In the *Dialogue* Galileo pays lip service to these arguments, for reasons to be explained below; his serious critical analysis is in the "Letter to Castelli," in the "Letter to the Grand Duchess Christina," and in "Considerations on the Copernican Opinion" (Finocchiaro 1989, 49–54 and 70–118). Other statements of such objections are found in Cardinal Bellarmine's letter to Foscarini (Finocchiaro 1989, 67–69) and in Oregius (1629, 194–95), quoted in Sosio (1970, 548–49n.1) and translated in Finocchiaro (1980, 10).

could have created a world in which the evidence suggests a moving earth despite its being motionless.

In summary, the idea updated by Copernicus was vulnerable to many counterarguments and much counterevidence. The earth's motion seemed epistemologically absurd because it flatly contradicted direct sense experience, and thus undermined the normal procedure in the search for truth. It seemed empirically untrue because it had astronomical consequences that were not seen to happen. It seemed physically impossible because it had consequences that contradicted the most incontrovertible mechanical phenomena, and because it directly violated many basic principles of the available physics. And it seemed religiously heretical or suspect because it conflicted with the words of the Bible and with the biblical interpretations of the Church Fathers and could be taken to undermine belief in an omnipotent God.

Copernicus was aware of many of these difficulties.[24] He realized that his novel argument did not conclusively prove the earth's motion, and that there were many counterarguments of apparently greater strength. I believe that these were the main reasons why he delayed publication of his book until he was almost on his death bed.

3.3. In light of the many objections, a common response to Copernicanism was to regard the earth's motion as a mere instrument of calculation and prediction rather than as a description of physical reality. This view may be called the *instrumentalist* interpretation of Copernicanism and was popularized by an anonymous foreword preceding Copernicus's own preface in the printed book. This foreword was written and inserted without his approval or knowledge by one of the editors supervising the book's publication—Andreas Osiander. It is unlikely that Copernicus would have endorsed this interpretation since it is clear from the book that although he was aware of the difficulties, he treated the earth's motion as a description of physical reality and not as a mere instrument of calculation and prediction; in short, Copernicus subscribed to a *realist* interpretation of the geokinetic theory, in accordance with the doctrine of epistemological realism.

The unauthorized foreword soon became public knowledge among experts, but many scholars adopted the instrumentalist interpretation as the only way out of the difficulties.[25]

24. See Book One of his great work, Copernicus (1992).

25. For more details on the instrumentalist response to Copernicus and related issues, see Westman (1972; 1975a, 285–345; 1975b; 1994).

One different response was that of Danish astronomer Tycho Brahe. He decided to collect new data by means of systematic naked-eye observations and the construction of new instruments. The scope, range, accuracy, and precision of his observations were unprecedented. On the basis of his observational data, Tycho constructed a new theory different from both the Ptolemaic and the Copernican ones. In the Tychonic system, the earth was still motionless at the center of the universe; the stellar sphere still had the westward diurnal motion around the earth; and the sun still had the eastward annual motion around the earth. But the planets revolved in orbits centered at the sun, so that the system was to that extent heliocentric; however, the sun carried the whole solar system around the motionless earth. Moreover, the nested planetary spheres were abolished since they no longer fit properly in the new arrangement; for example, some of the orbits of the heavenly bodies intersected, so that the spheres would have had to interpenetrate one another.

One of Tycho's assistants inherited his data and analyzed them in a deeper and more systematic and sophisticated manner. He was the German mathematician and astronomer Johannes Kepler. Rejecting Tycho's compromise, Kepler was firmly committed to Copernicanism and spent his life analyzing Tycho's data in ways that would support Copernicanism. He succeeded in this effort, although it also led to a revision of Copernicanism. In fact, Kepler discovered that the planets revolve around the sun in elliptical rather than circular orbits, with the sun located at one of the two foci of these ellipses. Unfortunately, Galileo never did pay the proper attention to Kepler's writings and so either was unaware of or neglected Kepler's ellipses.

3.4. Let us now turn to Galileo's attitude toward Copernicus's theory.[26] Galileo began university teaching in 1589, almost fifty years after Copernicus's death. In his official position as professor of mathematics, his duties included teaching astronomy and physics as well as mathematics. Although acquainted with the Copernican theory, he did not regard it as sufficiently well established to teach it in his courses; instead he covered traditional geostatic astronomy. Nor was he directly pursuing the geokinetic theory in his research. Rather his research consisted of investigations into the laws in accordance with which bodies move. Here his work was original and revolutionary, for he was critical of the traditional physics, and was attempting to construct a new theory of how

26. For an elaboration and documentation of this interpretation, see Finocchiaro (1988b); this also corresponds in large measure to Drake (1978).

bodies move. It did not take him long to realize that the physics he was building was very much in line with the geokinetic theory, in the sense that what he was discovering about the motion of bodies in general made it possible for the earth to move and rendered unlikely its rest at the center of the universe. In short, he soon realized that his physical research had important consequences in astronomy, namely, to strengthen the Copernican theory by removing the physical and mechanical objections to it and providing some new arguments in its favor.

There is evidence that he was aware of this reinforcement but was still dissatisfied with Copernicanism. For the other objections were still there, especially the empirical astronomical difficulties. It was only the invention of the telescope and the astronomical discoveries it made possible that removed most of them and paved the way for the removal of the others.

The invention of the telescope and Galileo's role in it is a fascinating story in its own right but is too long and complex to be told here.[27] Suffice it to say the following. This instrument was first invented by others in 1608, but in 1609 Galileo was able to improve its quality and magnification sufficiently to produce an astronomically useful instrument that could not be duplicated by others for quite some time. With this instrument, in the next few years he made several startling discoveries: for example, that like the earth, the moon has a rough surface covered with mountains and valleys and appears to be made of the same rocky, opaque, and nonluminous substance; that the planet Jupiter has four satellites, and thus constitutes a miniature planetary system; that Venus shows phases similar to those of the moon and changes its apparent diameter by a factor of six; that Mars shows changes in apparent diameter by a factor of eight and corresponding changes in brightness; and that there are dark spots on the surface of the sun, which appear and disappear at irregular intervals (like clouds on earth), but which move in such a way as to indicate that the sun rotates monthly on its own axis.

Some of these discoveries were published in 1610 in a book entitled *The Sidereal Messenger*; others were published in 1613 in a book entitled *Sunspot Letters*. *The Sidereal Messenger* allowed Galileo to leave the position of professor of mathematics at the University of Padua, which he had held for eighteen years. He went back to his native Tuscany to be

27. For a general account, see Drake (1957, 1–88; 1978, 134–76); for an account stressing the scientific and methodological issues, see Ronchi (1958); cf. Crombie (1967), Feyerabend (1975), King (1955), Rosen (1947), and Van Helden (1984; 1994).

under the patronage of its ruling grand duke and to devote full time to research and writing. Galileo requested that his official title should include the word philosopher as well as mathematician, his reason being that he claimed "to have spent more years studying philosophy than months studying pure mathematics."[28] His request was granted, and so for the rest of his life he held the position of Philosopher and Chief Mathematician to the Grand Duke of Tuscany.

The new telescopic evidence led Galileo to a reassessment of the status of Copernicanism, for it removed most of the empirical astronomical objections against the earth's motion and added new arguments in its favor. Thus, he now felt not only that the geokinetic theory was simpler and more coherent than the geostatic theory (as Copernicus had shown), not only that it was more physically and mechanically adequate (as he himself had been discovering in his twenty years of university research), but also that it was empirically superior in astronomy (as the telescope now revealed). But he had not yet published anything of his new physics, and the epistemological and especially the theological objections had not yet been dealt with; so the case in favor of the earth's motion was still not conclusive.

Thus, his new attitude toward the geokinetic theory was one of direct pursuit and tentative acceptance, by contrast with the indirect pursuit and contextually divided loyalty of the pre-telescopic situation.[29] But in Galileo's publications of this period we find yet no explicit acceptance of or committed belief in the earth's motion. We do find a more favorable attitude expressed in his private correspondence, and a stronger endorsement in the *Sunspot Letters* of 1613 than in *The Sidereal Messenger* of 1610. But that is as one would expect.

Besides realizing that the pro-Copernican arguments were still not absolutely conclusive, Galileo must have also perceived the potentially explosive character of the religious objections. In fact, at first he did not

28. Favaro (10:353). This is very important. I would want to stress the lesson regarding scientific methodology and the question of the relationship between science and philosophy. The assertion may also be partially true and may be an indirect admission of his studies of Aristotelian logic and epistemology in his early career; this point has been emphasized by Wallace (1984; 1992a; 1992b). It is also true, although Galileo did not include it in his request, that the social position of philosophers was higher than that of mathematicians, and so the request was part of his strategy for the social legitimation of the new science; this has been stressed by Biagioli (1993).

29. This also contrasts with the attitude of a true believer and of zealous commitment, which one finds attributed to Galileo in such works as Feyerabend (1975), Koestler (1959), and Langford (1966).

get involved despite the fact that his book of 1610 had been attacked by several authors on biblical grounds, among others. Eventually, however, he was dragged into the theological discussion. What happened was the following.

## 4. The Galileo Affair[30]

4.1. In the years following the publication of *The Sidereal Messenger* in 1610, as support for the geostatic theory continued to dwindle, conservatives began relying more and more heavily on biblical and theological arguments; this was done as much by philosophers and scientists as by theologians and churchmen. These discussions became so frequent and widespread that the ducal family must have started to wonder whether they had a heretic in their employment. Thus in December 1613 the Grand Duchess Dowager Christina confronted one of Galileo's friends and followers named Benedetto Castelli (1578–1643), who had succeeded him in the chair of mathematics at the University of Pisa; she presented him with the biblical objection to the earth's motion. This was done in an informal, gracious, and friendly manner, and as much out of genuine curiosity as out of worry. Castelli answered in a way that satisfied both the duchess and Galileo when Castelli informed him of the incident. But Galileo felt the need to write a very long letter to his former pupil containing a detailed refutation of the biblical objection; recall that this objection argued that the geokinetic theory must be wrong because many biblical passages state or imply that the geostatic theory is right.

In this letter, Galileo suggested that the objection has three fatal flaws. First, it attempts to prove a conclusion (the earth's rest) on the basis of a premise (the Bible's commitment to the geostatic system) which can only be ascertained with a knowledge of that conclusion in the first place; in fact, the interpretation of the Bible is a serious business, and normally the proper meaning of its statements about natural phenomena can be determined only after we know what is true in nature; thus, the business of biblical interpretation is dependent on physical investigation, and to base a controversial physical conclusion on the Bible

30. This account is adapted from Finocchiaro (1989), which also contains the essential documents. Other useful accounts and documents may be found in Biagioli (1993, 313–52), Blackwell (1991; 1994), Feldhay (1995), Langford (1966), Redondi (1987), and Santillana (1955a).

is to put the cart before the horse. Second, the biblical objection is a non sequitur, since the Bible is an authority only in matters of faith and morals, not in scientific ones, and thus its saying something about a natural phenomenon does not make it so, and therefore its statements do not constitute valid reasons for drawing corresponding scientific conclusions. Finally, it is questionable whether the earth's motion really contradicts the Bible, and an analysis of the Joshua passage shows that it cannot be easily interpreted in accordance with the geostatic theory, but that it accords better with the geokinetic view, especially as improved by Galileo's own discoveries; the biblical objection is therefore groundless, aside from its other faults.

Though unpublished, Galileo's letter to Castelli began circulating widely, and copies were made. Some copies came into the hands of traditionalists, who soon counterattacked. In December 1614, at a church in Florence, a Dominican friar named Caccini preached a Sunday sermon against mathematicians in general and Galileo in particular, on the grounds that their beliefs and practices contradicted the Bible and were thus heretical. In February 1615 another Dominican, named Lorini, filed a written complaint against Galileo with the Inquisition in Rome, enclosing his letter to Castelli as incriminating evidence. Then in March of the same year, Caccini made a personal appearance before the Roman Inquisition; in his deposition he charged Galileo with suspicion of heresy, based not only on the content of the letter to Castelli, but also on the book of *Sunspot Letters*, and on hearsay evidence of a general sort and of a more specific type, involving two other individuals. The Inquisition responded by ordering an examination of these two individuals and of the two mentioned writings.

In the meantime, Galileo was writing for advice and support to many friends and patrons who either were clergymen or had clerical connections. He had no way of knowing about the details of the Inquisition proceedings, which were a well-kept secret; but Caccini's original sermon had been public, and also he was able to learn about Lorini's initial complaint. Galileo also received the unexpected support of a Neapolitan clergyman named Paolo Antonio Foscarini (1580–1616), who published a book arguing in detail for the specific thesis that the earth's motion is compatible with the Bible. Finally, in December 1615, after a long delay due to illness, Galileo went to Rome of his own initiative to try to clear his name and prevent the condemnation of Copernicanism.

The results of the Inquisition investigations were as follows. The consultant who examined the letter to Castelli reported that in its essence it did not deviate from Catholic doctrine. The cross-examination of the two witnesses exonerated Galileo since the hearsay evidence of his utterance of heresies was found to be baseless. And the examination of his *Sunspot Letters* apparently failed to reveal any explicit assertion of the earth's motion or other heretical assertion. But the Inquisition felt it necessary to consult its experts for an opinion on the status of Copernicanism.

On 24 February 1616 a committee of eleven consultants reported unanimously that Copernicanism was philosophically and scientifically untenable, and theologically heretical. If we want to understand how this opinion came about, we must recall all the traditional scientific and philosophical arguments against the earth's motion; and we must view the judgment of heresy in the light of the biblical objection combined with Catholic Counter-Reformation rejection of new and individualistic interpretations of the Bible. But the Inquisition must have had some misgivings about this opinion, for it issued no formal condemnation. Instead two milder consequences followed.

One was to give Galileo a private warning to stop believing, supporting, and defending the truth of the earth's motion. The warning was conveyed to him within a few days by Cardinal Robert Bellarmine, the most influential and highly respected theologian and churchman of the time, with whom Galileo was on very good terms, despite their philosophical and scientific differences. The exact content, form, and circumstances of this warning are not completely known; but they are extremely complex and controversial. Moreover, as we shall see, the occurrence and propriety of the later Inquisition proceedings in 1633 hinge on the nature of this warning.

The other development was a public decree issued by the Congregation of the Index, the department of the Church in charge of book censorship, one of whose main tasks was to compile an index or list of prohibited books. In March 1616 this Congregation published a decree containing three main points. First, it prohibited completely and condemned Foscarini's book, the work that had tried to show that the earth's motion is compatible with the Bible. Second, it suspended Copernicus's book, pending correction and revision. Third, the decree ordered analogous censures for analogous books. Galileo was not mentioned at all.

The message in this decree of the Index was confusing enough; even more confusing were the circumstances of Bellarmine's private warning to Galileo (which I have not discussed). Moreover, at this time Galileo began receiving letters from friends in Venice and Pisa saying that there were rumors in those cities to the effect that he had been personally put on trial, condemned, forced to recant, and given appropriate penalties by the Inquisition. Having shown these letters to Cardinal Bellarmine, Galileo was able to convince him to write a brief and clear statement of what had happened and of how Galileo was affected. Thus in a document half a page long, the most authoritative churchman of his time declared the following: Galileo had been neither tried nor otherwise condemned, but rather he had been personally notified of the decree of the Index and of the fact that in view of this decree the truth of the earth's motion could be neither held nor defended nor supported.

With this certificate in his possession, Galileo left Rome soon thereafter. But before we, too, leave this first phase of the affair, let us recall the three different documents that will play a crucial role later: the private, oral warning given to Galileo by Cardinal Bellarmine in the name of the Inquisition; the public decree of the Index; and Bellarmine's certificate to Galileo. The propriety or impropriety of his subsequent behavior will depend on which one of these three items is stressed.

4.2. For the next several years Galileo did refrain from defending or explicitly discussing the geokinetic theory. The event that put an end to the interlude took place in 1623, when the old pope died and Cardinal Maffeo Barberini was elected Pope Urban VIII. Urban was a well-educated Florentine, and in 1616 he had been instrumental in preventing the outright condemnation of Copernicanism by the Inquisition. He was also a great admirer of Galileo and employed one of Galileo's closest acquaintances as personal secretary. Furthermore, at about this time, Galileo's book on the comets entitled *The Assayer* was being published in Rome, and so it was decided to dedicate the book to the new pope. Urban appreciated the gesture and liked the book very much. Finally, as soon as circumstances allowed, in the spring of 1624, Galileo went to Rome to pay his respects to the pontiff. Galileo stayed about six weeks and was warmly received by Church officials in general and the pope in particular, and the latter granted him weekly audiences.

The details of these six conversations are unknown. But there is evidence that Urban did not think Copernicanism was a heresy. Nor did he think it had been declared a heresy by the Church in 1616, for he interpreted the decree to mean that the earth's motion was a dangerous doctrine whose study and discussion required special care. He thought that the theory could never be proved necessarily true because the earth's motion is not directly perceivable, and so all its supporting arguments must be indirect ones in which observed phenomena are explained as effects of the earth's motion; but any such effects could always be produced by other causes, and such possibility cannot be denied because the denial would amount to wanting to limit God's power to do otherwise; as mentioned above, this was a version of the divine omnipotence objection.[31] This argument, together with his interpretation of the Index's decree of 1616, must have reinforced his liberal inclination that as long as one exercised the proper care, there was nothing wrong with the hypothetical discussion of Copernicanism.

At any rate, Galileo must have gotten some such impression during his six conversations with Urban, for upon his return to Florence he began working on a book. This was in part the work on the system of the world which he had conceived at the time of his first telescopic discoveries, but it now acquired a new form and new dimensions in view of all that he had learned and experienced since. After a number of delays in its writing, licensing, and printing the book was finally published in Florence in February 1632 with the title *Dialogue on the Two Chief World Systems, Ptolemaic and Copernican*.

The author had done many things to avoid trouble, to ensure compliance with the many restrictions under which he was operating, and to satisfy the various censors who issued him permissions to print. To emphasize the hypothetical character of the discussion, he had originally entitled it *Dialogue on the Tides* and structured it accordingly.[32] But the censors decided to try to make the book look like a vindication of the Index's decree of 1616; this corresponded partly to the pope's alleged wishes and partly to one of Galileo's several aims. The book's preface, whose compilation must be regarded as a joint effort by the author and the censors, claimed that the work was being published to prove to non-Catholics that Catholics knew all the arguments and evidence about the scientific issue, and thus their decision to believe in the geostatic theory was motivated by religious reasons, not by scientific ignorance. It went on to add that the scientific arguments seemed to favor the geokinetic

31. The argument is referred to in selection 1 and stated (without criticism) in selection 16; but some indirect discussion of it may be found in selection 4.

32. I realize, of course, that my interpretation here goes against the prevailing view that Galileo regarded the tidal argument as conclusive, that he therefore wanted to advertise this in the book's title, and that the Church did not want to endorse this argument; for an exposition of this view, see Shea (1972, 172–89). For some criticism, see Finocchiaro (1980, 16–18 and 76–78); cf. also Finocchiaro (1988b), Drake (1986b), and MacLachlan (1990).

theory, but that they were inconclusive, and thus the earth's motion remained a hypothesis.

Galileo also complied with the explicit request to end the book with a statement of the pope's favorite argument. Moreover, to make sure he would not be seen as holding, supporting, or defending the geokinetic thesis (which he had been forbidden to do), Galileo did two things. He wrote the book as a dialogue among three speakers: one defending the geostatic side, another taking the Copernican view, and the third being an uncommitted observer who listens to both sides and accepts the arguments that seem to survive critical scrutiny. And in many places throughout the book, usually at the end of a particular topic, the Copernican spokesman utters the qualification that the purpose of the discussion is to inform and enlighten, not to decide the issue, which is a task reserved for the proper authorities. Finally, it should be stressed that Galileo had obtained written permissions to print the book, first from the proper Church officials in Rome (when the plan was to publish it there), and then from the proper officials in Florence (when a number of external circumstances dictated that it be printed in the Tuscan capital).

Given all these precautions, how could anything go wrong? The book was well received in scientific circles, but rumors and complaints began circulating in Rome.

The most serious complaint involved a document found in the file of Inquisition proceedings of 1615–1616. It reads like a report of what took place when Cardinal Bellarmine, on behalf of the Inquisition, gave Galileo the private warning to abandon his geokinetic views. The cardinal had died in 1621; so he was no longer available to clarify matters. The document states that in February 1616 Galileo had not only been ordered to stop holding or defending the geokinetic thesis, but had also been prohibited from discussing it in any way whatsoever, either orally or in writing. The charge, then, was that his book of 1632 clearly violated this special injunction, since whatever else the book did and however else it might be described, it undeniably contained a discussion of the earth's motion. To be sure, the document does not bear Galileo's signature, and so it was of questionable legal validity; under different circumstances such a judicial technicality might have been taken seriously. But at the time the politics of the Thirty Years War had rendered the pope too vulnerable to defend Galileo; moreover, too many other difficulties were being raised about the book.

One of these was that the book only paid lip service to the stipulation about a hypothetical discussion, which represented Urban's compromise;

in reality, the book allegedly treated the earth's motion not as a hypothesis, but in a factual, unconditional, and realistic manner. This was a more or less legitimate complaint on the part of the pope, but the truth of the matter is that the concept of hypothesis was ambiguous and had not been sufficiently clarified in that historical context. By hypothetical treatment, Urban meant a discussion that would treat the earth's motion merely as an instrument of prediction and calculation, rather than as a potentially true description of reality; however, Galileo took a hypothesis to be an assumption about physical reality, which accounts for what is known, and which may be true, although it has not yet been proved to be true.

Third, there was the problem that the *Dialogue* was actually a defense of the geokinetic theory. Despite the dialogue form, despite the repeated disclaimers that no assertion of Copernicanism was being intended, despite the inconclusive character of the Copernican arguments, and despite the presentation of the anti-Copernican arguments, it was readily apparent that the geostatic arguments were being criticized and that the Copernican ones were being portrayed favorably; and to do this is one way of arguing in favor of a conclusion, which in turn is one way of defending a conclusion.

There were many other complaints. They were so numerous, and some were so serious that the pope might have been forced to take action even under normal circumstances. But Urban was himself in political trouble due to his behavior in the Thirty Years War. Thus in the summer of 1632 sales of the book were stopped, unsold copies were confiscated, and a special commission was appointed to investigate the matter. The pope did not immediately send the case to the Inquisition, but he took the unusual step of appointing a special commission first. This three-member panel issued its report in September 1632, and it listed as areas of concern all of the above mentioned problems. In fact, it is from the report that we learn about these complaints, which had been accumulating since the book's publication. In view of the report the pope felt he had no choice but to forward the case to the Inquisition. So Galileo was summoned to Rome to stand trial.

The entire autumn was taken up by attempts on the part of Galileo and the Tuscan government to prevent the inevitable. The Tuscan government got involved partly because of Galileo's position as philosopher and chief mathematician to the Grand Duke, partly because the book contained a dedication to the Grand Duke, and partly because the Grand Duke had been instrumental in getting the book finally printed in Florence. At the end of December the Inquisition sent Galileo an ultimatum: if he did not come to Rome of his own accord, they would send some officers to arrest him and bring him to Rome in chains. On 20 January 1633, after making a last will and testament, Galileo began the journey. When he arrived in Rome three weeks later, he was not placed under arrest by the Inquisition but was allowed to lodge at the Tuscan embassy; he was ordered not to socialize and to keep himself in seclusion until he was called for interrogations.

These were slow in coming; the first hearing was held on April 12. To various questions about the *Dialogue* and the events of 1616, Galileo replied as follows. He admitted receiving a warning from Cardinal Bellarmine in February 1616, and described this as an oral warning that the geokinetic theory could be neither held nor defended, but only discussed hypothetically. He denied having received a special injunction not to discuss the earth's motion in any way whatsoever, and he introduced Bellarmine's certificate as evidence. He also claimed that the book did not hold or defend the earth's motion, but rather showed that the arguments in its favor were not conclusive, and so it did not violate Bellarmine's warning.

This was a strong and practicable line of defense. In particular, just as the special injunction was news to Galileo, so Bellarmine's certificate must have surprised and disoriented the inquisitors. Thus it took another three weeks before they finally decided on the next step. In the meantime Galileo was detained at the headquarters of the Inquisition but allowed to lodge in the chief prosecutor's apartment. What the inquisitors finally decided was essentially what today we would call out-of-court plea bargaining. That is, they would not press the most serious charge (violation of the special injunction), but Galileo would have to plead guilty to the lesser charge of having inadvertently transgressed the order not to support or defend Copernicanism, in regard to which his defense was weakest.

The deal was worked out as follows. The Inquisition asked three consultants to determine whether Galileo's *Dialogue* held, supported, or defended the geokinetic theory; in separate reports all three concluded that the book clearly supported and defended the doctrine, and came close to holding it. Then the executive secretary of the Inquisition talked privately with Galileo to try to arrange the deal, and after lengthy discussions he succeeded. Galileo requested and obtained a few days to think of a dignified way of pleading guilty to the lesser charge. Thus, on April 30, the defendant appeared before the Inquisition for the second time and signed a deposition stating the following. Ever since the first hearing he had reflected about whether, without meaning to, he might have done anything wrong; it dawned on him to reread his book, which he had not done for the past three years since completing the manuscript. He was surprised by what he found, for the book gave the reader the impression that the author was defending the geokinetic theory, even though this had not been his intention. To explain how this could have happened, he attributed it to vanity, literary flamboyance, and an excessive desire to appear clever by making the weaker side look stronger. He was deeply sorry for this transgression and was ready to make amends.

After this deposition, Galileo was allowed to return to the Tuscan embassy. The trial might have ended here, but it did not conclude for another six weeks. Obviously the pope would have to approve the final disposition of the case. So a report summarizing the events from 1615 onward was compiled for him. Reading it did not resolve Urban's doubts about Galileo's intention, and so he directed that the defendant be interrogated under the verbal threat of torture, in order to determine his intention. The pope added that even if his intention were found to have been pure, Galileo had to make a public abjuration and had to be held under formal arrest at the pleasure of the Inquisition, and his book had to be banned.

On June 21 Galileo was subjected to the interrogation under the formal threat of torture. The result was favorable, in the sense that, even under such a formal threat, he denied any malicious intention, and showed his readiness to die rather than admit that. The following day, at a public ceremony in the convent of Santa Maria sopra Minerva in Rome, he was read the sentence and then recited the formal abjuration.

The sentence found Galileo "vehemently suspected of heresy";[33] it banned the book; and it condemned him to formal arrest. The phrase "vehement suspicion of heresy" was a technical legal term that meant more than it may convey to modern ears. In effect the Inquisition dealt with three types of religious crimes in descending order of seriousness: formal heresy, vehement suspicion of heresy, and slight suspicion of heresy. So Galileo was convicted of the second most serious offense. Two distinct heresies were mentioned: the cosmological thesis that the earth rotates daily on its axis and orbits the sun yearly, and the methodological principle that one may believe and defend as probable a thesis contrary to the Bible.[34]

33. Finocchiaro (1989, 291).

34. Insufficient attention has been paid to the double character of the heresy attributed to Galileo and to the probabilistic character of the proscribed methodological principle; for more details see section 5.4 of this introduction, and cf. Finocchiaro (1986).

It took another six months before Galileo was allowed to return home, to remain under arrest in his own house. He was first sent back to the Tuscan embassy, where he stayed for another ten days; then for about five months he was under house arrest at the residence of the archbishop of Siena, who proved to be a very congenial and sympathetic host.

## 5. Historical Aftermath and Philosophical Prelude

5.1. The Inquisition's sentence of 1633 ended the Galileo Affair insofar as it extinguished the original controversy, but it ignited a new dispute that continues to our own time. This second Galileo Affair (so to speak) hinges on a number of questions *about* the original one: what exactly happened at various junctures in the period 1613–1633; what really were the issues; what were the causes; who was to blame; could the scandal have been avoided; what useful lessons, if any, can be derived. This is not the proper place to relate the story of this second affair, even though its fascination rivals that of the original one; but, a few details are in order.

First, before his death in 1642, Galileo himself won a curious and ironic kind of revenge. For when his condition of house arrest forced him to stay away from all controversy, he went back to his studies of mechanics and the laws of motion; he had carried out the relevant work earlier during his university career, but that research had been interrupted by his telescopic discoveries. The suffering he had experienced in his dealings with the Inquisition renewed the determination of his indomitable spirit. Thus, in 1638 he published in Holland the *Two New Sciences*, a work on the foundations of mechanics and engineering and his most important book from a purely scientific point of view. Although the forbidden topic of the earth's motion was not so much as mentioned, the laws of motion elaborated therein helped to provide later (by way of Isaac Newton) a more effective proof of that phenomenon than any he himself had been able to formulate. Here is one of those ironies of history, where the temporary victor of a particular battle creates conditions that pave the way for his eventually losing the war.

Moreover, it was inevitable that Galileo's contemporaries would start spreading rumors and constructing myths, besides formulating interpretations and evaluations. For example, in the *Areopagitica* John Milton recalled his meeting with him in Florence by saying that "there it was that

I found the famous Galileo, grown old a prisoner to the Inquisition, for thinking in astronomy otherwise than the Franciscan and Dominican licensers thought."

Upon his death, Galileo's body was buried at the Church of Santa Croce in Florence in a grave without inscription. In 1734 the Inquisition agreed to a Florentine request to erect a mausoleum for him in that church. In 1744 the Inquisition allowed the *Dialogue* to be published in an edition of his collected works, although it was accompanied by several qualifications. In 1822 the Inquisition decided to allow the publication of books discussing the earth's motion in accordance with modern science, and so the 1835 edition of the Index of Prohibited Books for the first time omitted the *Dialogue* from the list.

The latest significant twist was added in November 1979 by the first compatriot of Copernicus to ever be pope and the first non-Italian to be so since the Galileo Affair. The occasion was a speech to the Pontifical Academy of Sciences commemorating the birth of Albert Einstein (1879–1955), the German-born physicist widely regarded as the greatest scientist of the twentieth century. John Paul II seemed to admit that the Church had committed an injustice when he stated that Galileo "had to suffer a great deal—we cannot conceal the fact—at the hands of men and organisms of the Church."[35] Then the pope boldly tried to reverse the traditional view which sees a conflict between science and religion; he argued that "in this affair the agreements between religion and science are more numerous and above all more important than the incomprehensions which led to the bitter and painful conflict that continued in the course of the following centuries."[36]

It is doubtful whether these words will put an end to the second Galileo Affair, but they must put an end to our discussion of it. For it is time to return to the connection of Galileo's book to the other two historical events mentioned earlier.

Regarding the Copernican Revolution, the year Galileo died, Isaac Newton was born. Newton was able to use the discoveries and ideas of Copernicus, Kepler, Galileo, Descartes, Huygens, and others. In 1687 the transition from a geostatic to a geokinetic worldview reached a climax when Newton published the *Mathematical Principles of Natural Philosophy*. In it the Copernican worldview was refined, revised, and elaborated in terms of four basic principles which became the foundation of

35. John Paul II (1979).

36. Ibid. This theme has been elaborated in Coyne et al. (1985).

modern science: the law of inertia, the law of force and acceleration, the law of action and reaction, and the law of universal gravitation. This provided a theoretical framework which explained a great number and variety of old and new facts and implied that the earth moves with axial rotation and orbital revolution. Thus, the basic tenets of Copernicanism were established beyond any reasonable doubt.

Even so, Newton did not really provide a *direct* experimental proof of the earth's motion. This came later with such observations as the aberration of starlight by James Bradley in 1729, the annual stellar parallax by Friedrich W. Bessel in 1838, and Jean Foucault's pendulum in 1851. However, I would argue that these more direct proofs merely confirmed what had already been established by Newton; they make sense only within his system, and anyone who was not convinced by it would not have been convinced by them. Newton's *Mathematical Principles* is thus the proper climax of the Copernican Revolution.

Regarding the Scientific Revolution, recall that when we use this label to refer to the developments from Copernicus to Newton, we are stressing that the transition from a geostatic to a geokinetic view involved the discovery of some of the most fundamental laws of nature. I have already mentioned inertia, force, and action-and-reaction, which are now called Newton's laws of motion, or more simply the laws of motion or Newton's laws; I have also mentioned universal gravitation. There are others, such as the principle of the superposition of motions, the law that in free fall the distance increases as the square of the time, and the fact that projectiles near the surface of the earth follow a trajectory having the shape of a parabola.

The connection is that the *Dialogue* contains inklings, intuitions, approximations, applications, and more or less explicit formulations of many of these laws. This is the sort of content which scientists tend to find in the book. For example, in the *Mathematical Principles* Newton attributed to Galileo a knowledge of the law of inertia, the law of force, the principle of superposition, and the law of squares.[37] Now, it is almost certain that Newton had read the *Dialogue* before publishing his work but did not read Galileo's *Two New Sciences* until afterward.[38] These two facts imply that Newton interpreted the *Dialogue* as containing formulations of these laws.

37. Newton (1934, 21–22).
38. I. B. Cohen (1967); cf. Herivel (1965, 35–41).

Some of Newton's readings are questionable; this is especially true of the law of force and acceleration. A more controversial case involves the law of inertia.[39] However, more important than the degree of their historical-textual accuracy is the fact that such readings have been advanced by Newton, as well as by other protagonists of the Scientific Revolution and other scientists in general; for scientists' judgments are the elements of the historical link between Galileo's book and the Scientific Revolution.

Thus, one may read the *Dialogue* to learn about the Scientific Revolution, the Copernican Revolution, and the Galileo Affair. Since these are historical events that can be concretely delimited, such readings may be called historical readings. Moreover, since the history in question is the history of science and intellectual history, one could also speak of historical-scientific and historical-intellectual readings, or more loosely of scientific and intellectual readings. However, in the process of such readings, there emerge interpretive and evaluative questions whose resolution requires a more abstract type of analysis that may be called philosophical. Before discussing the details of such philosophical readings (see the appendix), let us examine how one is led to them directly from the historical readings.

5.2. Suppose we begin with the question of whether the *Dialogue* contains a formulation of the law of inertia. Let us consider this law in the version given by Newton: "Every body continues in its state of rest, or of uniform motion in a right line, unless it is compelled to change that state by forces impressed upon it."[40] Now, we can indeed find a passage in the *Dialogue* (selection 8, p. [173]) asserting that a terrestrial body will remain

39. There are passages in the *Dialogue* suggesting that circular motion is natural and rectilinear motion is not; they are omitted below, but see Galileo (1967, 19–32 and 162–67) and Favaro (7:43–57, 188–93). These passages have led some readers to attribute to Galileo a notion of "circular inertia," which is far from the rectilinear inertia that defines the nature of a body in classical physics; see, for example, Franklin (1976, 58–62 and 84–87), Koyré (1966, 205–90; 1978, 154–236), Shea (1972, 116–38), and Shapere (1974, 87–121). I would argue that here we are dealing with two senses of "natural," and so there is no conflict; when Galileo says that circular motion is natural he means that it can be empirically actual and everlasting; when the law of inertia says that rectilinear motion is natural, the meaning is that such motion is an inherent tendency (which has to be combined with other kinetic tendencies to yield an empirically actual phenomenon); so I am inclined to find many approximations of the law of inertia in the book. My interpretation is along the lines of Chalmers and Nicholas (1983), Coffa (1968), Drake (1970, 240–78; 1978, 126–32), Finocchiaro (1980, 33–34, 87–92, and 349–53), McMullin (1967, 3–51), and Sosio (1970, l-li).

40. Newton (1934, 13).

motionless or conserve its horizontal motion forever unless disturbed by some accidental or external impediment. Then one obvious task would be to compare and contrast these two statements to determine whether they are equivalent (which does not seem to be the case); whether the Galilean statement is an approximation to, or special case of, the Newtonian one (which is what I hold); or whether the two claims are actually inconsistent (as some interpreters have argued on the grounds that Galileo's "horizontal motion" refers to motion along the terrestrial circumference and so is circular rather than rectilinear).[41] Such comparison and contrast are a logical exercise, namely, an exercise in the analysis of the inferential interrelationships between the two assertions. For the issues are whether each implies the other, whether the first implies the second but not conversely, or whether the second implies the negation of the first. Such logical exercises involve critical reasoning, for a key element of critical reasoning is the analysis and evaluation of inferences and logical relationships. This in turn means that the determination of the historical link between the *Dialogue* and the Scientific Revolution requires critical reasoning on the part of the reader and the historian.

Let us now assume we have resolved affirmatively the question about inertia. Or we could use as an example some other principle that is undeniably formulated in the book, such as the law of squares.[42] Then we could ask what Galileo's rationale was, what reasons led him to formulate it, or why he accepted it. A distinct but related question would ask what method or procedure Galileo followed in arriving at the law. These questions are important partly because it is important not only to reach the right conclusions, but to do so for the right reasons or as a result of following the right procedures or methodological principles. For example, the writings of Descartes contain a formulation of the inertial principle that is in some ways more explicit and general than Galileo's; and his influence on Newton in this regard is more direct and substantive. However, Descartes arrived at that principle largely on the basis of theological and metaphysical premises about the unchanging nature of God; now, despite the intrinsic interest of such a justification, the argument has never impressed working scientists, probably because they find such reasoning and such a procedure questionable.[43]

41. See, for example, Franklin (1976, 58–62 and 84–87), Koyré (1966, 205–90; 1978, 154–236), Shea (1972, 116–38), and Shapere (1974, 87–121).

42. Favaro (7:244–60); Galileo (1967, 218–33).

43. This is a complex and controversial issue; see, for example, Shea (1991).

Another important matter would be to determine how Galileo applied the inertial principle, what use he put it to, what problems he solved with it, or what further conclusions he derived from it. This is another way in which this principle is contextualized and related to other ideas. Thus, it is important to realize that he used his inertial principle to help him refute two anti-Copernican objections. To criticize the ship analogy objection (selection 8) he argued that, on a ship moving forward, a rock dropped from the top of the mast will fall at the foot of the mast, as it does on a motionless ship. And in his criticism of the objection from the extruding power of whirling (selection 9), he establishes that when a body is subject to whirling circular motion and is then released, it will have a tendency to move along the straight line tangent to the point of release. Such applications of a law of nature are instances of critical reasoning.

In summary, from the viewpoint of the Scientific Revolution and modern science in general, we want to know whether and to what extent various basic laws of nature are stated or suggested in the *Dialogue*. But these laws cannot be understood in isolation, abstracted from the context in which they were formulated. We also must know the reasoning that led to acceptance of the laws, the reasoning whereby the laws were applied to solve other problems and to derive further consequences, and the procedures or methodological principles followed. Once we get involved in analyzing and evaluating such reasoning and such methodological principles, we are involved in critical reasoning and in methodological reflection.

5.3. Analogous points can be made in regard to the Copernican Revolution, taking the latter specifically in the sense of the transition from the geostatic to the geokinetic worldview. As we have seen, this process began with Copernicus's new and important (though inconclusive) argument in support of the geokinetic idea and ended with Newton's conclusive (though theoretical and thus indirect) proof. To understand the place of the *Dialogue* in this development, we need to understand, first, what Copernicus's argument was, whether or not it was conclusive, and why it was inconclusive (if it was indeed inconclusive). Then the same determinations need to be made about the argument in the *Dialogue*; to be more exact, since this work contains many distinct arguments, such determinations have to be made about all these arguments. Moreover, one would want to know Galileo's intentions, namely, whether his arguments were meant to be conclusive. Finally, for any argument which is not conclusive, one needs to examine further whether is strong, weak, or worthless. All this involves critical reasoning in an essential way.

However, critical reasoning is even more deeply relevant to Galileo's book. For the point just made shows that the reader has to engage in critical reasoning in order to analyze and evaluate its content. Now, it so happens that Galileo himself is explicitly engaged in critical reasoning because the book presents not merely his own arguments favorable to the earth's motion, but also a critical analysis of the anti-Copernican arguments. In short, the reader needs to have some appreciation of critical reasoning merely to understand what Galileo is doing, let alone to analyze and evaluate it.

In my opinion, the overall structure of the *Dialogue* is that of an argument supporting the conclusion that the earth moves (with axial rotation and orbital revolution); the overall argument in the book is not conclusive but strong; and Galileo did not regard it as conclusive, but did regard it as somewhat stronger than it is. Now, these assertions are really generalizations because the book contains many distinct arguments supporting Copernican conclusions, and not all these arguments are equally strong; some are very strong, some of medium strength, some weak but not worthless, and a few close to worthless; most of them have a noteworthy degree of strength.

This view is not universally shared.[44] Some readers claim that some of the book's arguments are intended to be conclusive, for example the one based on explaining the apparent motion of sunspots in terms of the earth's motion, and especially the one based on explaining the tides as the effect of the same motion; such claims are documented by means of passages where Galileo seems to attribute conclusiveness to these arguments.[45] My reply is that such textual evidence shows merely that he regarded these arguments as very strong, the tidal argument as the strongest. For such textual evidence must be taken in context, and the crucial feature of the context is that each argument is only one of several being advanced to support the same conclusion; normally when more than one argument is given to support the same conclusion, it is done because none is conclusive by itself but rather each is meant to reinforce the others. That is, if Galileo really believed that either the sunspot argument or the tidal argument conclusively established the earth's

44. Langford (1966) and Shea (1972) provide good examples of alternative readings; McMullin (1967, 3–51) contains an interpretation that overlaps with mine.

45. For the sunspot argument, which is omitted below, see Galileo (1967, 344–56, especially 344–45) or Favaro (7:372–83, especially 372); for the tidal argument, see selection 15, especially p. [450].

motion, he would not have felt that he needed the other to prove the same point. There is another crucial feature of the context: Galileo is clearly aware that he has not refuted the anti-Copernican argument based on the absence of an annual stellar parallax, but has merely outlined a research program which might (or might not) yield the evidence needed to refute it (selection 14); but to claim that he considered any one pro-Copernican argument (or even all collectively) as conclusive is tantamount to claiming that he had forgotten about the important counterevidence of the missing parallax.

Another controversial issue arises because the *Dialogue* does not explicitly discuss the Tychonic system. As mentioned above, Tycho held that the five planets revolve in heliocentric orbits, but that the sun revolves in a geocentric annual orbit, carrying these planets along with it in the process, while they all also share the universal diurnal motion around the motionless earth. Such neglect would represent both logical and methodological flaws.

Logically, the criticism starts by pointing out that, although Galileo's arguments show the Copernican system to be superior to the Ptolemaic one, they do not show the former to be any better than the Tychonic system; for the Copernican and the Tychonic systems were observationally equivalent and the available evidence could be explained equally well by either; for example, the phases of Venus require only a heliocentric orbit by this planet, but imply nothing about whether the annual motion belongs to the earth or to the sun-Venus pair. Here, one flaw might be that of attacking a straw man, namely, the Ptolemaic system, which by 1632 was generally rejected by astronomers, rather than criticizing the Tychonic system, which was widely accepted and indeed was a serious rival to Copernicanism. Another flaw might be the fallacy of affirming the consequent, namely, the fallacy of arguing in the form "P is true; if Q then P; therefore, Q is true"; here, Q would stand for Copernicanism, and P would stand for any phenomenon (such as the phases of Venus) which can be explained by means of Copernican causes.[46]

Methodologically, Galileo's apparent neglect of Tycho would be at least a lapse of objectivity, for objectivity requires the consideration of all available relevant evidence and all available alternative theories. Another methodological flaw might be the appeal to ignorance, for Galileo directs his book to a lay audience of nonexperts, ignorant of the Tychonic

46. For such an interpretation, see Gingerich (1992, 111–12; 1993).

alternative, rather than to technical astronomers who would have been acquainted with it.

To answer such criticism, I would argue that, although Galileo does not discuss Tycho's alternative explicitly, he does so implicitly, and hence he is not really neglecting it. For the Tychonic theory does after all share a crucial common element with the Ptolemaic system: both hold the earth to be motionless at the center of the universe; and in both systems, the diurnal motion belongs to the whole universe except the earth, and the annual motion belongs to the sun. Thus, all the Galilean arguments for the earth's motion and against the geostatic thesis undermine both Tycho and Ptolemy. This is the case, for example, with the tidal argument (selection 15) and with the probable arguments for the earth's diurnal rotation (selection 6).

Another important point is that Galileo is clearly aware that not every argument supporting some part of Copernicanism necessarily supports the key geokinetic idea. For example, when he discusses the phases of Venus (selection 11, especially pp. [349–54]), he makes several things perfectly clear: that this phenomenon proves only the heliocentricity of Venus's orbit; that it does not even prove the general heliocentric thesis for all (five) planets; that distinct arguments are needed to support the heliocentric character of the orbits of Mercury, Mars, Jupiter, and Saturn; that an additional argument is needed to decide the issue of whether the annual motion belongs to the earth or the sun; and that the additional argument for the earth's annual motion is based on analogy and so is merely probable and much weaker than the argument for planetary heliocentrism.

However, we need not elaborate, for my main aim here is not to decide such controversies but to show that a proper understanding of the role of the *Dialogue* in the Copernican Revolution requires an appreciation of critical reasoning. My account is only one way of articulating the book's connection with critical reasoning; even if it is rejected, other accounts are no less dependent on critical reasoning than my own.

Finally, we need to discuss a different point about the connection between the *Dialogue* and the Copernican Revolution. It concerns the activity of methodological reflection rather than critical reasoning as such.

We have already seen that one response to Copernicus was the instrumentalist interpretation of the geokinetic hypothesis. This raised the general methodological question of whether scientific theories should be interpreted realistically or instrumentalistically. We have also seen that both Copernicus and Galileo were epistemological realists. However, that does not settle the issue, not only because one may want to criticize them, but also because in the twentieth century, the problem reemerged in an especially striking manner in connection with the interpretation of quantum physics.[47]

A related point is that direct sensory observation favored the geostatic view; the Copernicans did not claim that they could see, feel, or otherwise perceive or sense the earth's motion; like the Aristotelians, they felt the earth to be at rest and saw the heavenly bodies move around us; even today, direct naked-eye observation reveals the same thing. As the argument from the deception of the senses suggested, Copernicanism contradicted a basic epistemological principle, namely, that under normal conditions the human senses are reliable, that they tell us what is really happening, that normal observation reveals reality to us. Thus, the Copernican Revolution forced mankind to come to grips with the fact that this is not always so. Should the principle be rejected? If so, what would take the place of the senses in the search for truth? Should the principle be revised? If so, how? Assuming one would not want to go from the extreme that the senses are always reliable to the opposite extreme that they are never reliable, one might want to say that the senses are sometimes reliable and sometimes unreliable; but then how does one decide which is the case in a specific situation?

Another way of looking at the methodology of the situation is to say that Copernicanism postulated the existence of an unobservable process, the earth's motion. Even the telescope did not render this visible, as it did the lunar mountains, Venus's phases, and sunspots. The question then arose whether it is proper in scientific inquiry to use unobservable entities. This issue was all the more problematic because the introduction of entities that are not directly observable had few precedents in the history of science, and so this step was full of uncertainties and required great intellectual courage.

Galileo's introduction of the telescope raises analogous issues. The telescope enabled one to revise in a constructive manner the empiricist principle that the human senses are normally reliable. For one could now say that they together with instruments which enhance their power are normally reliable; instead of saying simply that normal observation reveals reality to us, one could say that normal observation together with instrumentally aided observation reveals reality to us. But

47. For more details about the instrumentalism controversy, see Duhem (1969), Feyerabend (1964), Finocchiaro (1992b), Gardner (1983), Jardine (1984), and Westman (1972).

the introduction of artificial instruments in the search for truth was problematic. The following objections against the telescope had to be dealt with.

Some people questioned its methodological legitimacy, arguing that there was no place in scientific inquiry for instruments which make us perceive things that cannot be perceived without them, and that we should use only the natural senses operating under normal conditions; obviously this objection could not be dismissed, as we can appreciate today if we compare the situation then with the recent issue of whether psychedelic drugs put users in contact with a deeper level of reality or merely make them see things that are not there. Others questioned its legitimacy in principle, on theological grounds; that is, the use of artificial instruments might be taken to imply that the natural human senses are flawed; but they were created by God and so cannot fall short of their purpose, which is to enable humans to glimpse reality; in short, if God had wanted us to see the lunar mountains, for example, He would have given us eyes sharp enough to detect them. Still others questioned the reliability of the telescope by pointing out that Galileo had not provided a completely adequate scientific explanation of how and why it worked, based on the theory of optics. Moreover, all empirical checks involved terrestrial observation, and there was not even one instance of a test showing that it was truthful in the observation of heavenly phenomena. Finally, the practical operation of the instrument required that one learn how to use it and how to avoid aberrant and deviant observations, which were very mysterious at the time and which are known today to stem from impurities of the lenses, improper lens shape, and other features of poor design.[48]

Thus, it is easy to see how the transition from the geostatic to the geokinetic view raised issues such as the following: What is the role of observation in the quest for knowledge? What is the role of unobservable entities in the search for truth? What is the role of artificial instruments in inquiry? What is the role of scientific hypotheses, and should they be interpreted realistically or instrumentalistically? These are questions of principle, involving the nature of knowledge and the proper procedure to follow. It is not surprising, then, that the *Dialogue* is full of discussions addressing such issues. Such methodological reflection is related to but distinct from the critical reasoning mentioned earlier. Two elements

48. For more details about the telescope controversy, see Crombie (1967), Feyerabend (1975), King (1955), Ronchi (1958), Rosen (1947), and Van Helden (1984; 1994).

of both are analysis and evaluation; but whereas critical reasoning is concerned with analysis and evaluation of arguments, methodological reflection is concerned with analysis and evaluation of general procedural principles for the discovery of truth and the acquisition of knowledge.

5.4. Let us now go on to explore how the understanding of the historical link between the *Dialogue* and the Galileo Affair also leads us to critical reasoning and methodological reflection. Let us start with methodology. We have already seen that the heresy attributed to Galileo by the sentence of 1633 has two parts: the view that the earth moves and the view that one may believe and defend as probable, opinions contrary to the Bible. The former is, of course, the key cosmological tenet of Copernicanism, but the latter is really a methodological principle. It is a methodological principle in my sense of the term because the principle obviously refers to opinions about physical reality and the context is obviously one of the search for truth about the physical world. Thus, the second alleged heresy is the methodological principle that in scientific inquiry it is legitimate to believe and defend as probable physical theories contrary to the Bible.

This is certainly a principle that Galileo accepted. In the *Dialogue* he does not explicitly articulate it,[49] but he is clearly practicing it. The book unquestionably holds and defends as (at least) probable the geokinetic idea, which the Church held to be contrary to the Bible.

This methodological principle involved the issue of whether biblical assertions are *decisive* in scientific inquiry. A related but distinct issue, at the opposite end of the spectrum, was that of whether biblical assertions are *relevant*; whether they carry *any* weight. Earlier, in the "Letter to Castelli" (1613) and in the "Letter to the Grand Duchess Christina" (1615), Galileo had explicitly elaborated the principle that biblical assertions are irrelevant; he had done so in order to refute the biblical objection to Copernicanism. And that, as we have seen, was what started the affair in 1613.

Although this second principle is stronger than the first, it should not be equated with a total rejection of biblical authority in general. In fact, Galileo did accept biblical authority in matters of faith and morals, and so his rejection of the authority of the Bible is partial.

49. The preface (selection 1) and selection 12 do explicitly discuss the issue but do so briefly. At any rate, as already mentioned, the preface cannot be taken at face value; and, the discussion in selection 12 is too indirect, ambiguous, and evasive. Such passages are best treated from the viewpoint of rhetoric, as explained in the appendix (3).

Analogous issues involved the question of the scientific import of religious authorities other than the Bible. Galileo was also inclined to reject them. He did this explicitly in the above mentioned letters, although not in the *Dialogue*. For example, as suggested by the objection from the consensus of the Church Fathers, some regarded their collective wisdom to carry weight in science. Galileo's attitude toward them was analogous to his attitude toward the Bible: he was willing to accept their authority on questions of faith and morals but not on questions of physical truth. A similar issue arose for the relationship between theology and physical science: some people regarded theology as the queen of the sciences in the sense that theologians had the right to dictate what physical theories scientists should or should not accept, whereas Galileo denied theology any such right.[50]

The issues and Galileo's attitude were more complex in regard to the authority of the Catholic Church and its official institutions such as the pope, the Inquisition, and Sacred Councils. Because at that time the Church held great political and economic power, it could dictate to a scientist like Galileo whether or not he could research a given topic, publish a given work, or retract his views, and what the content and form of a published book should be. And Galileo seemed to admit and submit to such power. However, as he also added,[51] this is not power to make certain views true or false; it is only power to declare them heretical. It is not power to make someone think internally in a certain way; it is rather power to make someone utter certain words or externally display a certain behavior. Some methodological issues raised here are the questions of who has the right to decide what topics should be studied and how results should be disseminated (especially if such studies and such dissemination need financial backing). One may well agree with Galileo that in physical inquiry, statements by religious authorities carry no weight in regard to the merits of the issue, but this is not to say that such statements have no role in regard to questions of research agenda and result dissemination. In fact, his behavior implicitly recognized that the Church does have a role in regard to the practicalities of the search for truth, and it implicitly recognized the principle that nonscientific authorities have a role to play in the practical conduct of scientific inquiry.

50. For more details, see "Galileo's Letter to the Grand Duchess Christina," in Finocchiaro (1989, 87–118) and cf. the analysis in Finocchiaro (1986).

51. "Letter to the Grand Duchess Christina," in Finocchiaro (1989, 114).

Galileo's attitude toward the Church is complicated in another way and thus raises another type of methodological question. In denying the scientific authority of the Bible and other religious entities, he may be taken as advocating the separation of church and science. Such separation is a simple and clear, if controversial, principle about their relationship. However, it is not the only alternative to the traditional idea of the dominance of religion over science. This traditional idea may be taken to embody the principle that assertions about physical reality by religious authorities are decisive. The separation of church and science may be taken to express the principle that such assertions are irrelevant. However, one may formulate a third principle different from both. This would state that physical claims by religious authorities are neither decisive nor irrelevant, but are worthy of consideration and subject to comparative evaluation along with other types of arguments, evidence, and reasons; religious considerations would have some weight, not enough to undermine conclusive observational or theoretical proofs, but enough to outweigh observational or theoretical arguments if these were merely probable but not conclusive.

This principle is the one suggested by Galileo's preface to the *Dialogue*. There he states that if Catholics choose to believe in the geostatic view, they do not do so out of scientific ignorance, but because the scientific arguments in favor of the earth's motion (while good) are inconclusive, and they attach greater weight to the religious considerations favoring the geostatic view. Because of the restrictions under which he was operating, it is questionable whether this suggestion represents anything more than lip service to ecclesiastical authority; certainly he did not put this principle into practice in either this book or his other works or activities. Moreover, one could question the general viability and self-coherence of this principle. For it suggests the picture of a methodology according to which no type of consideration would be excluded in principle, but all would be considered and evaluated on their own specific merits; now, although this may seem admirably open-minded, it may be too indiscriminate; the other side of the coin of the readiness to examine each type of consideration on its own merits is the unwillingness to formulate guidelines for the conduct of inquiry which might represent the lessons of past experience.

However, this is not the place to decide such issues since my main purpose is to introduce them. Accordingly, a final refinement is necessary in unearthing the methodology of the Galileo Affair; it involves the question of the role of authority in general. We have already seen that, in regard to religious authority, the *Dialogue* does not advance explicit criticism, but contains only implicit criticism (by actually arguing for a proposition contrary to the Bible), explicit lip service (preface), and ambiguous evasion (selection 12). But there is a scientific authority which is repeatedly, directly, and explicitly criticized in the book, namely, Aristotle. Now, when Galileo's attitude toward the science-religion question and toward the methodological role of religious authorities is examined in the context of his critique of Aristotle, one begins to see that the problem with religious authorities may be due more to their being authorities in the first place than to their being religious. In short, another crucial methodological issue is the question of the role of authority in the search for truth, meaning authority in general or the authority of other scientists.

Recall that the biblical objection reduces to the argument that the earth must be standing still because it is so stated or implied in the Bible. Galileo's main criticism was that the Bible is not a scientific authority, and so the objection has no force. Consider now another objection that was in the mind of many traditionalists: the earth is motionless because Aristotle said so.[52] This objection cannot be criticized in the same way as the biblical argument, for Aristotle was a scientific authority, and Galileo himself did not want to deny that. The key flaw with the new objection is that it is superficial or incomplete—it does not say why Aristotle asserted that the earth is motionless. Galileo's point is that scientific authorities should not be merely quoted regarding the conclusions they reached, but they should be studied so that we understand their evidence and reasoning; their arguments are more important than their conclusions.

One reason for this is that, if we understand their arguments, we will be able to tell how they would argue in other situations. For example, on a related topic, Aristotle argued that the heavens are unchangeable because no changes have ever been observed in the heavens. Then Galileo could plausibly say (selection 3) that this manner of reasoning is such that, if Aristotle were alive nowadays, he would conclude that the heavens *are* changeable, on the strength of the new observational evidence of novas and sunspots.

However, appealing to the authority's reasoning may not be the end of the story and may raise other difficulties. For example, if one were discussing the specific geocentric thesis, a traditionalist might start by saying

52. Selection 5 may be read as in part a discussion of this objection.

that the earth is located at the center of the universe because Aristotle said so. Then he might see the need to elaborate by reporting that Aristotle believed this because of the following argument (selection 2): the natural motion of heavy bodies is contrary to that of light bodies; light bodies move naturally straight upward; straight upward means toward the circumference of the universe; therefore, the natural motion of heavy bodies is toward the center of the universe; but the natural motion of heavy bodies is also seen to be toward the center of the earth; therefore, the center of the earth and of the universe coincide. However, this argument could be criticized as committing the fallacy of begging the question. An Aristotelian might retort that the argument cannot commit such a fallacy because Aristotle was the founder of the science of logic; here the proponent is appealing to the logical authority of Aristotle in order to defend the logical worth of this argument. Is this a legitimate use of authority? Galileo answered negatively. Again, he did not deny the logical authority of Aristotle but added that this authority extends to logical theory, or the theory of reasoning, which is not the same as logical practice, or actual reasoning. That is, as an undisputed authority in the science of logic, Aristotle's authority has weight in the domain of reasoning about reasoning, but here we are dealing with reasoning about natural phenomena; thus the charge of begging the question cannot be dismissed by appealing to his authority.

From such considerations one might conclude that, while it would be too extreme to hold that there is no place for authority in science, its role is problematic. Its role is more as a source of evidence and reasoning than of mere claims; but the authority's argument is not sacrosanct; it must be evaluated, and so investigators cannot avoid doing their own thinking. Religious authorities also have the drawback that by their very nature they normally do not provide scientific evidence and reasoning; if and when they can, they are not just religious authorities, but scientific authorities as well.

Whether or not one accepts such conclusions, the corresponding methodological questions certainly arise directly out of the Galileo Affair. The last one of these methodological questions also leads us into critical reasoning. For, if one asks whether authority has a role to play in science, our discussion suggests that we could join Galileo in answering that authorities can play a role if the investigator is willing to analyze and evaluate their arguments. Let us then go on and discuss more directly other ways in which this episode involves the analysis and evaluation of arguments.

The condemnation of 1633 was precipitated by the book's publication in 1632, but was also the result of several restrictive norms in terms of which it was being judged. Thus, a critical understanding requires that we ascertain the origin, meaning, application, and validity of these norms; and we should bear in mind that validity may have to be judged from many different viewpoints, such as religious, legal, moral, and methodological.

For example, the special injunction forbade Galileo to even discuss the earth's motion. Its origin, meaning, and application are clear: the *Dialogue* certainly discusses this topic and so violates the letter of the document claiming to be a report of Cardinal Bellarmine's warning and bearing the date of 26 February 1616. However, its validity is just as clearly questionable, even from a purely legal point of view, because it lacks Galileo's signature.

Consider next the stipulation that Galileo's discussion was supposed to be hypothetical. The origin of this restriction lay in the conversations conducted during the six papal audiences Galileo was granted in 1624 when he visited Rome to pay homage to the newly elected Pope Urban VIII. The validity of this norm is relatively uncontroversial for us today, at least given a particular conception of the notion of hypothesis; that is, there was no conclusive proof of the earth's motion, and so this thesis should not have been presented as a fact, but as a hypothesis, more or less probable, possibly false but likely true, and capable of some degree of support. But this was not the only meaning of the concept of hypothesis, and the pope himself was inclined toward an instrumentalist conception, according to which a hypothesis is not a description of physical reality but a mere instrument of calculation. Thus, the application of this norm is problematic, and we get different results depending on how we construe a hypothesis. The book's discussion is clearly realist and not instrumentalist, and so it violated the instrumentalist interpretation of the restriction; but then the validity of the norm is questionable because instrumentalism cannot be regarded as unquestionable. However, on the probabilist or fallibilist interpretation of a hypothesis, we would have an arguably valid norm, but I would question whether there is any violation because (as mentioned above) I would argue that the book's arguments were intended as very strong though not conclusive.

The most important norm in the present context is the one ordering Galileo not to support or defend the geokinetic hypothesis. The only thing clear about this norm is its origin: it is mentioned in almost all the relevant documents.[53] Moreover, Galileo did not deny knowledge of it. However, it is less clear to what extent he accepted this restriction: after the proceedings of 1615–1616 he paid lip service to it and felt obliged to act as if he were obeying it, for he did not want to openly defy the Church; but it is equally certain that he had some misgivings about this stipulation.

What did the stipulation mean? It seems to have meant that Galileo was not supposed to present evidence and reasons in favor of the earth's motion or to answer objections against it. This involves the presentation of favorable arguments and the criticism of counterarguments, which in turn is an essential element of critical reasoning. In short, Galileo was being forbidden to engage in critical reasoning about the earth's motion.

The application of this norm is then straightforward: the *Dialogue* clearly violates it. In favor of the earth's motion it presents arguments based on the greater simplicity of axial terrestrial rotation vis-à-vis geocentric universal revolution, on the relationship between size of orbits and periods, on the heliocentrism of planetary motions, on the explanation of retrograde planetary motion, on the motion of sunspots, and on the explanation of tides. And, to support the same conclusion, it tries to refute the many traditional and new anti-Copernican arguments.

Was the stipulation valid, so that its violation by Galileo may be somehow held against him? Given my interpretation that he was being forbidden from engaging in critical reasoning, this question is really asking why critical reasoning is valuable. Methodologically, one could justify critical reasoning (and consequently criticize this restriction on Galileo) by arguing that critical reasoning is conducive, indeed indispensable, to discovering the truth. Here it is important to bear in mind that freedom to think critically is especially important when one is dealing with hypotheses (in the fallibilist sense), namely, with propositions

53. The relevant documents usually speak of prohibition and include the words *tenere* and *defendere*. These Latin words are best taken to mean "to hold, support, and defend" a belief. This interpretation has three elements: accepting or believing an idea; justifying the idea positively (with supporting reasons and evidence); and justifying it defensively (by criticizing objections, counterevidence, and counterarguments). The Latin terms should not be taken to mean simply to support and defend, or simply to hold and defend. As Morpurgo-Tagliabue (1981) has argued, the key point is that the Latin *tenere* can mean simply to hold a belief as well as to hold and support a belief; to this we should add that *defendere* can mean simply to justify defensively or to justify both defensively and positively. In *The Galileo Affair* (Finocchiaro 1989) I usually translated *tenere* and *defendere* as "to hold and defend," but this should be understood in the broad sense that includes all three of the above mentioned elements.

that are not yet susceptible of being conclusively proven or refuted; in such cases, the process of critical reasoning helps us to construct the arguments that at some later stage may be reformulated or gathered together so as to make a conclusive case. Therefore, this restriction on Galileo could not be justified by saying that at that time there was no conclusive proof of the earth's motion, for far from making critical reasoning inappropriate, the lack of conclusive proof is the normal precondition for it. In short, critical reasoning would have been largely inappropriate if the geokinetic view could have been conclusively proved, and it was fruitful given that that view was not conclusively proved.

Morally, the freedom to engage in critical reasoning could be derived from the basic human rights of free speech, thought, expression, and inquiry; thus, it might be held to be inalienable. However, it is also true that in 1616 Galileo had agreed to refrain from arguing in favor of the earth's motion, and this might be taken to imply his moral obligation to so refrain. Here we would have to examine the circumstances under which that agreement was extracted from him. These circumstances were far from ideal, and so we might plausibly argue that his agreement was involuntary, or at least more involuntary than voluntary. This point is reinforced by his subsequent behavior, which amounted to a series of attempts to circumvent the implications of a strict interpretation of that promise. This behavior was motivated not only by the belief that the original stipulation was ultimately wrong, but also by the belief that lip service to it was sufficient.

There is one final norm that should be considered, namely, the prohibition on believing in the earth's motion. All of the relevant documents contain some version of this restriction on Galileo. It is distinct from, but related to, the previous stipulations. What does it amount to? Note that we are talking about a scientific belief, which could be defined as one formed as a result of a systematic search for the truth, namely, as a result of arguments and evidence. Moreover, let us be clear that we are talking about the belief that the earth moves, as distinct from the belief that the earth *probably* moves. Finally, believing in the earth's motion is supposed to be something different from supporting and defending this idea. All this suggests that the issue of whether Galileo believed in the earth's motion is equivalent to the question of whether he regarded his whole argument as conclusive. I have already stated that I think he regarded the argument as very strong, but not conclusive, although others would make the stronger attribution. However, the important point is that to prohibit Galileo from believing in the earth's motion amounted to prohibiting him from considering the arguments for this proposition to be conclusive. This prohibition in turn corresponds to one of the meanings of the requirement of hypothetical discussion, the fallibilist meaning. That is, by requiring a hypothetical discussion, the pope probably intended to require an instrumentalist discussion and exclude a realist one, rather than to require a fallibilist discussion and exclude a demonstrativist one; but Galileo was also being subject to the latter requirement via the language of belief prohibition.

If this is the meaning of this restriction, then it was a legitimate one since (as already mentioned) the pro-Copernican arguments were not conclusive. The main issue is then the question of application, which involves the question of whether the book's arguments are or are intended to be conclusive. As we have seen, this is a controversial question, but there is no doubt that its resolution requires analysis and evaluation of these arguments.

To summarize, at the end of the Galileo Affair he was tried and condemned because of at least four alleged transgressions committed in the *Dialogue*. He was not supposed to have even treated of the earth's motion; if he did treat of it, his treatment was supposed to have been instrumentalist rather than realist; should his treatment be realist, he was not supposed argue in favor of the idea; and if he did so argue, he was not supposed to present a conclusive case. He clearly committed the first three of these transgressions, but it is controversial whether he committed the fourth one. In any case, to merely understand that he committed the third violation (that the book does indeed support and defend the earth's motion), and to determine whether he committed the fourth transgression (whether the argument was or was meant to be conclusive), both require critical reasoning; once again, we need to see Galileo as being himself engaged in critical reasoning, and we need to be ourselves engaged in critical reasoning about his book.

5.5. This connection between critical reasoning and the Galileo Affair is analogous to the connections explored earlier when we considered how the *Dialogue* relates to the Scientific Revolution and the Copernican Revolution. We have also seen that the book's connection to all three episodes raises several distinct though overlapping methodological questions. Critical reasoning and methodological reflection may be regarded as philosophical notions in the sense that philosophers are especially inclined to scrutinize them; in particular, critical reasoning may be handily labeled a logical notion insofar as logic or logical theory is traditionally defined as the study of reasoning. Thus, besides being read historically to learn about these three developments, the *Dialogue* may also be read philosophically, namely, as an introduction to critical reasoning and methodological reflection.

However, this talk of philosophical readings should not be misunderstood, any more than should the talk of historical readings. To prevent such misunderstanding, we need to distinguish between readings which are of primary importance and readings whose importance is secondary.

First, let us clarify that there is no good reason why a historian should object to a logical or methodological reading of the Galilean text. For, as we have seen, these terms refer to the activities of critical reasoning and methodological reflection; these activities are widely practiced and potentially relevant to all human beings and scholarly disciplines; and issues relating to these activities arise directly out of the three historical developments in question. We are *not* talking about purely theoretical and abstract technical questions comprehensible and relevant only to experts specializing in the branches of professional philosophy known as systematic epistemology or systematic methodology; nor are we talking about metaphysical problems such as the place of values in a world of facts, which in themselves are easily comprehensible and universally relevant, but whose connection with Galileo's book is not directly and immediately apparent and requires intricate argumentation.[54] Such investigations have their own place and function, and they may be conceived as differing only in degree from the philosophical readings attempted in the present context; however, the present context is one of a more primary, elementary, and immediate type of philosophical reading.[55]

Analogously, there is no good reason why a philosopher should be skeptical of the historical readings sketched and suggested here. For we are talking about the history of the Copernican Revolution, Scientific Revolution, and Galileo Affair whose significance, as we have seen, is such that philosophers can neglect them only at their own loss. The historical readings of the *Dialogue* I have discussed do *not* involve merely erudite curiosities and antiquarian relics. Admittedly, the historical dimension of Galileo's book does include some of these and is not restricted to the key developments on which I have focused. For example,

54. An example of metaphysical analysis is the classic work by Burtt (1954); examples of systematic epistemology are Gardner (1983, 218–32), Rosenkrantz (1977, 135–61), and Thomason (1994a).

55. The issue here is analogous to that discussed later in the appendix (1.6 and 2.1), in regard to the difference between critical reasoning and logical theory and between methodological reflection and systematic methodology.

in the book Galileo is also settling scores (and not always fairly) with some old and new enemies, such as Jesuit astronomer Christopher Scheiner and philosophy professor Scipione Chiaramonti; thus, the work is in part a chapter in the history of Galileo's disputes with these people. However, the general human and cultural relevance of such disputes is relatively minor at best, and the corresponding issues are relatively complex; thus, they are best left to a subsequent stage of study, after the more fundamental historical readings stressed here. This is especially true because the instructiveness of the lessons of those disputes, if any, is likely to be in the domain of rhetoric, rather than in science and philosophy. But this brings up the topic of rhetoric.

I stated earlier in this introduction (1.2) that the *Dialogue* also has an important rhetorical dimension, important in the sense of primary importance. Thus, a rhetorical reading is also valuable. But, for a number of reasons, this will be postponed till the appendix (3).

One reason is that, however important the rhetorical point of view is, it is no more a part of philosophy than it is of history, or conversely, it is as much a part of history as it is of philosophy; thus, it is best elaborated after the details of the philosophical readings, just as the exposition of these is more appropriate after the historical details.

Furthermore, in a sense the rhetorical dimension constitutes a viewpoint distinct from, but related to, the historical and philosophical orientations. But as already suggested when the notion was first mentioned, rhetoric seems to subsume more disparate ideas and activities than either history or philosophy, and so its discussion at this point would be relatively confusing.

It is also true that rhetorical readings are dependent on philosophical analyses and historical interpretations in a way in which both of the latter are not dependent on the former. For example, in passages such as Galileo's preface (selection 1) and the book's ending (selection 16), the historical facts underlying the *Dialogue* suggest naturally that a rhetorical reading is necessary, rhetorical at least in the sense of pertaining to the question of whether he means what he says and is saying what he means. Similarly, whenever a reader of the text believes that Galileo has committed a particularly egregious error in reasoning, or has expressed an especially implausible methodological claim, then the appropriateness of rhetorical considerations suggests itself, in the sense that the questions arise as to whether the logical error or methodological expression was intentional and what his real purpose was; but in both cases the rhetorical reading is based on a philosophical evaluation, which in turn is based on an interpretation of the argument or methodological expression.[56]

Finally, I stated when I initially mentioned the notion of rhetoric above that there are difficulties and tensions in the relationship between the rhetorical reading and other interpretations from the point of view of science, critical reasoning, and methodology. Partly as a result of such undeniable tensions, a tendency has developed lately that stresses the rhetorical dimension of the *Dialogue*, but does so in order to diminish its scientific, logical, and methodological value.[57] I believe this is one-sided and reductionistic; it prevents the appreciation of aspects of the book which are clearly present in the text. Instead I wish to practice and encourage a more judicious approach that gives some relative autonomy to the various primary readings (historical, scientific, logical, methodological, *and* rhetorical) without pretending that any one of these is all-important and that the others are reducible to it. It is partly in accordance with this judgment that the discussion of rhetoric is postponed till later.

56. Accordingly, note that the discussion of rhetoric below (appendix, section 3) will be motivated in part by Galileo's methodological assertion that rhetoric has no role to play in scientific inquiry, which I believe contradicts both the reality of the situation as well as Galileo's own practice; the issue, however, is complicated by the many meanings of the term *rhetoric*. Similarly, Feyerabend (1975) is a leading exponent of the rhetorical interpretation and bases it in large measure on a critical analysis of Galileo's discussion of the vertical fall argument (selection 8); hence, it is not surprising that his rhetorical conclusion needs to be qualified to the extent that his reconstruction of the Galilean argument needs to be qualified; cf. Finocchiaro (1980, 192–200). And Hill (1984) advances a rhetorical interpretation of the passage on the extruding power of whirling largely based on his critique of Galileo's physical, mathematical, and logical errors.

57. See, for example, Gingerich (1993, 50) and Moss (1993, 257–300). I seem to have contributed to this tendency myself (Finocchiaro 1980) since many readers attached too much importance to the subtitle of my earlier book (*Rhetorical Foundations of Logic and Scientific Method*) and interpreted me as advocating a one-dimensional and reductionistic interpretation of science in general and of the *Dialogue* in particular; however, more perspicacious readers understood that my book's content was such that it would have been more accurate to speak of "logical foundations of rhetoric and scientific method" (Lupoli 1986). In that earlier book I had a reason for speaking of "rhetorical foundations," which is partly methodological and partly rhetorical, namely, to emphasize the critique of logicism, formalism, and positivism. Now that the scholarly and cultural pendulum seems to have swung to the other extreme (of seeing rhetoric everywhere, or nothing but rhetoric), I wish to counteract this tendency by trying to de-emphasize the rhetorical reading of Galileo's book, and so I relegate discussion of rhetoric to the end.

# Selections from Galileo's *Dialogue*

# Outline of Selections

- 4b. Differences between the moon and the earth
- 4c. The limitations of human knowledge
- 4d. The powers of human understanding

Second Day

5. Independent-Mindedness and the Role of Aristotle's Authority
   - 5a. The case of the Aristotelian who witnessed an anatomical dissection
   - 5b. The case of reading the invention of the telescope into an Aristotelian text
   - 5c. The case of the opportunistic interpreter of Aristotle's doctrine on the soul
   - 5d. The proper use of Aristotle's authority
6. Diurnal Rotation, Simplicity, and Probability
   - 6a. The principle of the relativity of motion
   - 6b. The simplicity argument for terrestrial rotation
   - 6c. Eight reasons why terrestrial rotation is simpler
   - 6d. The principle of simplicity
7. The Case against Terrestrial Rotation and the Value of Critical Reasoning
   - 7a. Aristotle's original five arguments against the earth's motion
   - 7b. Five more recent arguments
   - 7c. The greater open-mindedness of Copernicans as compared to Ptolemaics
   - 7d. Rational-mindedness
   - 7e. Four other anti-Copernican arguments
8. Vertical Fall, Superposition of Motions, and the Role of Experiments
   - 8a. The anti-Copernican argument from actual vertical fall
   - 8b. The anti-Copernican argument from apparent vertical fall
   - 8c. The ship experiment
   - 8d. The principle of conservation of motion
   - 8e. The role of experiments
9. Extruding Power and the Role of Mathematics
   - 9a. The anti-Copernican argument from the extruding power of whirling
   - 9b. A comparison of tangential extrusion with downward fall
   - 9c. The mathematical impossibility of extrusion
   - 9d. The role of mathematics
   - 9e. The cause of extrusion

10. The Deception of the Senses and the Relativity of Motion
    - 10a. The anti-Copernican argument from the deception of the senses
    - 10b. The principle of the relativity of motion
    - 10c. The deception of reason

Third Day

11. Heliocentrism and the Role of the Telescope
    - 11a. The center of the universe
    - 11b. The heliocentrism of planetary motions
    - 11c. The annual motion
    - 11d. The indispensability of the telescope
    - 11e. The traditional anti-Copernican arguments from Mars, Venus, and the moon
12. The Role of the Bible
13. Stellar Dimensions and the Concept of Size
    - 13a. The stellar dimensions argument against Copernicanism
    - 13b. A low estimate of actual stellar dimensions
    - 13c. Precautions in observing apparent stellar diameters
    - 13d. Reformulation of the argument
    - 13e. Higher estimates of stellar distances
    - 13f. The concept of size
14. Stellar Parallax and the Fruitfulness of a Research Program
    - 14a. The anti-Copernican argument from the appearance of fixed stars
    - 14b. Annual changes in apparent magnitude and angular separation
    - 14c. Annual changes in angular elevation above the ecliptic
    - 14d. Individual differences in the annual changes in stellar appearances
    - 14e. The methodology of theoretical explanation
    - 14f. A research program for observing stellar parallaxes

Fourth Day

15. The Cause of the Tides and the Inescapability of Error
    - 15a. The problem
    - 15b. Criticism of alternative theories
    - 15c. Acceleration causes tidal-like motions
    - 15d. The primary cause of the tides
    - 15e. Concomitant causes of the tides
    - 15f. Secondary tidal effects and additional concomitant causes

16. Ending
    - 16a. Final judgment on the tidal argument
    - 16b. Final judgment on the overall pro-Copernican case
    - 16c. The divine omnipotence argument
    - 16d. The limitations of human knowledge

# To the Discerning Reader

## [1. Preface][1]

[29] Some years ago there was published in Rome a salutary edict[2] which, to prevent the dangerous scandals of the present age, imposed opportune silence upon the Pythagorean opinion of the earth's motion. There were some who rashly asserted that that decree was the offspring of extremely ill-informed passion and not of judicious examination; one also heard complaints that consultants who are totally ignorant of astronomical observations should not cut the wings of speculative

1. This preface was compiled jointly by Galileo and a Vatican secretary named Niccolò Riccardi, who had the title of Master of the Sacred Palace and was in charge of evaluating books published in Rome. It corresponds verbatim to the latest version sent by him to the Florentine inquisitor on 19 July 1631, except that the word *mobility* in the second sentence of the third paragraph here had been changed to *immobility* in that copy. See Favaro (19:328–30), Finocchiaro (1989, 213–14), the introduction (4.2), and the appendix (3.2).

2. Here Galileo refers to the anti-Copernican decree published by the Congregation of the Index on 5 March 1616. Although the exact content of this decree is difficult to interpret and controversial, the Congregation temporarily banned Copernicus's book *On the Revolutions* until it was suitably revised, and condemned and totally prohibited certain interpretations of Copernicanism. The revisions to Copernicus were published by the same Congregation in 1620 and consisted of the deletion or rephrasing of about a dozen passages containing religious references or suggesting a realistic interpretation or categorical acceptance of the geokinetic thesis. The interpretations proscribed by the decree were in fact those which regarded the earth's motion as a description of physical reality, as an established fact, or as compatible with the Bible. See Finocchiaro (1989, 148–50 and 200–202).

intellects by means of an immediate prohibition.[3] Upon noticing the audacity of such complaints, my zeal could not remain silent. Being fully informed about that most prudent decision, I thought it appropriate to appear publicly on the world scene as a sincere witness of the truth. For at that time I had been present in Rome; I had had not only audiences but also endorsements by the most eminent prelates of that Court; nor did the publication of that decree follow without some prior knowledge on my part. Thus it is my intention in the present work to show to foreign nations that we in Italy, and especially in Rome, know as much about this subject as transalpine[4] diligence can have ever imagined. Furthermore, by collecting together all my own speculations on the Copernican system, I intend to make it known that an awareness of them all preceded the Roman censorship, and that from these parts emerge not only dogmas for the salvation of the soul, but also ingenious discoveries for the delight of the mind.[5]

To this end I have in the discussion taken the Copernican point of view, proceeding in the manner of a pure mathematical hypothesis and striving in every contrived way [30] to present it as superior to the viewpoint of the earth being motionless, though not absolutely but relative to how this is defended by some who claim to be Peripatetics; however, they are Peripatetics only in name since they do not walk around but are satisfied with worshipping shadows, and they do not philosophize with their own judgment but only with the memory of a few ill-understood principles.[6]

3. One of these critics was Galileo himself, as we can see from his "Letter to the Grand Duchess Christina" (Finocchiaro 1989, 87–118); thus, the account that follows should not be taken literally.

4. *Transalpine* literally means "beyond the Alps," a range of mountains marking the northern border of the Italian peninsula and separating it from north-central Europe. Italians traditionally referred to other Europeans, especially in north-central Europe, as "those who inhabit the region beyond the Alps."

5. Here Galileo states the book's aim, saying that it is to give a justification of the anti-Copernican decree of 1616. It is clear that he did not want his book to be interpreted as a violation of that decree and an act of defiance toward the Church; but it is unclear that it was necessary or plausible to make it look as if the book were providing a service to the Church by showing that the decree was not the result of scientific ignorance.

6. This paragraph describes the book's procedure or approach. Two main features are involved. The first is that of treating the Copernican view as a purely mathematical hypothesis; this was meant to correspond to Pope Urban VIII's favorite interpretation of Copernicanism, which was a standard way of avoiding conflict with religion; in short, the anti-Copernican decree could be interpreted as not prohibiting the hypothetical discussion of the earth's motion. However, as explained in the introduction (3.3, 4.2, and 5.4) and the appendix (2.6), here lay an unresolved ambiguity between the instrumentalist and the fallibilist conception of a hypothesis; it turns out that the book's discussion *is* hypothetical in the fallibilist but not in the instrumentalist sense, so that while Galileo could plausibly

Three principal points will be treated. First, I shall attempt to show that all experiments feasible on the earth are insufficient to prove its mobility[7] but can be adapted indifferently to a moving as well as to a motionless earth; and I hope that many observations unknown to antiquity will be disclosed here. Second, I shall examine celestial phenomena, strengthening the Copernican hypothesis as if it should emerge absolutely victorious and adding new speculations; these, however, are advanced for the sake of astronomical convenience and not for the purpose of imposing necessity on nature.[8] Third, I shall propose an ingenious fancy. Many years ago I had occasion to say that the unsolved problem

---

claim that he had abided by the pope's interpretation of the anti-Copernican decree, his critics could also plausibly charge that he had violated it. The second feature of the book's approach is the comparison of the two theories in the light of their supporting arguments; Copernicanism is shown to be better than the geostatic view in the sense that the Copernican arguments are shown to be stronger than the geostatic arguments. This comparison qualifies and relativizes the discussion of Copernicanism in the sense of leaving the door open to the question of how intrinsically strong it is from an absolute point of view; for example, it is possible for the Copernican arguments to be better without being conclusive, and so no categorical commitment to the truth of Copernicanism is implied; and it is possible that the scientific arguments for Copernicanism are better, while others favor the opposite view, in which case one would have to weigh scientific and nonscientific arguments.

7. The word *mobility* conveys the wrong emphasis from the viewpoint of Copernicanism; that is, it sounds odd for a Copernican to want to stress that terrestrial experiments cannot prove the earth's *mobility*; it would be more proper to stress that they cannot prove its *immobility*. On the other hand, an Aristotelian would want to phrase the point in terms of *mobility* (being unprovable by terrestrial evidence). That is, as literally phrased, this sentence seems to take the point of view of a critic rather than an advocate of Copernicanism. Was this a mere slip of the pen on Galileo's part, or does it have a deeper significance? It was probably not a slip of the pen because, as mentioned above, except for this word, the whole preface corresponds exactly to the text sent by the Master of the Sacred Palace to the Florentine inquisitor; in the latter copy the word *mobility* has been corrected to *immobility* by the insertion of the prefix *im* between the lines (see Favaro 19:328–30 and Pagano 1984, 110–12). As for the second question, it should be noted that during the trial of 1633, in his first deposition Galileo claimed among other things that "I do not think that by writing this book I was contradicting at all the injunction given me not to hold, defend, or teach the said opinion, but rather that I was refuting it" (Finocchiaro 1989, 260), and that "with the said book I had neither held nor defended the opinion of the earth's motion and sun's stability; on the contrary, in the said book I show the contrary of Copernicus's opinion, and show that Copernicus's reasons are invalid and inconclusive" (Finocchiaro 1989, 261–62). This was Galileo's weakest claim, which he did not repeat later; but it corresponds to the anti-Copernican phraseology of this sentence in the published preface.

8. Notice, once again, Galileo's attempt to suggest that the book is a hypothetical discussion. The talk of astronomical convenience is a specific suggestion of hypothetical instrumentalism, whereas the denial of a desire to necessitate nature suggests hypothetical fallibilism.

of the tides could receive some light if the earth's motion were granted.[9] Flying from mouth to mouth, this assertion of mine has found charitable people who adopt it as a child of their own intellect. Now, so that no foreigner can ever appear who, strengthened by our own weapons, would blame us for our insufficient attention to such an important phenomenon, I decided to disclose those probable arguments which would render it plausible,[10] given that the earth were in motion.[11] I hope these considerations will show the world that if other nations have navigated

9. See Galileo's "Discourse on the Tides," written in 1616 and conceived two decades earlier (Finocchiaro 1989, 119–33). It is incorporated into the Fourth Day of the *Dialogue* and corresponds to selection 15.

10. Galileo's word is *persuasibile*, which could be rendered literally as "persuasible" and stresses the element of persuasion. For a discussion of the possible rhetorical significance of this word, see Moss (1993, 10).

11. Though what is stated in this preface must be taken with a rhetorical grain of salt, this is an explicit statement that Galileo views the tidal argument to be merely plausible rather than conclusive, and this judgment corresponds to the many such qualifications he makes in the "Discourse on the Tides," which was written in 1616 just before the anti-Copernican decree. The anti-Copernican decree prohibited the *defense* of Copernicanism; thus, from the viewpoint of this decree, one may ask whether it would have been better to claim plausibility or to claim conclusiveness for the tidal argument. It seems to me that a plausible argument could more easily be interpreted as a defense than a conclusive demonstration could, for both defenses and plausible reasoning are relevant in a situation in which the issues cannot be decisively resolved, whereas if a conclusive demonstration is really conclusive, then the conclusion is established as true and there is less or no need for "defending" it. That is, perhaps Galileo took a bigger risk of violating the anti-Copernican decree by presenting the tidal argument as merely plausible than if he had presented it as conclusive.

Similarly, one may ask whether Galileo's qualified claim in the previous paragraph—that the Copernican arguments are better than the geostatic arguments—amounts to a defense of Copernicanism. It could be claimed that to defend a thesis is to give arguments supporting it and to criticize arguments opposing it; that the book shows that the Copernican arguments are better by giving arguments supporting Copernican conclusions and refuting geostatic arguments; and that therefore the book is a defense of Copernicanism. The Inquisition consultants at the trial of 1633 engaged in this plausible reasoning (Finocchiaro 1989, 262–76). But Galileo must have had in mind the following equally plausible argument: in order to discuss opposing arguments, one must do more than just present them; one must analyze them and evaluate them for their correctness and for the strength of the support given to their conclusions; as long as the analysis and evaluation are conducted in an objective manner, there is nothing wrong even if such a procedure yields the result that one side is better supported than the other. It is as if there were two kinds of defense: a pejorative kind consisting of supporting a thesis by fallacious and bad arguments and irrelevant evidence, and a nonpejorative kind consisting of a critical examination of all relevant arguments followed by the judgment that they support one thesis rather than another; the anti-Copernican decree could be taken to prohibit the first kind of defense but not the second, and then the issue of its violation would reduce to the question of whether Galileo's own analyses and evaluations are objective, fair, and correct.

more, we have not speculated less, and that to assert the earth's rest and take the contrary solely as a mathematical whim does not derive from ignorance of others' thinking but, among other things, from those reasons provided by piety, religion, acknowledgment of divine omnipotence, and awareness of the weakness of the human mind.[12]

Furthermore, I thought it would be very appropriate to explain these ideas in dialogue form; for it is not restricted to the rigorous observation of mathematical laws, and so it also allows digressions which are sometimes no less interesting than the main topic.

Many years ago in the marvelous city of Venice I had several occasions to engage in conversation with Giovanfrancesco Sagredo,[13] a man of most illustrious family and of sharpest mind. From Florence we were visited by [31] Filippo Salviati,[14] whose least glory was purity of blood and magnificence of riches; his sublime intellect fed on no delight more avidly than on refined speculations. I often found myself discussing these subjects with these two men and with the participation of a Peripatetic philosopher,[15] who seemed to have no greater obstacle to the understanding of the truth than the fame he had acquired in Aristotelian interpretation.

12. Here Galileo suggests that Catholics choose to believe that the geostatic thesis is true, not out of scientific ignorance, but for religious reasons. But he has just finished saying that the scientific arguments favor the geokinetic thesis, and so the geostatic belief is held despite the scientific counterevidence; thus, he seems to be committed to saying that religious reasons can outweigh scientific arguments. However, since he also claims that the Copernican arguments are not conclusive, he need only be committed to saying that this primacy of religion over science holds only when the scientific arguments are inconclusive. Further refinements would be needed if one were to take into account the variety of religious reasons. For example, the "acknowledgment of divine omnipotence" refers to the pope's favorite argument, which is explicitly stated at the end of the book (selection 16); and some versions of this argument are indeed cogent and unanswerable. Biblical arguments are perhaps being mentioned under the label of "piety"; but perhaps they are not explicitly mentioned for a good reason, namely, that Galileo still accepts his argument in the "Letter to the Grand Duchess," showing that when a scientific proposition is capable of being conclusively proved (even if it has not yet been proved) biblical assertions carry no weight at all; for an elaboration of this interpretation, see Finocchiaro (1986).

13. Sagredo (1571–1620) was a diplomat and public official of the Republic of Venice; he became Galileo's best friend when (1592–1610) Galileo lived in Padua, a part of the Republic and the site of its state university.

14. Salviati (1582–1614) was a wealthy Florentine nobleman whose interest in science and philosophy earned him membership in the Lincean Academy in 1612; he was one of Galileo's closest friends in Florence.

15. This person may have been Cesare Cremonini (1550–1631), professor of philosophy at the University of Padua and a colleague with whom Galileo was on friendly terms despite their intellectual opposition.

Now, since Venice and Florence have been deprived of those two great lights by their very premature death at the brightest time of their life, I have decided to prolong their existence, as much as my meager abilities allow, by reviving them in these pages of mine and using them as interlocutors in the present controversy. There will also be a place for the good Peripatetic, to whom, because of his excessive fondness of Simplicius's[16] commentaries, it seemed right to give the name of his revered author,[17] without mentioning his own. Those two great souls will always be revered in my heart; may they receive with favor this public monument of my undying friendship, and may they assist me, through my memory of their eloquence, to explain to posterity the aforementioned speculations.

These gentlemen had casually engaged in various sporadic discussions, and, as a result, in their minds their thirst for learning had been aroused rather than quenched. Thus they made the wise decision to spend a few days together during which, having put aside every other business, they would attend to reflecting more systematically about God's wonders in heaven and on earth. They met at the palace of the most illustrious Sagredo, and after the proper but short greetings, Salviati began as follows.

16. Simplicius was a Greek philosopher who lived in the sixth century A.D. and is best known as one of the greatest commentators on Aristotle.

17. Simplicio is the Italian name of the Aristotelian commentator called Simplicius in English; it is also an Italian word with the connotation of *simpleton*. This ambiguity added spice to the book but compounded Galileo's problems after its publication. I have used the Italian *Simplicio* for the name of this character.

# First Day

. . . [1]

## [2. Natural Motion and Aristotle's Logic]

[57] . . . SIMPLICIO. Aristotle was someone who did not expect more than is appropriate from his intellect, even though it was very sharp; thus in his philosophizing he judged that sensible experience should have priority over any theory constructed by the human intellect,[2] and said that those who denied their senses deserved to be punished by being deprived of them. Now, who is so blind as not to see that

1. The passages omitted here correspond to Favaro 7:32.1–57.4 and Galileo 1967, 9–32. Galileo begins with a discussion of the three-dimensionality of the world and the role of mathematics in physical inquiry. Then he gives a critical analysis of Aristotle's general theory of natural motion; his view is that straight and circular motions are not two distinct instances of simple natural motion but rather two different stages of natural motion: straight motion can be acquired naturally but cannot naturally continue forever, whereas circular motion can naturally continue forever but cannot be acquired naturally without prior straight motion. This view provides a theoretical, a priori criticism of the Aristotelian law of natural motion. On the other hand, selection 2 contains an empirical critique of the same law, embedded in the context of criticizing an anti-Copernican argument which used that law as a key premise. For more details, see Finocchiaro (1980, 33–34, 79–84, 106–7, and 346–53).

2. Cf. Aristotle, *On the Generation of Animals*, III, 10, 750b27 and 760b31. This sentence expresses a crucial methodological idea, which I call the fundamental principle of empiricism. For more details, see the appendix (2.4), Brown (1977; 1987), Duhem (1954), Franklin (1986), and Hacking (1983).

the parts of the element earth and of the element water, being heavy bodies, move naturally downwards, namely, toward the center of the universe, assigned by nature herself as the goal and end of straight downward motion; and similarly that fire and air move straight toward the concave side of the lunar orb,[3] as the natural end of upward motion? Since this can be seen clearly, and since we are sure that the same holds for the whole as for the parts,[4] must we not conclude that it is true and manifest that the natural motion of the earth is straight toward the center, and of fire straight away from the center?[5]

SALVIATI. By virtue of this argument of yours, the most you can pretend that one should grant you is the following:[6] because the parts of the earth return spontaneously (and hence naturally) to their place by means of straight motion when they are removed from their whole (namely, from the place where they rest naturally), that is when they are displaced into a wrong and disorderly arrangement, therefore (assuming that the same holds for the whole as for the parts)[7] one could infer that if the terrestrial globe were violently[8] removed from the place assigned to it by nature then it would return there by a straight line. As I said, this is the most one could grant you, giving you all sorts of advantages. However, if one were to examine these issues rigorously, one would first deny that, when returning to their whole, the parts of the earth move in a straight line rather than a circular or other mixed one; and you would have a hard time demonstrating the contrary, as you will plainly see in the answers[9] to the particular reasons and evidence advanced by Ptolemy and Aristotle. Secondly, one could say that the parts of the earth move not in order to go to the center of the world but to go to and unite with their

3. In this context, Aristotelian natural philosophy conceived the lunar orb as a spherical shell consisting of aether and surrounding the earth at the distance of the moon, in which the moon was embedded and whose monthly rotation carried the moon in its eastward revolution around the earth; the concave side of the lunar orb was the side of the spherical shell facing the earth.

4. Here and elsewhere in the book, which is in Italian, Galileo writes this principle in italics and in Latin: *eadem est ratio totius et partium*; cf. Aristotle, *On the Heavens*, I, 3, 270a11–12.

5. The last several sentences (from "Now, who is so blind . . . ") express an important anti-Copernican argument, which I shall call the natural motion objection. Cf. Aristotle, *On the Heavens*, II, 14, 296b8–22.

6. Some of the points in Salviati's following answer to Simplicio echo some of the views found in Copernicus, *On the Revolutions*, I, 8–9 (1992, 15–18).

7. In the original, *eadem sit ratio totius et partium*.

8. That is, by some external force.

9. Cf. selection 8 and the passage in Favaro (7:190–93) or Galileo (1967, 164–67).

whole, and that this explains why they have a natural tendency to go toward the center [58] of the terrestrial globe, by which tendency they try to form and conserve it; and then what other whole and what other center would you find in the world toward which the entire terrestrial globe would try to return after having been removed from there, in accordance with the principle that whatever holds for the parts would hold for the whole?[10] Add to this that neither Aristotle nor you will ever be able to prove that the earth is de facto at the center of the universe; instead, if we can assign a center to the universe, we will rather find the sun to be located there, as you will understand in due course.[11]

Now, from the unanimous cooperation of all the parts of the earth to form a whole, it follows that they contribute to that goal from all directions with equal tendency, and that in order to unite together as much as possible they adopt a spherical shape. Therefore, why should we not believe that the moon, sun, and other heavenly bodies have a round shape simply because of a unanimous instinct and a natural convergence of all their component parts? And if any of them by some violence were ever to be separated from its whole, is it not reasonable to believe that it would return there spontaneously and by natural instinct, and accordingly to conclude that straight motion belongs equally to all bodies?[12]

SIMPLICIO. If you want to deny not only the principles of the sciences but also plain experience and the senses themselves, there is no doubt

10. Here Galileo suggests this paradox: if we say that when displaced, the parts of the earth move spontaneously toward the center of the whole *earth*, and if we accept the principle that whatever holds for the parts also holds for the whole, then we would have to conclude that if the whole earth were displaced it would move spontaneously toward *its own* center. But it is an intrinsic impossibility for the whole earth to move toward its own center. This represents an elegant counterexample to the principle that whatever holds for the parts also holds for the whole, thus yielding a conclusive refutation of it. Note that this counterexample is talking about the center of the earth and not of the universe; in fact, it is *not* inherently impossible for the whole earth to move toward the center of the universe, any more than it is impossible for a terrestrial body to move toward the earth's center.

11. See selection 11.

12. Here Galileo suggests that both terrestrial and heavenly bodies spontaneously undergo straight motion in certain circumstances, namely, that when a body is displaced from the whole of which it is a part, it tries to return to the whole by means of straight motion. This suggestion is part of his attempt to undermine the earth-heaven dichotomy, which was a fundamental doctrine of Aristotelian natural philosophy; it divided the universe into two radically different regions, largely on the basis of natural straight motion as belonging to terrestrial bodies and natural circular motion as belonging to heavenly bodies. In the omitted passage that precedes this selection, he elaborates this claim about the universality of straight motion and also an analogous conclusion about circular motion, namely, that circular motion belongs equally to both terrestrial and heavenly bodies in the sense that it can naturally continue forever in both domains.

that you can never be convinced or dissuaded from any opinion you accept; and I shall keep quiet because one cannot dispute against those who deny the principles,[13] rather than because I am persuaded by your reasons. Now, let us take the things you just mentioned, such as your casting doubt even on whether the motion of falling bodies is straight or not; how can you reasonably deny that the parts of the earth (namely, the heaviest substances) fall toward the center with straight motion given that, when released from a very high tower whose walls are very straight and built plumb, they graze it (so to speak) and hit the ground exactly at the same point as a plummet hanging by a string tied where they were released? Is not this argument more than sufficient to show that such motion is straight and toward the center? In the second place, you cast doubt on whether the parts of the earth move in order to go where Aristotle claimed (toward the center of the universe), as if he [59] had not conclusively demonstrated it by means of contrary motions, when he argues as follows: the motion of heavy bodies is contrary to that of light ones; but the motion of light bodies is seen to be directly upwards, namely, toward the circumference of the universe; therefore, the motion of heavy bodies is directly toward the center of the universe, and it happens accidentally that it is toward the center of the earth because the latter coincides with the former.[14] Finally, it is useless to try to determine the motion of a part of the lunar or solar globe that has been separated from its whole because one is looking for what would follow as a consequence of an impossibility; in fact, as Aristotle also demonstrates, heavenly bodies are inert, impenetrable, and indivisible, so that the case cannot happen; and even if it could happen and the separated part returned to its whole, it would return neither as light nor as heavy, for Aristotle himself proves that heavenly bodies are neither heavy nor light.

13. This rule about the limits of discussion is written in italics in Latin in Galileo's original text: *contra negantes principia non est disputandum*. Cf. Aristotle, *Physics*, I, 2, 185a1–3; VIII, 3, 253b1–5.

14. This passage (after the colon) expresses an important Aristotelian argument, which I call the basic geocentric argument; cf. Aristotle, *On the Heavens*, II, 14, 296b8–26; IV, 4, 311a15–312a21. This argument is connected with the natural motion objection (considered earlier in this passage) insofar as they share the Aristotelian law of natural motion for heavy bodies; this proposition, which was used as an unsupported premise in the earlier argument, is now provided with an empirical justification. The same proposition could be supported by the Aristotelians by means of a theoretical, a priori argument; but Galileo has already dealt with that theoretical justification in the omitted passage that immediately precedes the present selection. For an account of this and more details, see Finocchiaro (1980, 353–56).

SALVIATI. As I just said, you will hear how reasonably I doubt whether falling bodies move in a straight and perpendicular line when I examine this particular argument.[15] Regarding the second point, I am surprised that you should need an explanation of Aristotle's paralogism, which is intrinsically so evident, and that you do not see that Aristotle assumes what is in question. Note then that . . . [16]

SIMPLICIO. Please, Salviati, speak of Aristotle with more respect. How can you ever convince anyone that he could have committed a serious error like assuming as known what is in question, given that he was the first, only, and admirable explainer of syllogistic forms, demonstration, fallacies, the methods for exposing sophisms and paralogisms, and in short the whole of logic? Gentlemen, one must first understand him perfectly, and then try to impugn him.[17]

SALVIATI. Simplicio, we are here talking to each other in a friendly manner to inquire into the truth. I will never hold it against you that you show me my errors; when I do not penetrate Aristotle's mind you should freely rebuke me, and I will be grateful to you for it. In the meantime, allow me to state my difficulties and also to respond to your last remarks. I say that, as you know very well, logic is the organ[18] of philosophizing; but, just as it can happen that an artisan is excellent in making [60] organs but unlearned in playing them, so one can be a great logician but little able to use logic; similarly, there are many who know by heart the whole of poetics but are then incapable of writing even a few verses, and others who know all the precepts of Da Vinci[19] but are then

15. Cf. selection 8 and the passage in Favaro (7:190–93) or Galileo (1967, 164–67).

16. This ellipsis occurs in the original text. Galileo wants to convey the impression that Simplicio impatiently interrupts Salviati at this point.

17. Aristotle is indeed universally recognized as the founder of the science of logic; his achievement in this field was such that no other comparable theoretical breakthrough was made until the end of the nineteenth century, although there were numerous minor advances in the two intervening millennia. But as Salviati goes on to suggest, logic is the theory of reasoning, and being good in the general theory of reasoning is not the same thing as being good in the practice of reasoning on a particular subject. Since Galileo's concern is the latter, he can question Aristotle's reasoning about physical phenomena without questioning his theoretical analysis of reasoning.

18. Salviati's argument in this paragraph involves a play on words: here the term *organ* means instrument, in the sense of a mental instrument; in the next clause, it refers to a particular musical instrument; but the term also refers to Aristotle's writings on logic and methodology, entitled *Organon*.

19. Leonardo Da Vinci (1452–1519) is one of the greatest painters who ever lived, best known for the paintings *Mona Lisa* and *The Last Supper*; he is also known as the author of a *Treatise on Painting* (*Trattato della Pittura*). And he is regarded as one of the greatest examples of the versatility of human talent since he was also a scientist, engineer, architect, musician, and sculptor.

unable to paint a stool. One does not learn how to play the organ from those who make organs but from those who can play them; one learns poetry by constantly reading poets; one learns how to paint by constantly drawing and painting; and one learns the art of demonstrating by reading books full of demonstrations, which are exclusively the books of mathematicians and not those of logicians.[20]

Now, returning to the argument, I say that what Aristotle sees in regard to the motion of light bodies is that fire begins its motion at any place on the surface of the terrestrial globe, moves away from it in a straight line, and rises higher; this is truly motion toward a circumference greater than the earth's, and Aristotle himself makes it move toward the concave side of the moon's orb. But one cannot claim that this circumference is that of the universe, or concentric with it, so that motion toward the former is also motion toward the circumference of the universe; this cannot be claimed unless one first supposes that the earth's center (from which we see rising light bodies move away) is the same as the center of the universe, which is equivalent to saying that the terrestrial globe is located at the center of the universe; but this is what we question and what Aristotle intends to prove. And you tell me that this is not an obvious paralogism?[21]

SAGREDO. This Aristotelian argument seems to me flawed and inconclusive also in another way even if one were to agree that the circumference toward which fire moves in a straight line is the one that encloses the universe. For, given any point whatever inside a circle and not only the center, every body starting from there and moving in a straight line toward any part whatever will undoubtedly go toward the circumference and will arrive there by continuing its motion, so that it will be very true to say that it moves toward the circumference; but it will not be true that a body moving along the same lines in the opposite direction goes toward the center, unless the selected point is the center itself or the motion is along a line which goes through the selected point and the center. Thus, [61] to say "fire moves straight and goes toward the circumference of the universe, therefore the parts of the earth which move along the same lines

20. This is one of the many expressions of Galileo's high regard for mathematics, the point here being that mathematics is the art of rigorous proof, and so it is the best thing to study if one wants to learn how to do rigorous proofs; cf. especially selection 9 but also selections 4, 13, and 14. The reason why studying logic is not equally effective is that it is the theory of proof, which must be applied to yield actual proofs; but the application is not necessarily as good as the theory. Thus he is not really disparaging logic but rather giving the distinction between logic and mathematics as a special case of the more general distinction between theory and practice. For more details, see Barone (1972), Finocchiaro (1980), and Wallace (1992a; 1992b).

21. The fallacy is presumably an instance of begging the question.

in the opposite direction go toward the center of the universe" does not prove anything unless one first assumes that the lines of fire, when extended, pass through the center of the universe; now, because we know for certain that they do pass through the center of the terrestrial globe (being perpendicular to its surface and not inclined), therefore the argument needs to assume that the earth's center is the same as the center of the universe, or at least that the parts of fire and of the earth ascend and descend exactly along a single line which passes through the center of the universe; and this is false and repugnant to experience, which shows that the parts of fire do not ascend by a single line but by an infinite number of them drawn from the earth's center toward all parts of the universe and always perpendicular to the surface of the terrestrial globe.

SALVIATI. Sagredo, yours is a very clever way of leading Aristotle into the same difficulty and showing clearly the error, though you add another fault. We know the earth to be spherical,[22] and so we are certain it has its center; given that the motions of its parts are all perpendicular to the earth's surface, it is necessary to say that these parts move toward that center; we understand how, by moving toward the earth's center, they move toward their whole and their universal mother; and then we are supposed to be so good as to let ourselves be persuaded that their natural instinct is not to go toward the earth's center but toward that of the universe, concerning which we do not know where it is or whether it exists, and which is merely an imaginary point and a nothing without any property, even if it does exist.[23]

Let me now respond to Simplicio's last point, to the effect that it is useless to discuss whether the parts of the sun, moon, and other heavenly bodies would naturally return to their whole if separated from it, the reason being that the case is impossible since it is clear from Aristotle's demonstrations that heavenly bodies are inert, impenetrable, indivisible, etc. I answer that none of the conditions on which Aristotle bases the distinction between heavenly bodies and elemental ones has any solidity other than what

22. Note that the Copernican controversy was about the location and the motion of the earth and had nothing to do with its shape, which both sides agreed was a sphere. Whatever the belief of uneducated people may have been, among scientists and philosophers the issue of the earth's shape had been settled since the time of the ancient Greeks.

23. This agnosticism about the center of the universe is a recurring cosmological theme in the *Dialogue* and in Galileo's work in general. His heliocentrism concerns the issue of the center of planetary revolutions, not that of the center of the whole universe. His caution is due in part to religious concern, in the sense that the issue of the center of the universe is related to the question of the infinity of the universe; in regard to the latter, belief in this infinity had been one of the reasons leading to the trial and execution of Giordano Bruno by the Inquisition in the year 1600. See also selection 11, [347]; Galileo's "Reply to Ingoli" (Finocchiaro 1989, 179); and his letter to Fortunio Liceti of 24 September 1639 (Favaro 18:106).

derives from the distinction between the natural motions of the two; hence, having denied that circular motion belongs only to heavenly bodies and having affirmed that it is appropriate for all naturally moving bodies,[24] we must [62] draw the necessary consequence that the attributes of generable or ingenerable, changeable or unchangeable, divisible or indivisible, etc., belong equally and commonly to all bodies, heavenly as well as elemental;[25] or perhaps Aristotle was wrong and mistaken in deducing from circular motion the attributes he assigned to heavenly bodies.

SIMPLICIO. This manner of philosophizing tends to subvert all natural philosophy and to throw into disorder and to upset the heavens, the earth, and the whole universe. But I believe the foundations of the Peripatetics are such that one need not fear that upon their ruins one can erect new sciences.

SALVIATI. Do not worry about the heavens or the earth, and do not be afraid for the subversion of these things or of philosophy. For, in regard to the heavens, it is useless that you should fear for what you yourself consider to be unchangeable and inert; and, as to the earth, we are trying to make it more noble and more perfect insofar as we strive to make it similar to the heavenly bodies and in a sense to place it in heaven, from which your philosophers have banished it. Philosophy itself cannot but benefit from our disputes because if our thoughts are true then we will make new gains, and if they are false then their refutation will confirm further the earlier doctrines.[26] You should rather worry about certain philosophers and try to help and support them, for science itself cannot but advance. Now, returning to our plan, please feel free to put forth what you remember in support of the great difference that Aristotle postulates between heavenly bodies and the elemental region, making the former ingenerable, indestructible, unchangeable, etc., and the latter degradable, changeable, etc.

. . .[27]

24. As I said earlier, this conclusion is elaborated in a passage (omitted here) just before this selection.

25. This unity of terrestrial and heavenly bodies is a recurring theme in the *Dialogue* and in Galileo's work in general; it corresponds to his rejection of the earth-heaven dichotomy and is explicitly discussed in selection 3.

26. This sentence expresses an important methodological principle, which seems to lie at the heart of the justification of such institutions as freedom of thought, speech, inquiry, expression, and the press.

27. The passage omitted between selections 2 and 3 corresponds to Favaro 7:62.28–71.33 and Galileo 1967, 38–47. It contains a criticism of Aristotle's argument for the earth-heaven dichotomy based on contrariety. This argument is the one that in the next selection (3, [73]) Simplicio calls the a priori argument; it is summarized in a note there. The criticism contains a very rich range of logical themes (cf. Finocchiaro 1980, 34, 108, and 357–72).

# [3. Heavenly Changes and Aristotle's Empiricism]

[71] . . . SIMPLICIO. Here are, to begin with, two very powerful demonstrations proving that the earth is very different from heavenly bodies. First, bodies which are generable, degradable, changeable, etc. are very different from those which are ingenerable, indestructible, unchangeable, etc.; the earth [72] is generable, degradable, changeable, etc. and heavenly bodies ingenerable, indestructible, unchangeable, etc.; therefore, the earth is very different from heavenly bodies.[28]

SAGREDO. With the first argument you bring back onto the table what was there earlier and has just been taken away.

SIMPLICIO. Not so fast, Sir! Hear the rest, and you will see how different it is from that one. In the earlier one the minor premise[29] was proved a priori,[30] but now I want to prove it a posteriori;[31] see whether this is the same. So I prove the minor premise, the major one being very obvious: sensible experience shows us how on earth there are constantly generations, decay, changes, etc.; none of these have ever been seen in

28. This argument does not have the form of a syllogism in the strict technical sense, but it is easily related to such a form.

29. In an argument with two premises and one conclusion, the minor premise is the one which contains the subject of the conclusion, whereas the major premise is the one which contains the predicate of the conclusion. For example, in an argument of the form "all A are B; x is A; so, x is B" the minor premise is "x is A" because it contains the conclusion's subject x, whereas the major premise is "all A are B" because it contains the conclusion's predicate B. Simplicio's argument is slightly more complicated, being of the form: "C are different from U; e is C and h is U; so, e is different from h"; applying the definition yields that the minor premise is "e is C and h is U," namely, the proposition that the earth is changeable and heaven is unchangeable.

30. In the omitted passage immediately preceding this selection, heavenly unchangeability is justified by means of the following argument: the heavens are unchangeable because change derives only from contrariety and contrariety derives ultimately from contrary motions; but there is no contrariety in the motions of the heavenly bodies because they are all circular and circular motion has no contrary. In calling this argument a priori, Simplicio means that it is relatively independent of sensory experience, namely, that its key premises do not depend directly on observation; this meaning yields a sufficient contrast with the next argument which he calls a posteriori. Sometimes the term *a priori* is used to mean "completely independent of sensory experience," in the sense of the epistemology of apriorism; but I do not think this is the relevant meaning here. In short, Simplicio means that the contrariety argument discussed earlier is a theoretical, as distinct from empirical, argument.

31. Here *a posteriori* means "based directly upon observation or sensory experience"; in fact, in the ensuing argument, the premises are clearly observation reports.

the heavens either with our senses, or according to tradition or the reports of the ancients; therefore, heaven is unchangeable etc., and the earth changeable etc., and thus different from heaven.[32] The second argument is based on an important and essential phenomenon, and it is this: bodies that by nature are dark and devoid of light are different from luminous and shining bodies; the earth is dark and without light, and heavenly bodies are resplendent and full of light; therefore, etc.[33] In order not to accumulate too large a pile, it is better to respond to these, and then I will advance others.

SALVIATI. As to the first, whose strength derives from experience, I desire you to explain to me more clearly the changes that you see occurring on the earth but not in heaven, and that make you call the earth changeable and the heavens not.

SIMPLICIO. On the earth I constantly see the generation and destruction of plants, trees, and animals, and the production of winds, rain, and storms; in short, this aspect of the earth is in a perpetual metamorphosis. None of these changes are seen in the heavenly bodies; their appearance and arrangement have been very exactly the same within human memory, without the generation of anything new or the destruction of anything old.

SALVIATI. However, since you quietly accept these observations that can be made or, to be more precise, have been made by vision, you must think that China and America are heavenly bodies; for, surely, you have never seen in them the changes you see here in Italy, and hence, as far as you are concerned, they are unchangeable.

32. The last three clauses (after the colon) express an important argument, which I shall call the observational argument for heavenly unchangeability. The rest of this selection is a multifaceted criticism of this argument.

33. This argument contains another Aristotelian reason for the earth-heaven dichotomy, an optical reason involving the question of whether a body gives off its own light. Note that Galileo leaves the conclusion of this second argument unstated, but that it is the same as the conclusion of the argument Simplicio gives in his first speech of this selection. This second argument is not further discussed in this or any other selection. But in another passage of the "First Day" (Favaro 7:95–124; Galileo 1967, 71–98), Galileo criticizes its minor premise by arguing that both the earth and moon (and the planets, though of course, not the sun) share the same optical properties in regard to luminosity, darkness, and reflecting power. Moreover, he is inclined to accept the major premise of this argument (that there is an important difference between luminous and dark bodies); indeed, this major premise together with the optical similarity between the earth and the moon (and planets) enables him elsewhere to formulate the following plausible but nonconclusive argument from analogy: that the earth probably moves because the moon and planets do and both the earth and the moon and planets are nonluminous, whereas the sun probably does not move in an orbital revolution because unlike the planets it gives off its own light; cf. Favaro (6:559–61; 7:291–92), Finocchiaro (1989, 196–97), and Galileo (1967, 266–68).

SIMPLICIO. Although I have not sensibly seen these changes in those places, we have unquestionable reports about them. Moreover, [73] because the same holds for the parts as for the whole, and because those countries are parts of the earth (just as ours is), it is necessary that those are changeable (as this one is).[34]

SALVIATI. And why have you not observed and seen them yourself with your own eyes, without having to simply believe the reports of others?

SIMPLICIO. Because those countries are so remote that our vision would be incapable of detecting such changes, aside from their not being exposed to our view.

SALVIATI. Now you see how you yourself have easily discovered the fallacy of your argument. For, if you say that the changes seen on the earth near us could not be detected when occurring in America due to the excessive distance, much less could you detect them when occurring on the moon, which is hundreds of times farther. And if you believe the changes in Mexico based on the news coming from there, what reports have come to you from the moon to make you conclude that there are no changes on it? Therefore, from your not seeing heavenly changes (which could not be seen due to the excessive distance even if there were any), and from your not having reports of them (as long as it is impossible to have them), you cannot infer that they do not exist, just as from seeing and hearing them on the earth you infer correctly that changes exist here.

SIMPLICIO. I can find some changes having occurred on the earth which are so great that, if such happened on the moon, they could very well be observed from down here. We have very ancient claims that formerly at the Strait of Gibraltar the mountains Abila and Calpe[35] were

34. Compare the argument expressed in this sentence with the natural motion objection of selection 2. Both appeal to the parts-whole principle, and here Galileo repeats the same Latin formulation stated earlier. But the argument here is from whole to parts, whereas the earlier one is from parts to whole. I have made this clear in my translation by correspondingly different renderings of the same Latin sentence, which literally reads "the same reasoning applies to the whole and to the parts" and so formally subsumes both directions of inference. Also compare the strength of the two inferences: the present one does not raise the paradox of the earlier one, nor does it seem to be fallacious in any simple or obvious sense. Note also that the rest of the discussion does not seem to take up this particular subargument.

35. The Strait of Gibraltar connects the Mediterranean Sea to the Atlantic Ocean and separates North Africa from Southern Europe; it is flanked by two mountains, Mt. Acha at Ceuta on the African side and the Rock of Gibraltar on the European side. The ancients called these mountains respectively Abila and Calpe, and jointly, the Pillars (or Gates) of Hercules. This story of the formation of the Mediterranean Sea was widespread in antiquity.

joined together along with some other smaller ones and kept out the ocean; but then, whatever the cause may have been, these mountains separated, leaving the way open for the seawater, which rushed in to such an extent as to form the whole Mediterranean Sea. Now, if we consider its great size and the difference in appearance between the surface of the water and that of the land when seen from afar, there is no doubt that such a change could very well have been observed by someone on the moon; analogously, similar changes on the moon should be visible to us inhabitants of the earth; but there is no record that such a thing has ever been seen; therefore, there is no reason left for saying that any of the heavenly bodies is changeable, etc.

SALVIATI. I would not dare say that such vast changes have occurred on the moon, but neither am I sure that there may not have been some. For such a change could show us only [74] some variation between the lighter and the darker parts of the moon; but I do not know that there have been on the earth any curious selenographers[36] who for a very long time have provided us with such exact lunar maps that we can be sure no change has ever occurred on the face of the moon; the most detailed description I find of its appearance is that some say it represents a human face, others that it is similar to a lion's muzzle, and still others that it represents Cain with a bundle of thorns on his shoulder.[37] Therefore, to say "the heavens are unchangeable because the changes noticeable in the earth are not seen to occur in the moon or other heavenly bodies" has no strength to prove anything.[38]

SAGREDO. I still have another qualm regarding this first one of Simplicio's arguments, and I should like it removed. So I ask him whether the earth was generable and degradable before the Mediterranean overflow, or whether it began then to be such.

SIMPLICIO. Undoubtedly it was generable and degradable before; but that change was so vast that it could have been observed even from the moon.

36. A selenographer is a specialist who studies the moon's surface. The term is similar to *geographer*: the Greek root *geo* refers to the earth; the Greek name *Selene* refers to the moon-goddess of ancient Greek mythology.

37. In the Bible, Cain is the eldest son of Adam and Eve who murdered his brother Abel out of jealousy. According to a legend to which Galileo refers here, God punished Cain by banishing him to the moon and forcing him to carry on his shoulder a bundle of thorns until the end of the world (Sosio 1970, 61n.2).

38. This completes the first line of criticism of the observational argument for heavenly unchangeability, and it is instructive to compare this criticism with the others. For such details, see the appendix (1.4).

SAGREDO. Now, if the earth was generable and degradable even before such a flood, why can the moon not be the same without a similar change? Why does one need for the case of the moon what is completely irrelevant for the case of the earth?

SALVIATI. A very shrewd question.[39] However, I suspect that Simplicio alters somewhat the meaning of the texts of Aristotle and the other Peripatetics; they say they hold the heavens to be unchangeable because one has never seen there the generation or destruction of any star, which is perhaps a smaller part of the heavens than a city is of the earth, and yet an infinite number of these have been destroyed in such a way that no signs of them have been left.

SAGREDO. I certainly felt otherwise, and believed that Simplicio concealed this interpretation of the text in order not to burden the Master and his followers with a more serious flaw. What an absurdity to say: "the heavenly region is unchangeable because in it there is no generation or destruction of stars." Is there perhaps someone who has seen the destruction of one terrestrial globe and the regeneration of another one? Is it not accepted by all philosophers that there are in the heavens very few stars which are smaller than the earth, but very many which are much much larger? [75] Therefore, the destruction of a star in the heavens is not a lesser event than the destruction of the whole terrestrial globe; and, if to correctly infer generation and destruction in the universe it is necessary to have the destruction and regeneration of such vast bodies as stars, you can forget about it completely because I assure you that one will never see the destruction of the terrestrial

39. Sagredo's and Simplicio's exchange in the last three paragraphs embodies a second, distinct line of criticism of the observational argument for heavenly unchangeability. Though it is brief and is not taken up again, it is extremely interesting. It raises questions about whether a general conclusion like heavenly unchangeability can ever be correctly justified on empirical, observational, a posteriori grounds. This in turn raises the general methodological problem of what (if anything) we learn from experience, and how we do so (if we do), given that there is such a difficulty about generalizing and extrapolating past or limited observations into the future or into wider domains. This is one formulation of what philosophers call the problem of the justification of induction, so-called because induction is sometimes defined as a process of reasoning consisting of arguments whose premises are statements of particular observations and whose conclusion is a generalization. This type of difficulty is also useful for appreciating that perhaps we should not expect a conclusion of such an argument to follow necessarily from its premises, but only to be supported with some degree of probability. But these are controversial issues which are far from settled; cf. L. J. Cohen (1989), Popper (1959), and Skyrms (1975).

globe or other whole body of the universe in such a way that it will be dissolved without any trace after having been visible for many centuries earlier.[40]

SALVIATI. However, to satisfy Simplicio in an overwhelming manner and remove him from error if possible, I say that in our age we have new phenomena and observations such that, if Aristotle were living nowadays, he would change his mind. This may be obviously gathered from his own manner of philosophizing. For he writes that he regards the heavens as unchangeable, etc. because no one has seen the generation of something new or the destruction of something old; so he implicitly conveys the impression that, if he had seen one of these phenomena, he would have concluded the opposite and given priority to sensible experience over natural theorizing (as is fitting);[41] if he had not wanted to value the senses, he would not have inferred the unchangeability from not sensibly seeing any change.

SIMPLICIO. Aristotle relied primarily on the a priori argument,[42] showing the necessity of heavenly unchangeability by means of natural, obvious, and clear principles; afterwards he established the same conclusion a posteriori by means of the senses and the traditions of the ancients.

SALVIATI. What you are describing is the method he used in writing his doctrine, but I do not think it is the method with which he investigated it; for I am firmly convinced that he first tried to ascertain the conclusion as much as possible by means of the senses, experience, and observations, and then he searched for ways of demonstrating it. In fact, this is more or less what one does in the demonstrative sciences.

40. In the last two paragraphs, Salviati and Sagredo advance a third criticism of the observational argument for heavenly unchangeability. This criticism involves an equivocation about *body* and *change*. The troublesome ambiguity is that *body* may mean a whole planet, star, or globe and a part of such a whole body; correspondingly, *change* may mean the creation or destruction of a whole and the creation or destruction of a part of such a body. But it is not easy to describe the alleged flaw more precisely. For more details, see the appendix (1.4) and Finocchiaro (1980, 358–59, 370–71, and especially 372–77).

41. This is another expression of what I have called the fundamental principle of empiricism; here not only is it being attributed to Aristotle, but it is apparently being endorsed by Galileo's spokesman. Cf. Aristotle, *On the Generation of Animals*, III, 10, 760b31; cf. also selection 2, [57].

42. Besides the observational argument for heavenly unchangeability, Aristotle had a theoretical justification, the so-called argument from contrariety, summarized in an earlier note to p. [72] of this selection.

This happens because, when the conclusion is true, by using the method of resolution[43] one easily finds some proposition already demonstrated or some self-evident principle; but when the conclusion is false, one can go on endlessly without ever arriving at a known truth, or else one may arrive at an obvious impossibility or absurdity. Have no doubt that, [76] long before Pythagoras found the demonstration for which he made a hecatomb,[44] he had become certain that in a right triangle the square of the side opposite the right angle is equal to the sum of the squares of the other two sides. The certainty of the conclusion helps a lot in finding the demonstration (as long as we understand we are still talking about the demonstrative sciences).[45] In any case, whatever Aristotle's procedure may have been, whether the a priori theorizing preceded the a posteriori senses or vice versa, the important point is that Aristotle himself places sensible experience before all theorizing (as

43. Here the method of resolution may be defined as a procedure used to find the propositions from which a given proposition follows; one starts by assuming that the given proposition is true and that there is a valid proof or argument of which it is the conclusion, and then seeks to find the premises of such an argument; it is contrasted with the method of composition, which is the reverse procedure where one finds the valid consequences of a given proposition or set of propositions. For example, at one point in his physical research Galileo discovered the empirical truth of the law of squares, namely, he became convinced that bodies falling freely from rest cover a distance which increases as the square of the time elapsed; then, for several years, he tried to formulate a proper principle which would enable him to derive this law; eventually he found this to be the principle of time proportionality—the proposition that the (instantaneous) speed of a falling body increases in direct proportion to the time elapsed from rest; here the method of resolution describes the procedure Galileo followed after his initial discovery. The notion has a long history and continues to generate controversy in regard to its exact meaning and role; moreover, it should be noted that the method of resolution is also known as the method of analysis, and the method of composition as the method of synthesis. For a flavor of these discussions, see Jardine (1976), Naylor (1990), Wallace (1983, 302; 1992b, 69–77 and 221), and Wisan (1978, 29).

44. In ancient Greece and Rome, a hecatomb was the slaughter of one hundred cattle as a ritual sacrifice to the gods. Pythagoras is alleged to have made such a sacrifice when he discovered the proof of the geometrical theorem bearing his name (stated in the next clause).

45. In this passage, Galileo makes an important methodological distinction, which has had considerable echo in modern science and philosophy. It is the idea that there are two distinct stages of inquiry, the context of exposition or justification and the context of investigation or discovery; the method of discovery usually involves empirical observational arguments and the method of justification usually involves theoretical a priori arguments; thus, the method of discovery typically does not coincide with the method of justification. He also gives an argument supporting the distinction. For more details, see Finocchiaro (1973, 229–38), Nickles (1980), Popper (1959), and Wallace (1992b).

we have said several times);[46] moreover, regarding the a priori arguments, we have already examined how far their strength goes.

Now, returning to the subject, I say that the things discovered in the heavens in our time are and have been such that they can give complete satisfaction to all philosophers. For, both on the individual bodies and in the general expanse of the heavens, we have seen and still see phenomena similar to those which we call generation and destruction when occurring near us: excellent astronomers have observed many comets generated and dissolved in the regions above the lunar orb, as well as the two new stars[47] of the years 1572 and 1604, located without any doubt much higher than any planet; and on the face of the sun itself by means of the telescope we see the production and decomposition of dense and dark substances very similar in appearance to clouds on the earth, and some of which are so large that they far exceed not only the Mediterranean basin but also all Africa and Asia. Now, Simplicio, what do you think Aristotle would say and do if he were to see these things?

SIMPLICIO. I do not know what would be done and said by Aristotle, who was master of the sciences; but I know well in part what his followers do and say, and what is proper for them to do and say in order not to be left without a guide, without protection, and without a head in philosophy. As for the comets, has not the *Anti-Tycho*[48] refuted those modern astronomers who wanted to make them heavenly bodies, refuting them with their own weapons (namely, by means of parallaxes[49] and cal-

46. Here Salviati distinguishes between what scientists actually do and what they explicitly say about how to proceed—between scientific practice and self-reflective articulation of scientific methodology; he suggests that the latter is more important. It is doubtful that this view of their relative importance is correct; many would instead agree with Einstein's (1934, 30) famous remark that "if you want to find out anything from the theoretical physicists about the methods they use, I advise you to stick closely to one principle: don't listen to their words, fix your attention on their deeds." But Salviati's general point is plausible, for the distinction as such says nothing about which is more important or how frequently the two coincide. For more details, see the appendix (2).

47. These were novas, indeed, supernovas.

48. Chiaramonti (1621), which criticizes Tycho Brahe's theory that comets are heavenly bodies and argues that they have a terrestrial origin. Galileo's criticism here is relatively mild because he too disliked Tycho and found the same erroneous theory of the terrestrial origin of comets attractive. Later (selection 10 and, more generally, Galileo 1967, 247–318) he criticizes Chiaramonti (1628), and that criticism is more extensive and severe.

49. A generally useful explanation of the nature of parallax is found in selection 14; the application there is not to comets and the new stars, but to stellar parallaxes deriving from the earth's annual motion.

culations turned around in all sorts of ways), and finally concluding in Aristotle's favor that they are all elemental? And with this demolished, which was the only ground of those who believe in the novelties, what else is left to them to remain standing?

SALVIATI. Be patient, Simplicio. What does this modern author say about the new stars of 1572 and 1604 and about the sunspots? [77] As for the comets, from my point of view I would have little difficulty in supposing them to be generated either below or above the moon; nor have I ever put much trust in Tycho's verbosity; nor do I feel any repugnance in believing that they are made of elemental matter, and that they can rise as much as they want without being stopped by the impenetrability[50] of the Peripatetic heavens, which I consider to be much thinner, softer, and more rarefied than our air. As for the parallax calculations, first I question whether comets are subject to this phenomenon,[51] and then the variability of the observations on which the calculations are based makes me equally suspicious of both opinions, especially since I feel the author of the *Anti-Tycho* modifies at will or considers fallacious those observations which contradict his purpose.

SIMPLICIO. As for the new stars, the author of the *Anti-Tycho* dismisses them very quickly in a few words;[52] he says that these modern new stars are perhaps not parts of the heavenly bodies, and that, if his opponents want to prove the existence of change and generation up there, they should demonstrate the occurrence of changes in the stars that have been described for a long time and that no one doubts are heavenly bodies (something they will never be able to do at all). As for the substances that some claim are generated and dissolved on the face of the sun, he makes no mention of them; so I infer he regards them as a

50. Some versions of the geostatic view claimed that the heavenly region around the central and motionless earth consisted of a series of nested spherical shells made of aether; in each was embedded a planet, and in the outermost one were attached the fixed stars; these distinct spherical shells rotated at different rates, and their rotation accounted for the revolution of the heavenly bodies around the earth. Aether was believed to be invisible and impenetrable, and so the heavenly region was not really empty. This arrangement made it impossible for a body to move through the various parts of the heavenly region, as some astronomers held comets to be able to do.

51. A reason might be the possibility that comets are optical phenomena like halos, mirages, and rainbows, which are not self-subsisting material bodies with a definite position. This is elaborated in Galileo's *Assayer*.

52. Cf. Chiaramonti (1621, 357).

fairy tale, or illusions[53] of the telescope, or effects produced in the air, or in short, anything but heavenly occurrences.

SALVIATI. But you, Simplicio, what have you thought of answering to the objection based on these annoying spots, which have come to mess up the heavens and even more so the Peripatetic philosophy? As its courageous defender, you must have found some escape or solution and must not deprive us of it.

SIMPLICIO. I have heard various opinions about this phenomenon.[54] Some say that they are stars which, like Venus and Mercury, turn around the sun along their individual orbits; that when they pass between us and the sun they appear dark to us; and that, being very numerous, it frequently happens that some of them appear bunched together and then separated from each other. Others think they are effects in the air; still others, illusions of the lenses; and others, other things. However, I am much inclined to believe, indeed I firmly hold, that they are a collection of numerous and various opaque bodies that come together accidentally; and so we often see that in a particular [78] spot one can count ten or more of the very small bodies, which have irregular shapes and look like snow flakes or flocks of wool or flies in motion; they change place relative to each other, sometimes congregating and sometimes coming apart between us and the sun, and they also move around the latter as their center. Nevertheless, we are not obliged to say that they are generated and dissolved, but only that sometimes they are hidden on the other side of the solar body and other times (though separated from the latter) they are not seen due to the immeasurable light of the sun; moreover, there is around the sun something like an onion made up of many layers,

53. The claim that the new telescopic observations might be optical illusions created by the new instrument is repeated several times in this selection, but is not explicitly discussed. When Galileo first published his telescopic discoveries (1610), this claim raised an important issue since it represented one way of defending the traditional view and rejecting the new evidence; indeed, at first there were legitimate questions about the methodological legitimacy of the telescope, the scientific explanation of its workings, its empirical reliability, and its practical operation and use; for more details, see the introduction (5.3), selection 11, Feyerabend (1975), Ronchi (1958), and Van Helden (1984; 1994). But by the time the *Dialogue* was published (1632), the controversy had largely been resolved in Galileo's favor, at least among the experts; indeed, as early as 1611 one of the Church's leading universities (the Jesuit Collegio Romano) had endorsed his telescopic discoveries, though not his interpretation of them; cf. Favaro (11:87–88 and 92–93).

54. The rest of this paragraph, except for the last sentence, is placed in quotation marks by Galileo, as if it were a direct quotation; but its content makes it clear that it is merely a paraphrase. The interpretation of sunspots as swarms of planets revolving around the sun was advanced by the Jesuit astronomer Christopher Scheiner.

each inside another, each in motion, and each sprinkled with some small spots; and, although their motion at first seemed variable and irregular, nevertheless, lately it has been observed that the exact same spots return after definite periods. This seems to me to be the most convenient escape found so far to account for such a phenomenon and at the same time retain the indestructibility and ingenerability of the heavens; and, if this were not sufficient, there will be no lack of loftier intellects who will find better explanations.

SALVIATI. If what is being discussed were a point of law or of other human studies in which there is neither truth nor falsehood, one could have great confidence in intellectual subtlety, verbal fluency, and superior writing ability, and hope that whoever excels in these qualities could make his reasoning appear and be judged better. But in the natural sciences, whose conclusions are true, necessary, and independent of the human will, one must take care not to engage in the defense of falsehood because a thousand Demostheneses[55] and a thousand Aristotles would be overcome by any average intellect who may have been fortunate enough to have discovered the truth. Therefore, Simplicio, give up your thought and your hope that there could be men so much more knowledgeable, learned, and well-read than we are as to be able to turn truth into falsehood, against nature.[56]

Now, since of all the opinions so far produced on the essence of these sunspots, you seem to regard as true the one you just expounded, it follows that all the others are false; but, to free you from this one (which is nevertheless a very false chimera), leaving aside countless other improbabilities, [79] I give you only two observations against it. The first one is

55. Demosthenes (384?-322 B.C.) was a classical Greek lawyer and statesman; he is regarded as one of the greatest orators who ever lived. He is best known for three speeches entitled *Philippics*, which tried (unsuccessfully) to persuade his fellow citizens to take effective measures to defend their country from King Philip II of Macedon (father of Alexander the Great).

56. In this paragraph Salviati expresses the methodological principle that rhetoric has no role to play in natural science; this view is widely shared by scientists and by many scholars who seek to understand science. But this is a beautiful occasion for applying the distinction between scientific practice and methodological reflection, elaborated elsewhere in this selection by Galileo himself; that is, just because he explicitly advanced an anti-rhetorical methodological principle, this does not mean that he actually proceeded in that manner. In fact, the examination of Galileo's works and especially the *Dialogue* shows that he constantly engages in rhetorical techniques; this does not mean that he is being unscientific, but rather that rhetoric has an important role to play in science. For more details, see the appendix (3), Brown (1988), Feyerabend (1975), Finocchiaro (1977a; 1980; 1990), Gross (1990), Kuhn (1970), Moss (1993), Pera (1994), and Prelli (1989).

that many such spots are seen being generated in the middle of the solar disc, and similarly many are seen being dissolved and vanishing far from the solar circumference; this is a necessary argument that they are generated and dissolved because, if they appeared only as a result of local motion[57] without being generated and dissolved, they would all be seen entering and leaving at the outer circumference. The other observation is for those who are not completely ignorant of the principles of perspective: from the changes in apparent shape and in apparent speed, one concludes necessarily that the spots are contiguous to the solar body; that while touching its surface, they move with it and upon it; and that they do not revolve along circles separated in any way from the solar body. This is proved by the fact that their motion appears very slow near the circumference of the solar disc and faster near the middle; and it is also proved by the fact that the shape of the spots appears very narrow near the circumference in comparison to what it appears near the middle, which happens because near the middle they are seen in their fullness and as they really are, whereas near the circumference they appear foreshortened due to the curvature of the spherical surface. For those who have been able to observe and analyze them diligently, both reductions (of the shape and motion) correspond precisely to what must appear if the spots are contiguous to the sun, and they hopelessly contradict motion in circles separated from the solar body (however small the distance); this has been amply demonstrated by our friend[58] in the *Sunspot Letters* addressed to Mr. Mark Welser.[59] From the same change of shape one gathers that none of the spots are stars or other bodies with a spherical shape; for, of all shapes, only the sphere is never seen foreshortened

57. In Aristotelian natural philosophy, local motion corresponded to what we call simply motion; the Aristotelian concept of motion was really synonymous with our notion of change and so subsumed any kind of change including generation, destruction, growth, decay, and qualitative change; thus, local motion was a special case of motion or change, in which what changes is place or location or position. Although the speaker here is Salviati who may be taken as representing Galileo, and although Galileo eventually gave up such a conceptual framework, the process was a slow one and some of the language continued to be used harmlessly in a way which did not signify a commitment to the Aristotelian concepts.

58. Namely, Galileo; besides "Academician," this is another description he uses in the book to refer to himself.

59. Welser was a German businessman, politician, and intellectual who (acting as the intermediary) initiated and sustained the correspondence between Galileo and the Jesuit astronomer Christopher Scheiner on the topic of sunspots. This correspondence began a dispute which occasioned the publication of Galileo's *Sunspot Letters* in 1613 and which each continued to discuss in other works.

and must always be represented as perfectly round; so, if any particular spot were a round body (as all stars are considered to be), it would appear equally round near the middle as well as near the edge of the solar disc; on the other hand, the fact that they appear so foreshortened and narrow near the edge but wide and large near the middle assures us that they are layers of small depth or thickness in relation to their length and width. In regard to the recent observation that after definite periods the exact same spots return, you should not believe it, Simplicio, and the person who told you this [80] wants to deceive you; in fact, notice that he has been silent about those which are generated and dissolved within the face of the sun far from the circumference, and that he has said nothing about the foreshortening, which proves conclusively their being contiguous to the sun. What truth there is in the return of the same spots, you can read about in the aforementioned *Letters*; that is, some spots occasionally last such a long time that they are not decomposed in the course of only one solar rotation, which occurs in less than a month.[60]

SIMPLICIO. Frankly, I have not made the long and diligent observations needed to properly master the facts of this matter; but I definitely want to make them, and then try myself to reconcile what experience reveals with what Aristotle demonstrates, for it is clear that two truths cannot contradict one another.[61]

SALVIATI. As long as you want to reconcile what the senses show you with the soundest doctrines of Aristotle, you will not have to work hard. That this is true may be seen as follows. Does not Aristotle say that one

60. The important physical conclusion advanced in this paragraph is that the spots are large thin layers of opaque matter contiguous to the sun and carried by it as it rotates on its axis; this claim is correct according to modern science. The argument supporting this conclusion is based on the observational evidence that the spots are seen to grow larger and move faster near the middle than near the circumference of the solar disc; this is a good example of a conclusive argument. Galileo can claim to have established this particular scientific conclusion beyond any reasonable doubt, without making the same claim about another, for example, the motion of the earth.

61. The claim that two truths cannot contradict one another is in a sense uncontroversial and definitionally true; that is, contradiction is defined as the relationship between two propositions such that if one is true then the other is false (for example, "the earth moves" contradicts "the earth stands still"); it follows immediately that if two propositions contradict one another, they cannot both be true; or by contraposition, if they are both true then they cannot contradict one another. But to assert this claim in a context like the present one is a way of expressing one's faith in the truthfulness of a particular author or authority, in this case Aristotle; in fact, the claim (in identical words) was often made in reference to the Bible to stress that biblical statements cannot contradict scientifically demonstrated truths. Cf. Galileo's "Letter to the Grand Duchess Christina" (Finocchiaro 1989, 96) and the footnote in selection 5, [134].

cannot confidently treat of heavenly phenomena because of the great distance?[62]

SIMPLICIO. He says so explicitly.

SALVIATI. Does he not affirm that what is given to us by experience and the senses must be put before any theory, however well founded it may seem?[63] And does he not say this decisively and without any hesitation?

SIMPLICIO. He does.

SALVIATI. Consider now these two propositions, which are both parts of Aristotle's doctrine: the last one, claiming that one must put the senses before theorizing; and the earlier one, which regards the heavens as unchangeable. The last one is much more solid and serious than the earlier one, and so it is more in accordance with Aristotle to philosophize by saying "the heavens are changeable because so the senses show me," than if you say "the heavens are unchangeable because theorizing so persuaded Aristotle."[64] Moreover, we can investigate heavenly phenomena much

62. Cf. Aristotle, *On the Heavens*, II, 12, 291b24ff.

63. Cf. Aristotle, *On the Generation of Animals*, III, 10, 760b31; cf. also selection 2, [57], where this principle is first introduced. In this restatement of the fundamental principle of empiricism, the priority of observation over theory is made stronger and more decisive since it is claimed to apply even when the theory seems well-founded. This version of the principle is less plausible because it may be taken to imply that if there is a conflict between an observation and a well-founded theory, then the theory should be abandoned; but for a theory to be well-founded means that it is supported by many observations; so if a particular conflicting observation is allowed to refute a well-founded theory that observation is regarded as more important than the other observations, and such priority of one observation over others may be unwarranted. Another issue here involves noting that the added clause speaks of the theory *seeming* to be well-founded and noting that such a theory may not be actually well-founded. See the appendix (2.4) for more details.

64. Here Salviati wants to show not only that changes can now be observed in the heavens and so the heavens are changeable, but also that this thesis of heavenly changeability is more in accordance with Aristotle than its alternative. The reason given for this greater correspondence is that in Salviati's time heavenly changeability was more in accordance with Aristotle's own fundamental principle of empiricism than was heavenly unchangeability. This technique presupposes that one be able to discriminate among the various statements found in a given author, and that one attach more importance to some than to others; for example, a methodological principle is normally more important than a particular substantive conclusion. Galileo uses the same technique in selection 11, where he distinguishes between Aristotle's definition of the center of the universe as the center of planetary revolutions and his thesis that the planets revolve around the earth. This technique is related to, but not identical with, the one used earlier in this selection ([75]) when Salviati says that in his time it was more in accordance with Aristotle's procedure to claim heavenly changeability because the Aristotelian procedure for arriving at a different conclusion (heavenly unchangeability) was the a posteriori method of discovery. There Galileo distinguished the procedure used from the result arrived at and gave more importance to the former than to the latter. That is, earlier he distinguished between Aristotle's words and deeds, whereas now he distinguishes between two kinds of words.

better than Aristotle could; for he admits that such knowledge is difficult for him on account of their distance from our senses, and so he implies that if one's senses could better represent those phenomena then one could philosophize about them with greater confidence; but, [81] by means of the telescope, we have brought them thirty or forty times closer to us than they were to Aristotle, so that we can perceive in the heavens a hundred things he could not see; among these are sunspots, which were absolutely invisible to him; therefore, we can treat of the heavens and of the sun more confidently than Aristotle could.

SAGREDO. I sympathize with Simplicio, and I see he is much moved by the strength of these very conclusive reasons; on the other hand, he is much confused and frightened by the fact that Aristotle has universally acquired great authority, that so many famous interpreters have labored to explain his meaning, and that other generally useful and needed sciences have based a large part of their reputation on Aristotle's influence. It is as if I hear him say: "On whom shall we rely to resolve our controversies if Aristotle is removed from his seat? Which other author shall we follow in the schools, academies, and universities? Which philosopher has written on all parts of natural philosophy, so systematically, and without leaving behind even one particular conclusion? Must we then leave the building in which are sheltered so many travelers? Must we destroy that sanctuary, that Prytaneum,[65] where so many scholars have taken refuge so comfortably, where one can acquire all knowledge of nature by merely turning a few pages, without being exposed to the adversities of the outdoors? Must we tear down that fortress where we can live safe from all enemy assaults?" I feel compassion toward him, no less than toward a man who would take a long time building a luxurious palace at great expense and with the help of hundreds of artisans; who would then see it run the risk of collapsing due to poor foundations; and who would try to save it from ruin by means of chains, props, buttresses, barbicans, and struts to avoid the great pain of seeing the fall of the walls adorned with beautiful paintings, of the columns supporting the stately balconies, of the golden stands, and of the doorposts, facades, and cornices made of expensive marble.

SALVIATI. Simplicio should not be afraid of such a downfall; for a much smaller charge I would undertake to insure him against such damage. There is no danger that such a great multitude of clever and shrewd

65. In ancient Greek city states, the Prytaneum was a public building where government meetings and official functions were held, and where distinguished citizens and visitors were maintained at state expense.

philosophers would allow themselves to be overcome by one or two noisemakers; indeed, without even using their pens against the latter, but by simply [82] being silent, they could make the latter the object of public contempt and scorn. It is inane to think of introducing a new philosophy by refuting this or that author; one must first learn to remake human brains and render them fit for distinguishing truth from falsehood, something that only God can do. But, where have we come, moving from one argument to another? I would not know how to get back to the main road without the guidance of your memory.

SIMPLICIO. I remember very well. We were discussing the answers of the *Anti-Tycho* to the objections against heavenly unchangeability; among these you included the one based on sunspots, which he had not mentioned. I think you then wanted to examine his answer to the one involving the new stars.[66]

SALVIATI. I now recall the rest. Continuing with the subject, it seems to me that in the answer of the *Anti-Tycho* there are some things that deserve criticism. First, consider that he cannot refuse to place the two new stars in the highest regions of the heavens, and that they lasted a long time and eventually vanished, but that they do not give him any trouble in maintaining heavenly unchangeability because they are not unquestionable parts of the heavens or changes occurring in the old stars; if so, for what purpose did he, with so much anxiety and anguish, take a position against the comets to banish them completely from the heavenly regions? Would it not have been sufficient for him to say the same thing about them as about the new stars, namely that, since they are not unquestionable parts of the heavens or changes occurring in any star, they present no difficulty for either the heavens or Aristotle's doctrine?

Secondly, I do not understand what goes on inside him; for, on the one hand, he admits that changes occurring in the stars would be destructive of the heavenly prerogatives (namely, indestructibility, etc.), the

66. Recall also that the *Anti-Tycho* was injected into the discussion after Salviati expressed his fourth criticism of the observational argument for heavenly unchangeability; this criticism rejects as false the crucial observational premise that no changes have ever been observed in the heavens, on the basis of the new evidence available about comets, novas, and sunspots; the criticism also claims that the same evidence which refutes this premise also refutes the conclusion about unchangeability and proves changeability instead. The *Anti-Tycho* was meant to undermine this new three-part evidence and thus to respond to this criticism. Injecting this book into the discussion no doubt gave Galileo the opportunity to criticize an author (Scipione Chiaramonti) who he felt deserved criticism; but it also represents an implicit recognition that the Ptolemaic view could not be demolished by simply appealing to new evidence because the alleged facts could not be established as facts in any simple manner but required critical reasoning.

reason for this being that the stars are heavenly bodies, as is clear from the unanimous consensus of everybody; on the other hand, nothing bothers him if the same changes should occur in the rest of the heavenly expanse outside individual stars. Does he think perhaps that heaven is not a heavenly body? I for my part should have thought that the stars are called heavenly bodies due to their being in the heavens and their being made of heavenly matter, and that therefore heaven is more heavenly than they are, just as one can say that nothing is more terrestrial than the earth or more fiery than fire itself.

Finally, there is the fact that he makes no mention of sunspots, in regard to which it has been conclusively proved that they are produced and dissolved, that they are [83] next to the solar body, and that they rotate with it or around it; this gives me a strong indication that maybe this author writes more to please others than for his own satisfaction. I say this because he shows himself to be competent in mathematics, and so it is impossible for him not to have been persuaded by the demonstrations that these phenomena are necessarily contiguous to the solar body and are generations and decompositions so large that none equally large ever occur on the earth. And if they are so large, so many, and so frequent in the solar globe itself (which may be reasonably regarded as among the most noble parts of the heavens), what reason is left capable of dissuading us that other such phenomena may occur on the other globes?

. . .[67]

## [4. Life on the Moon, and Human versus Divine Understanding]

[124] . . . SIMPLICIO. But do you perhaps believe that those large spots which are seen on the face of the moon are seas, and the brighter rest is land or some such thing?[68]

67. The passages omitted here correspond to Favaro 7:83.12–124.10 and Galileo 1967, 58–98. They contain, first, a critique of what I call the teleological argument for the earth-heaven dichotomy—that the heavens are unchangeable because heavenly changes would be superfluous and useless; the discussion is very brief but raises many interesting issues, as may be seen from Finocchiaro (1980, 35, 48–49, 320–25, 327–29, and 377–79). Then there is a lengthy discussion of the similarities between the earth and moon, with special attention to the roughness of the lunar surface (which was one of Galileo's telescopic discoveries) and to the earth's ability to reflect light; cf. Finocchiaro (1980, 35, 85–87, and 109–12).

68. The spots in question are visible with the naked eye and are the areas of the lunar disk that appear darker than the rest.

SALVIATI. What you ask marks the beginning of the differences I judge exist between the moon and the earth;[69] but, in regard to them, it is time to move quickly because we have lingered too long on the moon. So I say that, if in nature there were only one way in which a surface illuminated by the sun could be made to appear brighter than another, and if this were for one to be land and the other water, then it would be necessary to say that the moon's surface is part land and part water; but because we know of several conditions that can cause the same effect, and there may be others unknown to us, I would not dare claim that this rather than that occurs on the moon. We have already seen earlier how a bleached silver plate turns from white to dark when touched with a burnisher; how the moist part of the earth appears darker than the arid part; and how on mountain ridges the wooded parts appear duller than the bare and sterile parts. The last occurs because among [125] the trees there is a great deal of shadow, but the barren areas are entirely illuminated by the sun; and this mixture of shadows is so effective that in sculptured velvet the color of the cut silk appears much darker than that of the uncut one due to the shadows dispersed among the threads, and likewise plain velvet appears much duller than armozine[70] made of the same silk. It follows that, if on the moon there were things like very large forests, their appearance could produce the spots we see; such an effect would also follow if they were seas; finally, it is not repugnant to think that those spots may really be of a darker color than the rest, for this is the way snow makes mountains look brighter. What is seen clearly on the moon is that the darker parts are all plains, with few rocks and hills within them, though there are some; the brighter rest is all full of rocks, mountains, and small hills of round or other shape; in particular, around the spots there are very long ranges of mountains. We can be certain that the spots are flat surfaces from seeing how the line separating the illuminated from the dark

69. In the preceding passage omitted here (Favaro 7:87–124; Galileo 1967, 62–98), Galileo elaborates a number of similarities between the earth and the moon: solidity, spherical shape, mountainous surface, lack of intrinsic luminosity, reflecting power, opacity, reciprocal illumination, reciprocal eclipsing, reciprocal phases, and unevenness of apparent brightness. Here, he is willing to consider some important differences; this concern may be taken as a sign of his judiciousness—his desire to avoid one-sidedness and extremes.

70. A strong-corded, taffeta-like silk cloth originating from Hormuz in the Persian Gulf and now no longer used.

part makes an even cut when crossing the spots, but appears irregular and jagged when lying in the bright areas.[71] However, I do not know whether this flatness of the surface is by itself sufficient to produce the darkness, and I rather think not.[72]

Furthermore, I believe the moon to be very different from the earth: though I imagine it is not inert and dead, still I do not say that it has activity and life, much less that there is generation of plants, animals, and other things similar to ours; what I do say is that, if there are living things, they are very different and far removed from anything we can imagine. What moves me to believe this is my feeling that the matter of the lunar globe is not earth and water, which alone suffices to take away any generations and changes like ours; but, even if there were water and earth up there, in any case no plants and animals like ours would grow; there are two principal reasons for this.

The first is that the various apparent positions of the sun are so necessary for our generation that without them everything would be lacking. Now, the apparent positions of the sun in relation to the earth are very different from those in relation to the moon. In regard to the daily illumination, in most parts of the earth for every period of twenty-four hours we have some daylight and some night darkness; on the moon this effect has a period of one month.[73] Moreover, regarding the [126] annual falling and rising by which the sun produces the various seasons and the inequality of night and day, on the moon this is also completed in one month; and while with us the difference between the maximum and minimum elevation of the sun is about forty-seven degrees (which is equivalent to the distance between the

71. This is an observation possible only with the telescope.

72. The attitude displayed in this discussion of the cause of moon spots is a good example of epistemological modesty and tentativeness.

73. That is, the lunar cycle of night and day is one month. Galileo does not explain or justify this claim, perhaps because he regarded it as too well known and obvious. A derivation is the following: we know that the moon revolves eastward around the earth once a month, and that it always shows the same side toward the earth; the only way for the moon to keep the same hemisphere facing the earth is for the moon to rotate eastward on its axis once a month; that is, as the moon revolves in its orbit around the earth in a counterclockwise direction, it must be rotating on its own axis in the same direction just enough to compensate for the orbital displacement; now, disregarding the earth and considering the moon's monthly axial rotation in relation to the sun, any one point on the lunar surface will face the sun and experience daylight for about half a month, and it will be on the far side of the sun and experience darkness for the other half of the month.

two tropics),[74] for the moon it amounts merely to little more than ten degrees (which equals the sum of the maximum latitudes of the dragon[75] on the two sides of the ecliptic). Think now what would be the action of the sun inside the torrid zone[76] if its rays were striking it constantly for fifteen days, and you will easily understand that all trees, plants, and animals would be destroyed; and even if any living things were generated, they would be plants, trees, and animals very different from the ones we have here.

Secondly, I firmly hold that there is no rain on the moon. For if clouds were formed anywhere up there as they are on the earth, they would eventually hide some of the things we see on the moon with the telescope; in short, visibility would be subject to variation in some way. However, despite my long and diligent observations, I have never seen this effect; instead I have always noticed there a uniform and perfect serenity.

SAGREDO. To this one could reply that there could be very heavy dews or that it could rain there during nighttime, namely, when the sun does not shine on it.

SALVIATI. If from other evidence we had indications that in it there are generations like ours and the only thing lacking is rain, then we could find this or some other condition to substitute for the rain, as happens in Egypt with the flooding of the Nile. However, we do not find up there any phenomenon corresponding to ours among the many required to produce similar effects, and so there is no need to trouble ourselves to introduce only one merely on the grounds of its not being repugnant (rather than because we have positively observed it). Moreover, if I were

74. The tropics are the two circles parallel to the equator at a distance (latitude) of about 23½ degrees north and south, called respectively Tropic of Cancer and Tropic of Capricorn; the term applies both to the circles on the earth's surface and to the corresponding circles on the celestial (or stellar) sphere. Only between these two latitudes on the earth does the sun at noon appear directly overhead for part of the year; and on the celestial sphere these two latitudes are the limits of the sun's apparent north-and-south motion in the course of a year; from a Copernican point of view, these phenomena are consequences of the fact that the earth's axis of rotation is inclined 23½ degrees to the plane of its orbit around the sun. On the earth, the region between these two latitudes is hotter and undergoes less variation in temperature between summer and winter than other regions; for this reason it is called the torrid zone, but it is also called the tropical zone or the tropics.

75. By dragon (*dragone*) Galileo means the circle on the celestial sphere corresponding to the monthly eastward revolution of the moon around the earth; its plane intersects that of the ecliptic at an angle of about five degrees. He claims that the two intersecting circles form a shape similar to two snakes. Cf. Favaro (2:246; 7:91) and Sosio (1970, 83, 124, and 566).

76. The torrid zone is the region on the earth's surface around the equator, and more exactly from the latitude of 23½ degrees north to the latitude of 23½ degrees south; this region is so called because it is relatively hot.

asked what my initial impression and pure common sense suggest to me concerning the production up there of things similar or dissimilar to ours, I would always say "very dissimilar and wholly unimaginable for us"; for so require, it seems to me, the richness of nature and the omnipotence of the Creator and Ruler.[77]

SAGREDO. I have always felt it to be extremely rash to want to make human capacity the measure of what nature can and knows how to do, [127] whereas on the contrary there is no natural phenomenon (however minute it may be) that can be known in its entirety by the most speculative intellects.[78] Such a foolish pretension to understand everything can only originate from not ever having understood anything; for, if someone had tried only once to understand a single thing perfectly and had truly tasted how knowledge works, he would know that he does not understand an infinity of other conclusions.

SALVIATI. Your point is very well taken. To confirm it we have the experience of those who understand or have understood something; the wiser they are, the better they know and the more freely they admit how little they know. The wisest man of Greece, so pronounced by the oracle, said openly that he was aware he knew nothing.[79]

SIMPLICIO. Therefore, one should say that either the oracle or Socrates himself was lying since the former pronounced the latter most wise and the latter said he was aware of being most ignorant.[80]

SALVIATI. Neither the one nor the other follows, for both pronouncements can be true. The oracle judges Socrates most wise in relation to

77. The last five paragraphs may be reconstructed in terms of the following argument: there is no life on the moon (at least no life as we know it) because the lunar nights and days are very long (about two weeks each), because there are no significant seasons on the lunar surface, and because there are no rains or clouds on the moon. Each of these three premises connects with the conclusion by means of an intermediate proposition, and each is in turn supported by a subargument. If the connection between the absence of rains and clouds and the absence of life is taken to be the absence of water, this reconstruction enables us to integrate into this argument the preceding discussion of the moon spots; for we can then regard that discussion as an argument showing that it is unlikely that oceans are the cause of the moon spots. At this point, the discussion of the moon ends and an explicit methodological discussion begins, dealing with the nature, powers, and limitations of human knowledge.

78. This sentence is an expression of what I call the fundamental principle of epistemological modesty.

79. This refers to Socrates, a classic example of epistemological modesty. When the Delphic Oracle declared that he was the wisest man in the world, he found it hard to believe and undertook a lifelong search to decipher its meaning; eventually, he interpreted it to mean that he was most acutely aware of how little he knew and that true wisdom must include an awareness of the limitations of one's knowledge. Cf. Plato, *Apology*, 20d-23b.

80. The last clause is printed in italics in the original text.

other men, whose knowledge is limited; Socrates is aware of not knowing anything in relation to absolute wisdom, which is infinite; now, compared to the infinite, much, little, or nothing are the same because, for example, to arrive at an infinite number it is the same if we add thousands, tens, or zeros; therefore, Socrates knew well that his limited wisdom was nothing compared to infinite wisdom, which he lacked. However, among men one finds some knowledge, and this is not equally distributed among them; thus it was possible for Socrates to have a larger amount of it than others, and so for the oracle's answer to be true.[81]

SAGREDO. I feel I understand this point very well. Human beings, Simplicio, have the power to do things, but it is not equally shared by all. Undoubtedly, the power of an emperor is much greater than that of a private person, but the latter as well as the former are nothing in comparison to divine omnipotence. Some men understand agriculture better than others; but what does knowing how to plant a branch of a vine in a trench have to do with knowing how to make it take root, draw nourishment, and from this choose [128] something appropriate to make leaves, something to form tendrils, something for the grapes, something for the skin, and something for the seeds, all of which are works of the most wise nature? This is only one particular work among the countless ones in nature, and in it alone one recognizes an infinite wisdom; we may thus conclude that divine wisdom is infinitely infinite.[82]

81. This interpretation of Socrates indicates that Galileo's commitment to epistemological modesty should not be confused with skepticism, namely, the thesis that knowledge is impossible and human beings do not really know anything. I believe that skepticism is an injudicious exaggeration of modesty; it is an extreme position at one end of a spectrum having at the opposite end a view that is excessively optimistic, dogmatic, and arrogant. Thus epistemological modesty is a moderate and balanced mean between the extremes of skepticism and dogmatism. One commentator who has appreciated Galileo's remarks in this selection along the lines of epistemological modesty and balanced moderation is Sosio (1970, 123n.1 and 125n.1); cf. Biagioli (1993, 301–11).

82. This paragraph begins the elaboration of two contrasts: that between human and divine wisdom and that between human artifacts and natural products; each provides a reason to be epistemologically modest. The ensuing remarks on the nature of God are also relevant to the favorite anti-Copernican objection of Pope Urban VIII, which was based on God being all-knowing and all-powerful; as required by the book censors, that argument is stated (without criticism) at the end of the book (selection 16). Some scholars (for example, Wisan [1984b]) are inclined to think that these remarks constitute a preemptive strike against that final argument; such an interpretation would depend on an appropriate interpretation of these theological remarks; that is, for example, one would have to claim that this passage is trying to qualify the idea that God is all-powerful, or trying to extend the certainty of human knowledge beyond pure mathematics; the latter extension would conflict, I believe, with epistemological modesty.

SALVIATI. Here is another example. Would we not say that being able to find a beautiful statue in a piece of marble has exalted Buonarroti's[83] mind very far above the common minds of other men? This skill involves nothing but copying a single aspect and arrangement of the exterior and superficial members of a motionless man. But what is this in comparison to a man created by nature, consisting of so many internal as well as external organs, muscles, tendons, nerves, and bones, which are used for so many different movements? But what shall we say of the senses, the powers of the soul, and finally of the understanding? Can we not say, rightly, that the making of a statue yields by an infinite amount to the creation of a live man, indeed even to the creation of the vilest worm?

SAGREDO. And what do we believe the difference to be between the dove of Archytas[84] and a natural one?

SIMPLICIO. Either I am one of those men who lack understanding, or there is a clear contradiction in this account of yours. You regard the understanding of man, created by nature, as one of his major distinctions, indeed as the highest of all; but a little while ago you joined Socrates in saying that his understanding amounted to nothing; therefore, one must say that even nature did not understand how to create an intellect which understands.

SALVIATI. You make a very sharp objection. To respond one should refer to a philosophical distinction and say that understanding may be taken in two ways, namely, intensively or extensively. Extensively speaking, namely, in regard to the number of intelligible things (which is infinite), human understanding is like nothing; for, even if one understands one thousand propositions, one thousand in relation to infinity is like a zero. However, if we take the understanding intensively, insofar as this term implies that some proposition is understood perfectly, I say that the human intellect understands some propositions as perfectly and with as much absolute certainty as nature herself does. This is the case for the [129] pure mathematical sciences, namely, geometry and arithmetic; the divine intellect knows infinitely many more propositions because it knows them all; but for the few understood by the human intellect, I

83. Michelangelo Buonarroti (1475–1564) was an Italian architect, painter, and sculptor, one of the greatest who ever lived. He is best known for such statues as the *Pietà*, *David*, and *Moses* and for painting the ceiling of the Sistine Chapel.

84. Archytas of Tarentum (c. 400–350 B.C.) was a Greek scientist, philosopher, mathematician, and Pythagorean who influenced Euclid and contributed to acoustics and music theory. He is renowned for having built one of the most famous automatons in history: a wooden dove hanging from the end of a pivoted bar which was made to revolve by means of a jet of steam or compressed air.

believe our knowledge equals the divine one in regard to objective certainty, for it is capable of grasping their necessity, which seems to be the greatest possible assurance there is.[85]

SIMPLICIO. This seems to me to be a very serious and bold manner of speaking.[86]

SALVIATI. These are common assertions far from any shadow of rashness and boldness, and they do not detract at all from the majesty of divine wisdom; similarly, there is no diminution of His omnipotence in saying that God cannot bring it about that a fact is not a fact. However, Simplicio, I suspect you take exception to my words by interpreting them incorrectly. To explain myself better, I say that, in regard to the truth which we know by means of mathematical demonstrations, it is the same as the one known by the divine wisdom. However, I will readily admit that the manner by which God knows the infinity of propositions of which we know only a few is immeasurably more excellent than ours; the latter proceeds by discourse and by reasoning from one conclusion to another, whereas His has the character of simple intuition. For example, to acquire knowledge of some properties of a circle (which are infinitely many), we begin with one of the simplest ones, taking it as its definition, and then by reasoning we go on to another one, from this to a third, then to a fourth, and so on; on the other hand, the divine intellect understands the whole infinity of those properties by the simple apprehension of its essence and without temporal progression; to be sure, those infinitely many properties are potentially included in the

85. Here Salviati elaborates a comparison between the human and the divine intellect which is useful in steering clear of skepticism. The similarity involves a feature (or cluster of features) of mathematical knowledge that both man and God presumably share—the truths of arithmetic and geometry are certain, objective, absolute, and necessary. This is the traditional interpretation of the methodological status of mathematics, and there was no reason to question it until the advent of non-Euclidean geometry in the nineteenth century and the study of the foundations of mathematics in the twentieth century; nonetheless, it is a correct first approximation even today, although the situation is now known to be much more complicated. For more discussions of the methodology of mathematics, see especially selection 9, but also selections 2, 13, and 14, and cf. the appendix (2.5). For some valuable modern discussions, see Kitcher (1983) and Lakatos (1978).

86. Here Galileo shows he was aware that this comparison of divine and human understanding could be questioned from a theological viewpoint. The answer Salviati proceeds to develop in the next paragraph is both intrinsically plausible and in accordance with a respectable theological tradition; for example, it can be claimed to have a basis in the writings of St. Augustine (Santillana 1953, 115n.97). But this comparison caused trouble after the book's publication and was explicitly included in a list of complaints compiled by a special commission appointed by the pope to investigate the book's transgressions; see Finocchiaro (1989, 218–22).

definitions, and due to their infinity they are perhaps a single thing in their essence and in the divine mind. Even this is not completely foreign to the human intellect, though it is obscured by a deep and dense mist, which is partly thinned out and cleared up when we master some firmly demonstrated conclusions with such proficiency that we can speedily go through them. In short, in a triangle the square opposite the right angle equals the other two next to it;[87] but is not this fact equivalent to saying that two parallelograms constructed on a common base and between two parallel lines are equal to each other? And is not the latter fact ultimately the same as saying that two surfaces are equal if when superimposed [130] they do not extend beyond each other but are enclosed within the same boundaries? Now, these transitions for which our intellect requires time and a step-by-step procedure are made by the divine intellect in an instant (at the speed of light, as it were), which is to say that it has them all always present. Therefore, I conclude that the divine understanding surpasses ours by an infinite amount both in regard to the manner and to the number of things understood.[88] However, I do not devalue the latter so much that I regard it as absolutely nothing; instead, when I consider how many and how wonderful are the things understood, investigated, and done by men, I know and understand very well that the human mind is one of the most excellent works of God.

SAGREDO. In regard to what you are presently saying, I have many times considered within myself how great human ingenuity is; I have run through so many and such wonderful inventions made by men both in the arts and in the sciences; and I have reflected on my own intellect, which is very far from hoping to discover anything new or even to learn some things already known; all this makes me feel confounded with wonder, afflicted with despair, and almost depressed. When I look at one of the excellent statues, I ask myself: "When will you know how to remove the excess from a piece of marble in order to disclose the beautiful statue which lay hidden within it? When will you know how to mix colors, spread them on a canvass or wall, and thereby represent all visible

87. This is the Pythagorean theorem; in Euclidean geometry, its proof requires two steps using the two principles mentioned in the next two sentences.

88. Thus, even in regard to mathematical knowledge, there are important differences between human and divine understanding, besides there being a similarity. The similarity was that mathematical truths are known with certainty; the differences are that God knows infinitely many more truths than we do, and also that He knows them instantly and immediately without having to use the step-by-step reasoning process needed by human beings.

objects, as Michelangelo, Raphael, and Titian[89] were able to do?" If I think about what men have discovered by dividing musical intervals and formulating precepts and rules for manipulating them to give marvelous delight to the ear, when shall I cease to wonder? And what shall I say of the multitude and variety of instruments? When reading the best poets, what wonder fills those who carefully consider the creation of images and their expression! What shall we say of architecture? What of the art of navigation? But above all stupendous inventions, how sublime was the mind of whoever conceived a means of communicating his most private thoughts to any other person, though very distant in time or place; that is, a means of speaking with those who are in India, or those who have not been born yet or will not be born for another thousand or ten thousand years! And how easy it is, by various arrangements of twenty characters on [131] paper! Let this be the seal for all wonderful human inventions and the end of our reasoning today. Now that the hottest hours have passed, I think Salviati would like to go and enjoy some cool air in a boat. I expect both of you tomorrow in order to continue the discussions we have begun, etc.

89. These are three of the greatest artists of the Italian Renaissance. Michelangelo is the same Buonarroti mentioned above ([128]). Raphael (1483–1520) is best known for the grace, expressiveness, and harmony of his paintings. Titian (c. 1490–1576) is the greatest painter of the Venetian School and is best known for his skill and innovations in handling colors.

# Second Day

## [5. Independent-Mindedness and the Role of Aristotle's Authority]

[132] SALVIATI. Yesterday's digressions from the direct path of our main discussions were many, and so I do not know whether I can get back and proceed further without your help.

SAGREDO. I am not surprised that you are in a state of confusion, given that you have your mind full not only of what has been said but also of what remains; but I, who am a mere listener and know only the things I have heard, will perhaps be able to bring the argument back into line by briefly recalling them. As far as I can remember then, the gist of yesterday's discussions was the examination from its foundations of the question of which of the two following opinions is more probable and reasonable:[1] the one holding that the substance of the heavenly bodies is

1. This formulation of the issue in terms of probability and reasonableness is very important and is a recurring theme in the book. It may be contrasted with other formulations stating that the issue is either which of the two views is true or which of the two views can be proved conclusively to be true. Galileo's caution is especially revealing in view of the fact that the two opinions mentioned here are those discussed in the "First Day," namely, the earth-heaven dichotomy and the thesis of unification; and in regard to these, the evidence was almost conclusive in refuting the dichotomy and establishing unification. His caution is related to, but cannot be explained away by, the fact that he was operating under several ecclesiastic prohibitions, such as the anti-Copernican decree of 1616; for, from the viewpoint of these prohibitions it was problematic to claim even that Copernicanism was more probable than geocentrism; this may be seen from the fact that the qualification in Salviati's

ingenerable, indestructible, unchangeable, inert, in short, exempt from any but changes of place, and that therefore there is a fifth essence[2] very different from the familiar one of our elemental, generable, degradable, and changeable bodies; or else the other opinion which, taking this division of parts away from the world, believes that the earth enjoys the same perfections as the other constitutive bodies of the universe and is, in short, a movable and moving globe no less than the moon, Jupiter, Venus, or other planets. Lastly we drew many particular parallels between the earth and the moon, concentrating on the latter perhaps because we have greater and more sensible knowledge of it on account of its lesser distance. Having concluded that this second opinion is more likely than the other, I think that the next step is to begin [133] to examine whether the earth must be considered immovable, as most people have so far believed, or else movable, as some ancient philosophers and others more recently have held;[3] and if it is movable, we must ask what its motion may be.

SALVIATI. I understand already and recognize the direction of our path. However, before beginning to proceed further, I must say something about the last words you uttered—that we concluded that the opinion holding the earth to have the same properties as the heavenly bodies is more likely than the contrary one. For I have not concluded this, just as I am not about to conclude any other controversial proposition; instead I have meant to produce, for one side as well as for the other, those reasons and answers, questions and solutions which others have found so far, together with some that have come to my mind after long reflection, leaving the decision to the judgment of others.[4]

---

next speech applies explicitly to the claim that the Copernican position is "more likely than the contrary one"; and in the sentence that ended the trial of 1633, one of the heresies of which Galileo was found guilty was "that one may hold and defend as probable an opinion after it has been declared and defined contrary to Holy Scripture" (Finocchiaro 1989, 291); for further details, see Morpurgo-Tagliabue (1981) and Finocchiaro (1986; 1992b).

2. That is, a fifth element, besides the four terrestrial ones of earth, water, air, and fire; it was also called quintessence or aether.

3. The idea that the earth moves had been advanced in antiquity by the Pythagoreans and by Aristarchus. Copernicus's contribution in the sixteenth century may be interpreted as that of having worked out the details of a novel and significant argument in support of that ancient, discarded idea.

4. This disclaimer is only one of many similar ones interspersed in the text; it is part of Galileo's attempt to have his book interpreted as a harmless hypothetical discussion of the issues, rather than as an assertion or defense of Copernicanism, which would have violated the ecclesiastic restrictions placed upon him. Of course, his attempt was not entirely successful, as the trial of 1633 shows. For more details, see the introduction (4), the appendix (3.2), selection 1, and Finocchiaro (1980, 6–18).

SAGREDO. I let myself be carried away by my own feelings and made universal a conclusion that should have been left individual, thinking that others should feel what I felt within myself. Indeed I erred, especially since I do not know the opinion of Simplicio present here.

SIMPLICIO. I confess to you that I thought about yesterday's discussions the whole night, and I really find many beautiful, new, and forceful considerations. Nevertheless, I feel drawn much more by the authority of so many great writers, and in particular. . . .[5] You shake your head and sneer, Sagredo, as if I were saying a great absurdity.

SAGREDO. I merely sneer, but believe me that I am about to explode by trying to contain greater laughter; for you reminded me of a beautiful incident that I witnessed many years ago together with some other worthy friends of mine, whom I could still name.

SALVIATI. It will be good for you to tell us about it, so that perhaps Simplicio does not continue to believe that it was he who moved you to laughter.

SAGREDO. I am happy to do that. One day I was at the house of a highly respected physician in Venice; here various people met now and then, some to study, others for curiosity, in order to see anatomical dissections performed by an anatomist who was really no less learned than diligent and experienced. It happened that day that they were looking for the origin and [134] source of the nerves, concerning which there is a famous controversy between Galenist[6] and Peripatetic physicians. The anatomist showed how the great trunk of nerves started at the brain, passed through the nape of the neck, extended through the spine, and then branched out through the whole body, and how only a single strand as thin as a thread arrived at the heart. As he was doing this he turned to a gentleman, who he knew was a Peripatetic philosopher and for whose sake he had made the demonstration; the physician asked the philosopher whether he was satisfied and sure that the origin of the nerves is in the brain and not in the heart, and the latter answered after some reflection: "you have made me see this thing so clearly and palpably that one would be forced to admit it as true, if

5. The ellipsis is in Galileo's text as given in the National Edition (Favaro 7:133), which corrected the 1632 edition in a few places such as here.

6. The Galenists were followers of Galen (c. 130–c. 200), Greek physician to the Roman emperor Marcus Aurelius and writer of many treatises that made him the supreme authority on medicine until the sixteenth century. They held (correctly) that the nerves originate in the brain, whereas the Aristotelians held (incorrectly) that they originate in the heart; cf. Aristotle, *On the Generation of Animals*, V, 2, 781a20.

Aristotle's texts were not opposed in saying plainly that the nerves originate in the heart."

SIMPLICIO. Gentlemen, I want you to know that this dispute about the origin of the nerves is not as settled and decided as some believe.

SAGREDO. Nor will it ever be decided as long as one has similar opponents. At any rate what you say does not diminish at all the absurdity of the answer of the Peripatetic, who against such a sensible experience did not produce other experiences or reasons of Aristotle, but mere authority and the simple ipse dixit.[7]

SIMPLICIO. Aristotle has acquired such great authority only because of the strength of his arguments and the profundity of his discussions. However, you must understand him, and not only understand him, but also know his books so well that you have a complete picture of them and all his assertions always in mind. For he did not write for the common people, nor did he feel obliged to spin out his syllogisms by the

7. Galileo uses the Latin phrase "ipse dixit" in his Italian dialogue since the phrase has been adopted by other languages, including English. It derives from the traditional Latin version of the Bible, which uses the phrase in Genesis to refer to God's acts of creation. It literally means "he himself said it," referring to someone who is regarded as an authority; it is a way of appealing to authority in the course of a discussion. Some appeals to authority may be regarded as arguments where the conclusion is supported by the premise that an authority has asserted it; although logic textbooks often claim that all such arguments are fallacious, it is more correct to say that they are never conclusive or deductively valid, but only more or less probable or strong, and that their strength depends on whether the "authority" is really a genuine authority on the topic in question and if so on how reliable it is. A good example would be the argument that may be taken to be in the background of the discussion in this selection and that would also define a direct connection between this discussion and the Copernican controversy: the earth stands still at the center of the universe because Aristotle said so; another instructive example would be the so-called biblical objection against Copernicanism: the earth cannot move because it is so stated or implied in many passages of the Bible. Galileo's criticism of Aristotle's authority in this selection is as an indirect criticism of the former argument; on the other hand, the latter objection is in the background of the discussion in selection 12 and is explicitly analyzed and criticized in his "Letter to the Grand Duchess Christina" (Finocchiaro 1989, 87–118). To explore further the parallels between these two arguments, one should examine the implications of the fact that in both the biblical and the Aristotelian contexts, the traditionalists were fond of declaring, with identical words, that "two truths cannot contradict one another" (Finocchiaro 1989, 96; cf. footnote to selection 3, [80]); a critical examination of this principle could begin by noting that in one sense this is a tautological assertion, which is an immediate consequence of the definition of the concepts of *truth* and *contradiction*; but then one would have to go on and examine the rhetorical import of such an assertion and the unstated assumption underlying the stated claim. It is also instructive to compare the content, character, and import of the general criticism of Aristotle's authority in this selection with the criticism of his authority as a logician in selection 2, with the criticism of his scientific and methodological authority in selection 3, and with the appeal to Aristotle's definition of the center of the universe in selection 11; for an example of such a critical comparison, see the appendix (2.3).

well-known formal method; instead, using an informal procedure, he sometimes placed the proof of a proposition among passages that seem to deal with something else. Thus, you must have that whole picture and be able to combine this passage with that one and connect this text with another very far from it. There is no doubt that whoever has this skill will be able to draw from his books the demonstrations of all knowable things, since they contain everything.

SAGREDO. So, my dear Simplicio, you are not bothered by things being scattered here and there, and you think that by collecting [135] and combining various parts you can squeeze their juice. But then, what you and other learned philosophers do with Aristotle's texts, I will do with the verses of Virgil or Ovid,[8] by making patchworks of passages and explaining with them all the affairs of men and secrets of nature. But why even go to Virgil or other poets? I have a booklet much shorter than Aristotle or Ovid in which are contained all the sciences, and with very little study one can form a very complete picture of them: this is the alphabet. There is no doubt that whoever knows how to combine and order this and that vowel with this and that consonant will be able to get from them the truest answers to all questions and the teachings of all sciences and of all arts. In the same way a painter, given various simple colors placed separately on his palette, by combining a little of this with a little of that and that other, is able to draw men, plants, buildings, birds, fishes—in short, all visible objects—without having on his palette either eyes, or feathers, or scales, or leaves, or rocks; on the contrary, it is necessary that none of the things to be drawn nor any part of them be actually among the colors, which can serve to represent everything, for if there were, for example, feathers, they would not serve to depict anything but birds and bunches of feathers.

SALVIATI. There are still alive some gentlemen who were present when a professor teaching at a famous university, upon hearing descriptions of the telescope which he had not yet seen, said that the invention was taken from Aristotle. Having asked that a book be brought to him, he found a certain passage where Aristotle explains how it happens that from the bottom of a very deep well one can see the stars in heaven during the day,[9] and he said to the bystanders: "here is the well, which corresponds to the tube; here are the thick vapors, from which is taken the

8. These are two of the greatest poets of ancient Rome and of the Latin language. Virgil (70–19 B.C.) is best known for the *Aeneid*, a long epic poem relating the events leading to the founding of the city of Rome in the eighth century B.C. Ovid (43 B.C.–18 A.D.) is best known for the *Metamorphoses*.

9. Cf. Aristotle, *On the Generation of Animals*, V, 1, 780b21.

invention of lenses; and lastly here is the strengthening of vision as the rays pass through the denser and darker transparent medium."

SAGREDO. This way of all knowledge being contained in a book is very similar to that by which a piece of marble contains within itself a very beautiful statue, or a thousand of them for that matter; but the point is to be able to discover them. We can say that it is also similar to Joachim's[10] prophecies or to the answers given by heathen oracles, which are understood only after the occurrence of prophesied events.

[136] SALVIATI. And where do you leave the predictions of astrologers,[11] which after the event can be so clearly seen in the horoscope, or should we say in the configuration of the heavens?

SAGREDO. In the same vein alchemists,[12] driven by their melancholic humor,[13] find that all the greatest minds in the world have never written about anything except the process of making gold, but that, in order to say this without revealing it to the common people, they have contrived in various ways to conceal it under various covers. It is very amusing to listen to their comments on ancient poets, as they reveal the very impor-

10. Joachim of Floris (c. 1132–1202) was an Italian clergyman whose writings contain numerous vague and ambiguous prophecies, especially about the coming of a new age when the hierarchical structure of the Church and the separation between Christianity and other religions would no longer be needed.

11. Astrology is the pseudo-science that tries to predict human behavior and future events based on the positions and configurations of the heavenly bodies. It should not be confused with the science of astronomy. But until Galileo's time, astronomy and astrology were usually practiced by the same persons and the two terms were often used interchangeably. Galileo's criticism reinforced a preexisting trend critical of astrology that eventually resulted in the separation of the two. One of the most important earlier critics of astrology had been Nicole Oresme (c. 1320–1382), a French scholar and clergyman who also anticipated some of Galileo's critiques of the geostatic world view and some of his ideas on motion. One of the best examples of an astronomer-astrologer is Johannes Kepler (1571–1630), who discovered the three laws of planetary motion named after him. The relationship between astrology and astronomy is analogous to that between alchemy and chemistry.

12. Alchemy was the ancient quest to turn base metals (such as iron) into precious ones (such as gold). Though this quest was impossible and alchemy was largely a pseudoscience, some aspects of it (such as its emphasis on experimental tinkering) made alchemy into a forerunner of the modern science of chemistry. As with astrology, Galileo's dismissal of alchemy was rare for his time and shows his remarkably modern outlook. Not all of the founders of modern science had comparably "modern" intuitions. For example, Isaac Newton (1642–1727) spent considerable time and effort in alchemical investigations, even though he was one of the greatest scientists and mathematicians of all times.

13. According to a theory first formulated by Hippocrates (c. 460–c. 370 B.C.) and then developed by Galen, the melancholic humor is one of four bodily fluids (called "humors") whose individual content and mutual balance govern physical health and psychological well being. Personality disorders, deviant behavior, and eccentric temperament were usually explained by the excessive predominance of a particular humor.

tant mysteries lying hidden under those fables:[14] what is the meaning of the love affairs of the moon, of her coming down to earth for Endymion, and of her anger at Actaeon; when does Jupiter change himself into golden rain, and when into burning flames; and how many great secrets of the art are to be found in Mercury the interpreter, in Pluto's kidnappings, and in those golden boughs.

SIMPLICIO. I believe and to some extent know that the world is full of very extravagant brains, whose follies should not redound to the discredit of Aristotle. You seem sometimes to speak of him with too little respect,[15] but the mere antiquity and the great name he has acquired in the minds of so many outstanding men should suffice to make him respectable among all educated men.

SALVIATI. That is not the way it is, Simplicio. It is some of his excessively cowardly followers who are responsible for making us think less of him, or to be more exact, who would be so responsible should we want to applaud their triflings. Tell me, if you do not mind, are you so simpleminded that you do not understand that if Aristotle had been present to listen to the doctor who wanted to make him inventor of the telescope, he would have been more angry with him than with those who were laughing at the doctor and his interpretations? Do you have any doubt that if Aristotle were to see the new discoveries in the heavens, he would change his mind, revise his books, accept the more sensible doctrines, and cast away from himself those who are so weak-minded as to be very cowardly induced to want to uphold every one of his sayings? Do they not realize that if Aristotle were as they imagine him, he would be an intractable brain, an obstinate mind, a barbarous soul, a tyrannical will, someone who, regarding everybody else as a silly sheep, would want his decrees to be preferred [137] over the senses, experience, and nature herself? It is his followers who have given authority to Aristotle, and not he who has usurped or taken it. Since it is easier to hide under someone else's shield than to show oneself openly, they are afraid and do not dare to go away by a single step; rather than putting any changes in the heavens of Aristotle, they insolently deny those which they see in the heavens of nature.

14. The rest of this speech refers to figures of classical Greek and Roman mythology, and not to heavenly bodies. The moon was a goddess; Endymion was a young shepherd with whom she fell in love; and Actaeon was the unfortunate hunter who watched her bathe, for which she turned him into a stag, whereupon he was killed by his own dogs. Jupiter was the supreme god, Mercury was the messenger of the gods, and Pluto was the god who ruled the afterlife.

15. Cf. selection 2, [59].

SAGREDO. These people make me think of that sculptor who carved a large piece of marble into an image of Hercules[16] or of a thundering Jupiter (I forget which); with admirable skill he gave it so much liveliness and fierceness that it terrified anyone who looked at it; then he himself began to be afraid even though these qualities were the work of his own hands; and his terror was such that he no longer dared to face it with his chisel and mallet.

SALVIATI. I have often wondered how it can be that those who rigidly maintain everything Aristotle said do not notice how much damage they do to his reputation, how much discredit they bring him, and how much they diminish his authority instead of increasing it. For I often see them stubbornly wanting to defend propositions that I find palpably and manifestly false, and wanting to persuade me that this is what a true philosopher is supposed to do and what Aristotle himself would do; their behavior greatly undermines my belief that he may have philosophized correctly in regard to other conclusions less well known to me; on the other hand, if I saw them yielding and changing their mind in regard to the obvious truths, I would be inclined to believe that when they persisted they might have sound demonstrations which I did not know or understand.[17]

SAGREDO. Still, if they felt they were risking too much of Aristotle's reputation or their own by admitting not knowing this or that conclusion discovered by others, would it not be better to find it in his texts by combining them in accordance with the practice mentioned by Simplicio? For, if all knowledge is contained there, one must be able to find it.

SALVIATI. Sagredo, do not make fun of this advice, which you seem to propose in jest. In fact, not long ago a philosopher of great renown

16. In classical Greek mythology, Hercules was a hero renowned for his physical strength and courage.

17. This is another Galilean expression of epistemological modesty; the point is that it is better to sometimes admit that one is ignorant or mistaken, rather than to claim to know everything or to be always right. One reason for this is that one will then be taken more seriously. And one consequence is that the Aristotelians' lack of epistemological modesty actually harms the authority of their master. This plausible argument is analogous to the one Galileo advanced in his "Letter to the Grand Duchess" while criticizing the scientific authority of the Bible and adopting an argument out of St. Augustine's *De Genesi ad Litteram*; the criticism is directed at those who claim that biblical assertions about natural phenomena must be accepted (on pain of heresy); the point is that such zealots do the Bible a disservice by discrediting it in the eyes of people (especially potential converts) who know that physical truth is otherwise, for these people will then understandably be led to question the truth of the Bible where it really matters, namely, on questions of faith and morals; cf. Finocchiaro (1989, 94–95).

wrote a book on the soul which discussed Aristotle's opinion on whether or not it is immortal by presenting many passages that suggested a pernicious answer (these were passages discovered by himself in little known places rather than the ones quoted by Alexander[18] [138] because in these Aristotle allegedly did not even discuss this subject, let alone establish anything pertaining to it); when he was warned that he would have encountered difficulties in getting the printing license, he wrote back to his friend not to let this stop the application process because, if there were no other obstacles, he would have had no difficulty changing Aristotle's doctrine and supporting the contrary opinion with other assertions and passages also corresponding to Aristotle's mind.[19]

SAGREDO. Oh, what a scholar! I am at his command; he does not want to be duped by Aristotle, but wants to lead him by the nose and make him speak as he himself commands! See how important it is to know how to seize an opportunity.[20] One should not deal with Hercules when he is in a rage and overtaken by fury, but rather while he is playing

18. Alexander of Aphrodisias was a Greek philosopher who lived around 200 A.D. He is best known for his commentaries on Aristotle, especially the one dealing with the book *On the Soul*. He interpreted Aristotle as implying that a person's soul is not immortal and was consequently condemned at the Fifth Lateran Council in 1512 (Santillana 1953, 125n.8).

19. Regardless of the historical existence and identity of this "philosopher of great renown," this passage gives a good portrait of opportunism, in the pejorative sense, for the "pernicious answer" to the problem is the thesis that the soul is not immortal, which implies that death represents the end of personal existence; and the story has this philosopher adopt the religiously more orthodox answer in order to receive permission to publish his book. We may wonder whether such opportunism was any more prevalent among the Aristotelians in Galileo's time than among the adherents of any other dominant school of thought at any other time, for example today; we may also wonder whether it is possible to be opportunistic in a nonpejorative sense, for the character flaw in this example is perhaps the materialistic motivation for the change of mind, whereas it is possible for such a change to be motivated by a change in one's assessment of the evidence and arguments. For more details on methodological opportunism, see Albert Einstein's remarks in Schilpp (1951, 683–84), Feyerabend (1975), and Finocchiaro (1988a). There has been some speculation about whether this "philosopher of great renown" corresponds to an actual historical figure, and if so who he may have been. Fiorentino (1868, 326 and 336) suggests it may have been Zabarella (d. 1589); Strauss (1891, 513n.8) suggests Pendasio (d. 1603); Drake (1967, 475n; 1982, 579n.8) mentions the possibility of Cesare Cremonini, one of Galileo's colleagues at the University of Padua; and Sosio (1970, 138 n. 1) mentions Fortunio Liceti (1577–1657), with whom Galileo exchanged some correspondence.

20. Galileo's choice of this particular word (*tempo opportuno*, which literally means "opportune time") strengthens the remarks on opportunism in the previous note. This is also the effect of the wording in his marginal note on the story in Salviati's preceding speech; translated literally, that marginal note reads "opportune decision of a Peripatetic philosopher" (Favaro 7:138).

with Lydian maids.[21] Oh, the unbelievable cowardice of slavish minds! To make oneself spontaneously a slave, to accept decrees as inviolable, to be obliged to call oneself persuaded and convinced by arguments so effective and clearly conclusive that its proponents cannot decide even whether they are written for that purpose and are meant to prove that conclusion! But let us mention the greatest folly: that they themselves are still uncertain whether the same author holds the affirmative or negative side. Is not this like regarding a wooden statue as their oracle, resorting to it for answers, fearing it, revering it, and worshipping it?

SIMPLICIO. But, if one abandons Aristotle, who will be the guide in philosophy? Name some author.

SALVIATI. One needs a guide in an unknown and uncivilized country, but in a flat and open region only the blind need a guide; whoever is blind would do well to stay home, whereas anyone who has eyes in his head and in his mind should use them as a guide.[22] Not that I am thereby saying that one should not listen to Aristotle; on the contrary, I applaud his being examined and diligently studied[23] and only blame

21. In one of his many exploits, Hercules was condemned by the Delphic Oracle to be a servant to the queen of Lydia; she had him dress in women's clothes, live with her maids, spin wool like them, and make love to her.

22. This sentence makes it clear that the positive side of Galileo's negative criticism of Aristotle's authority is a plea for independence of mind. When formulated in terms of independent-mindedness, the methodological attitude advocated by Galileo may seem uninformative and uncontroversial, but it is very important. However, the examples and arguments in this selection show that to be independent is something more easily said than done. In fact, independence requires avoiding the uncritical acceptance of authority; for example, we must not give the views of the authority priority over our own experience, as the absurdity of the Peripatetic's attitude about the origin of nerves illustrates; and we must not act as if all knowledge had already been provided by the authority, as the absurdity of reading the invention of the telescope into Aristotle illustrates. But we must also avoid opportunism; that is, we must avoid being so self-centered and self-confident as to think we can make the authority stand for whatever we wish it to represent; this may be regarded as the lesson of the example of the unscrupulous author of the book on Aristotle's doctrine of the soul. And, as Salviati goes on to caution, we must not completely disregard an authority of Aristotle's caliber; such disregard would be a kind of solipsism. In short, independent-mindedness may be viewed as the balanced mean between the extremes of uncritical acceptance at one end and uncritical disregard at the other end.

23. It is important to realize that Galileo does not advocate the complete disregard of Aristotle, and that the target of his criticism is the uncritical Aristotelians and not Aristotle himself; Galileo objects to the abuse and misuse of Aristotle's authority and not to the proper recognition of it. This point is important from a methodological viewpoint, in order to understand that authority has a proper role in inquiry; the alternative would be an absurd methodological solipsism in which one would rely only on oneself and end up rediscovering the wheel. But the point is also important from a historical viewpoint, in order to understand properly the relationship of Galileo to Aristotle and Aristotelianism; for more details, see Wallace's important contributions.

submitting to him in such a way that one blindly subscribes to all his assertions and accepts them as unquestionable dictates, without searching for other reasons for them. This abuse carries with it another extreme impropriety, namely, that no one makes an effort any longer to try to understand the strength of his demonstrations. Is there anything more shameful in a public discussion [139] dealing with demonstrable conclusions than to see someone slyly appear with a textual passage (often written for some different purpose) and use it to shut the mouth of an opponent?[24] If you want to persist in this manner of studying, lay down the name of philosophers and call yourselves either historians or memory experts, for it is not right that those who never philosophize should usurp the honorable title of philosopher.[25]

However, we should get back to shore in order not to enter an infinite ocean from which we could not get out all day. So, Simplicio, come freely with reasons and demonstrations (yours or Aristotle's) and not with textual passages or mere authorities because our discussions are

24. The "extreme impropriety" to which Salviati objects here is the stress on what Aristotle said rather than on why he said it; the overemphasis on his claims and the neglect of his arguments; the failure to appreciate, understand, and evaluate his reasoning. This is a crucial distinction of general methodological importance; it should be added to, and compared with, the distinctions that emerged in previous selections between theory and practice and between method used and result reached. This distinction between claims and reasons is soon put into practice in the dialogue by undertaking the presentation, analysis, and evaluation of Aristotle's arguments.

25. This Galilean praise for philosophy is based on conceiving it as consisting essentially of reasoning and the analysis and evaluation of claims in the light of reasons. This conception of philosophy is widespread and fruitful; I believe it is essentially correct and can be supported partly by the ancient Greek etymological meaning of philosophy as *love of wisdom*, partly by analyzing the connotations of the verb *philosophize* or *philosophizing*, and partly by a comparative analysis and evaluation of this conception vis-à-vis others. This view of philosophy is a recurrent theme in Galileo, and here we must mention two revealing instances: after he resigned his professorship of mathematics in 1610 until his death, he held the title of "Philosopher and Chief Mathematician" to the grand duke of Tuscany; and the inclusion of the word *philosopher* in his title was something he initiated and insisted upon, claiming as a justification that he had spent more years studying philosophy than months studying pure mathematics (Favaro 10:353). Not all scholars would agree with this account of the connection between Galileo and philosophy; the best exponent of an antiphilosophical interpretation of Galileo is Drake (1957; 1976; 1980; 1981); but it is possible to accept many of Drake's positive characterizations of Galileo involving experimentation, measurement, and his tentativeness toward Copernicanism, without accepting the antiphilosophical conclusions; see, for example, Finocchiaro (1977b) and cf. MacLachlan (1990). In this passage, Galileo contrasts philosophers with historians and equates history with memory; this does not do full justice to the complexity of historical inquiry, in which there is a key role for the analysis and evaluation of documents and texts, and hence for reasoning; but the equation and contrast remain widespread views even today; for some relevant details, see Finocchiaro (1973) and Kuhn (1970).

about the sensible world and not about a world on paper. In yesterday's discussions the earth was drawn out of darkness and brought to light in the open heavens, and we showed that to want to number it among those bodies called heavenly is not so doomed and prostrate a proposition as to be left devoid of any vital energy; and so today we should examine how much probability there is in holding it fixed and completely motionless (referring to the globe as a whole) and how much likelihood there is in making it move with any motion (and if so what type this is).[26] I am undecided about this question, while Simplicio together with Aristotle is firmly on the side of immobility; because of this, he will present step-by-step the motives for their opinion, I will present the answers and arguments for the contrary side, and Sagredo will say what goes on in his mind and to which side he feels drawn.

SAGREDO. I am happy with this arrangement, but on the condition that I am free to introduce whatever simple common sense may suggest to me.

SALVIATI. Indeed, I beg you to do exactly that; for I think the various authors have left out few of the easier and (so to speak) cruder considerations, so that only some of the more subtle and esoteric ones may be wanting and lacking; but to investigate these, what subtlety can be more appropriate than that of Sagredo's intellect, which is most acute and penetrating?

SAGREDO. I may be all that Salviati says, but please, let us not start on another sort of ceremonial digression because right now I am a philosopher and have come to school and not to city hall.

## [6. Diurnal Rotation, Simplicity, and Probability]

SALVIATI. So let the beginning of our reflections be the following consideration: whatever motion is attributed to the earth, [140] it must remain completely imperceptible and seem nonexisting for us living there and sharing that motion, as long as we look only at terrestrial things; on the other hand, it is equally necessary that the same motion appear to us to be common to all other visible bodies and objects

26. Note that the language of probability and likelihood which had been previously used to formulate the issue of earth-heaven dichotomy versus unification is now used to formulate the question of earth's motion versus rest.

which are separated from the earth and so lack that motion.[27] Thus, the true method of investigating whether any motion may be attributed to the earth and, if so, what kind it may be is to consider and observe whether in the bodies separated from the earth one sees any appearance of motion belonging equally to all; for if a motion were seen, for example, only in the moon and had nothing to do with Venus or Jupiter or other stars, it could not belong to the earth in any way but only to the moon. Now, there is a motion which is very general and most important of all: it is the motion by which the sun, moon, other planets, and fixed stars (in short, the whole universe except only the earth) appear to us to move together from east to west in a period of twenty-four hours. In regard to this first phenomenon, this motion may belong either to the earth only or to the rest of the universe without the earth, for the same appearances would be seen in the one situation as in the other.[28] Aristotle and Ptolemy grasped this consideration, and so when they try to prove the earth to be motionless, they argue only against this diurnal motion; but Aristotle mentions something or other against another motion attributed to it by an ancient author, of which we shall speak in due course.[29]

27. This sentence expresses a version of the principle of the relativity of motion, a basic principle of classical physics (not to be confused with the principles of either special or general relativity in contemporary physics, introduced by Albert Einstein). One question is whether Galileo intends to apply the principle literally to all motion or only to uniform motion, and what would be meant by uniformity in the latter case. Another question is whether he gives arguments directly supporting this principle or takes it as an unsupported fundamental assumption useful in the analysis of other issues. A third question is whether he accepts this principle as unconditionally true or only as an approximation subject to exceptions, one of which might be the tides. A fourth question is whether and why the Aristotelians would disagree, in regard to which we should note Simplicio's next comment and the discussion it engenders ([142]). Finally, there is the question of how Galileo's relativity relates to that of Copernicus, who had a similar idea (*On the Revolutions*, I, 5; 1992, 11–12); this difference is usually defined by interpreting Copernicus as talking merely about optical relativity, but Galileo as focusing on mechanical relativity (Koyré 1966; Santillana 1953, 128n.11); cf. also Chalmers (1993).

28. That is, apparent diurnal motion may be explained in either of two ways: geostatically, by saying that apparent diurnal motion is real, namely, that all heavenly bodies actually revolve westward around a motionless earth every twenty-fours hours, as the Aristotelians claimed; or geokinetically, by saying that apparent diurnal motion is merely apparent and results from the earth's eastward rotation on its own axis daily, as the Copernicans held. Note that for the latter, the direction of the real motion is opposite to that of the apparent motion because this is required to bring about the observed effect.

29. This other motion is the annual motion which Galileo discusses in the Third Day, whereas the Second Day deals with the diurnal motion. The Aristotelian and Ptolemaic arguments are presented in selection 7, and textual references to their works may be found in the notes there.

SAGREDO. I understand very well the necessity of which you speak, but I have a difficulty which I do not know how to remove. Copernicus attributed to the earth another motion besides the diurnal one; so by the rule just stated, as regards the appearances, that other motion should remain imperceptible when we look at the earth but be visible in the whole rest of the universe; thus it seems one can necessarily conclude either that he clearly erred in attributing to the earth a motion which does not appear to be general in the heavens, or that if it is general then Ptolemy was equally wrong in not refuting it as he did the other.

[141] SALVIATI. Your difficulty is a very reasonable one; when we treat of the other motion you will see how much Copernicus's intellect surpassed Ptolemy's in cleverness and profundity, insofar as the former saw what the latter did not, namely, the wonderful accuracy with which this motion is reflected in all other heavenly bodies.[30] However, for now let us put off this aspect and return to the first point; in regard to this, I shall begin with the more general considerations and propose the reasons that seem to favor the earth's mobility, and then I shall listen to Simplicio for the opposite ones.

Firstly, let us consider the immense size of the stellar sphere in comparison to that of the terrestrial globe, which can fit inside the former many millions of times, and let us also think of the speed required for it to make one entire rotation in twenty-four hours; given these considerations, I cannot persuade myself that anyone can be found who would think it is more reasonable and credible for the celestial sphere to undergo rotation and the terrestrial globe to stand still.

SAGREDO. Let us assume that all phenomena which may be naturally dependent on these motions are such that the same consequences follow, without a difference, from one supposition as well as the other one;[31] if

30. Observationally speaking, at the time of Copernicus and Galileo, the earth's annual motion was "reflected" in the appearances of only the planets, not in those of the fixed stars. In fact, this lack of reflection constituted a key difficulty for Copernicanism; it was formalized in the objection from stellar parallax, which Galileo could not really refute because even his telescope did not reveal any stellar parallax. For more details, see selections 13 and 14.

31. This is a crucial premise in the geokinetic argument of this selection, which may be reconstructed as follows: the geokinetic hypothesis is more likely to be true than the geostatic one because (1) both can explain observed phenomena (such as apparent diurnal motion); (2) the geokinetic explanation is simpler; and (3) nature operates in the simplest possible ways. The principle of the relativity of motion serves as the main reason in the subargument supporting this crucial premise (1); the seven arguments listed in the following discussion are reasons why the geokinetic explanation is simpler, namely, seven subarguments supporting the second premise; and the third premise is a version of the principle of simplicity, which will soon emerge in the discussion.

this were so, my initial and general impression would be that whoever thought it more reasonable to make the whole universe move to keep the earth motionless was more unreasonable than someone who went up to the top of your cathedral to look at the city and its surroundings and demanded that they turn around him so that he would not have to bother turning his head. To overcome this absurdity and revise my impression, thus rendering this supposition more credible than the other one, the advantages deriving from it rather than the other would have to be great and many. But Aristotle, Ptolemy, and Simplicio must think that there are advantages in it; now, if these exist we should be told what they are, or else let it be admitted that there are not or cannot be any.

SALVIATI. Despite my having thought about it for a long time, I have been unable to find any difference, and so my finding seems to be that there cannot be any difference; hence I feel it is useless to continue searching for one. Let me explain. Motion exists as motion and acts as motion [142] in relation to things that lack it, but in regard to things that share it equally, it has no effect and behaves as if it did not exist.[32] Thus, for example, the goods loaded on a ship move insofar as they leave Venice, go by Corfu, Crete, and Cyprus, and arrive in Aleppo, and insofar as these places (Venice, Corfu, Crete, etc.) stay still and do not move with the ship; but for the bales, boxes, and packages loaded and stowed on the ship, the motion from Venice to Syria is as nothing and in no way alters their relationship among themselves or to the ship itself; this is so because this motion is common to all and shared equally by all; on the other hand, if in this cargo a bale is displaced from a box by a mere inch, this alone is for it a greater motion (in relation to the box) than the journey of two thousand miles made by them together.

SIMPLICIO. This doctrine is correct, sound, and entirely Peripatetic.

SALVIATI. I think it is even more ancient. Moreover, I suspect that, when Aristotle took it from some good school, he did not entirely grasp it, and that therefore he wrote it in altered form and so was the source of confusion with the help of those who want to support all his statements. I also suspect that, when he wrote that everything which moves, moves upon something unmoved, he engaged in an equivocation on the assertion that everything which moves moves in relation to

32. This sentence gives another interesting formulation of the principle of the relativity of motion.

something unmoved; the latter proposition suffers no difficulties, the former many.[33]

SAGREDO. Please, let us not break the thread, and let us proceed with the discussion we began.

SALVIATI. It is clear, then, that motion common to many movable things is idle and null in regard to their relationship among themselves (because nothing changes among them), and that it acts only in regard to the relationship between those movable things and others which lack that motion (for this is the relationship which changes).[34] We have also divided the universe into two parts, for which it is necessary that one of them is mobile and the other immobile; in regard to whatever may depend on this motion, to make the earth alone move is equivalent to making the rest of the universe move because the action for this motion lies only in the relationship between the heavenly bodies and the earth, and this is the only relationship that changes. Again, let us assume that, in order to bring about the same effect in the finest detail one can either have the earth alone moving with the whole rest of the universe stopped or have the earth alone still with the whole universe [143] moving by the same motion; if this assumption holds, who will believe that nature has chosen to let an immense number of very large bodies move at immeasurable speed to bring about what could be accomplished with the moderate motion of a single body around its own center?[35] Indeed, who will believe this, given that by common consent, nature does not do by means of many things what can be done by means of a few?[36]

33. Cf. Aristotle, *Physics*, VIII, 4–5, 254b7–258b9; and *On the Progression of Animals*, 2, 698b8ff. Santillana (1953, 129n.12) claims Aristotle got the idea from Democritus's atomism and judges Galileo was right in charging that Aristotle misunderstood the atomists. Democritus (c. 460–c. 370 B.C.) was a Greek thinker best known as one of the originators of the atomic theory of matter, according to which all things are made up of tiny indivisible particles, too small to be seen directly, but whose motions and properties account for all that is perceived. See also Wallace (1974, 269n.12).

34. Once again, this is the principle of relativity.

35. This subargument may be reconstructed as follows: the geokinetic explanation of diurnal motion is simpler than the geostatic explanation because (1) the number of moving bodies is much greater in the geostatic system than in the geokinetic one (thousands of stars versus one rotating earth); (2) the size of the bodies in motion is much larger in the former than in the latter (any star is bigger than the earth); and (3) the speeds of motion are much higher in the former than in the latter (traversing in twenty-four hours the circumference of the universe versus the terrestrial circumference).

36. This is the first of many statements of the principle of simplicity (or economy), which is a premise in the geokinetic argument of this selection; this argument may thus be called the simplicity argument for the earth's diurnal motion. Simplicity is an important and controversial methodological principle; but it is itself anything but simple; for example, note that here it is formulated as a general physical principle about natural operations.

SIMPLICIO. I do not understand very well how this very great motion is null for the sun, the moon, the other planets, and the innumerable array of fixed stars. How can you say it is nothing for the sun to pass from one meridian to another, rise above this horizon, set below that one, and bring day and night in turn; and also for the moon, other planets, and fixed stars to go through similar variations?

SALVIATI. All these variations you mention are nothing except in relation to the earth. To see that this is true, imagine that the earth is taken away: there will no longer be in the world any rising or setting of the sun or moon, any horizons or meridians, any days or nights; nor would their motion ever produce any changes among the moon, the sun, or any other stars whatever (be they fixed or wandering). In other words, to say that all these changes relate to the earth means that the sun appears first in China, then in Persia, and afterwards in Egypt, Greece, France, Spain, America, etc., and that the moon and the other heavenly bodies do the same. This phenomenon occurs in exactly the same way if, without involving such a large part of the universe, the terrestrial globe is made to turn on itself.

However, let us double the difficulty with another very great one. That is, if this great motion is attributed to the heavens, it is necessary to make it contrary to the particular motion of all the planetary orbs; each of these unquestionably has its own characteristic motion from west to east, at a very leisurely and moderate speed; but then one has to let this very rapid diurnal motion carry them off violently in the contrary direction, namely, from east to west. On the other hand, by making the earth turn on itself, the contrariety of motions is removed, and motions from west to east alone accommodate all appearances and satisfy them all completely.[37]

37. This paragraph gives a second reason why the geokinetic system is simpler than the geostatic one: all bodies move in the same direction (eastward) in the Copernican system, whereas in the Ptolemaic one, some motions (for example, the diurnal motion) are westward but others (for example, the annual motion of the sun) are eastward; that is, Copernicanism has fewer directions of motions (only one rather than two). The reason for this, which is not explained in this passage, is as follows. Observation reveals a mixture of directions of motion; for example, apparent diurnal motion is westward, but apparent annual motion is eastward. In the Ptolemaic system, since appearance corresponds to reality, the same mixture is postulated to exist in reality. But in the Copernican system, when apparent westward diurnal motion is explained by means of the earth's axial rotation, the direction has to be reversed, and the earth must be taken to rotate eastward; this is needed because we are replacing a motion in an orbit that encloses the earth with a motion of an earth that is not in an orbit about something else but is rather spinning around itself. On the other hand, Copernicanism does not have to reverse the direction of the annual motion because the eastward direction of the apparent annual motion of the sun corresponds to the order of the constellations of the zodiac, and given that the sun appears to move among them in this order, this appearance can result only by the earth orbiting in the *same*

SIMPLICIO. As for the contrariety of the motions, it matters little because Aristotle demonstrates that circular motions are not contrary to each other, and that theirs cannot be called true contrariety.

[144] SALVIATI. Does Aristotle demonstrate this, or does he merely assert it because it fits his purpose? If, as he himself states, contrary motions are those which reciprocally destroy each other, I do not see how two moving bodies which collide along a circular line would damage each other any less than if they were colliding along a straight line.

SAGREDO. Please, stop for a moment. Tell me, Simplicio, when two knights meet jousting in an open field, or when two whole fleets or armadas clash at sea breaking up and sinking each other, would you call such encounters contrary to one another?

SIMPLICIO. Let us call them contrary.

SAGREDO. How is it then that there is no contrariety for circular motions? For these occur on the surface of the land or the ocean, which (as you know) is spherical, and so they are circular after all. Do you know, Simplicio, which circular motions are not contrary to one another? They are those of two circles tangent to each other and such that the turning of one naturally makes the other one move in a different direction; but, if one is inside the other, it is impossible that their motions in different directions should not contrast with each other.

SALVIATI. In any case, whether the motions are contrary or not, these are verbal disputes. I know that in fact it is much simpler and more natural to explain everything by means of a single motion rather than by introducing two of them. If you do not want to call them contrary, call them opposite. Moreover, I am not saying that this introduction of opposite motions is impossible; nor am I claiming to be giving a necessary demonstration, but only inferring a greater probability.[38]

The unlikelihood is tripled by upsetting in a very disproportionate manner the ordered pattern we unquestionably see existing among those heavenly bodies whose revolution is not in doubt but most cer-

---

direction. In other words, to explain apparent annual motion, the two alternatives are either that the sun orbits the earth or that the earth orbits the sun; since either body is revolving around the other, in reality the same direction is required in both cases and this is the direction in which the sun appears to move along the zodiac. For more details, see Kuhn (1957, 160–65).

38. Note the probabilistic character of this argument: the Ptolemaic introduction of opposite motions is not impossible but improbable; it is improbable because the Ptolemaic system is less probable than the Copernican arrangement; the former is less probable because it is less simple; it is less simple because it has more directions of motion; and it has more of these because it has two directions whereas the latter system has just one.

tain.[39] The pattern is that when an orbit is larger, the revolution is completed in a longer period of time; and when smaller, in a shorter period.[40] Thus Saturn, which traces a greater circle than any other planet, completes it in thirty years; Mars in two; the moon goes through its much smaller orbit in just a month; and, in regard to the Medicean stars, we see no less sensibly that the one nearest Jupiter completes its revolution in a very short time (namely, about forty-two hours), the next one in three and one-half days, the third one in seven days, and the most remote one in sixteen. This very harmonious pattern is not changed in the least [145] as long as the motion of twenty-four hours is attributed to the terrestrial globe (rotating on itself). However, if one wants to keep the earth immobile, it is necessary first to go from the very short period of the moon to others correspondingly longer; that is, to that of Mars lasting two years, from there to the larger orbit of Jupiter requiring twelve years, and from this to the bigger one of Saturn with a period of thirty years; but then it is necessary to go to an incomparably greater orb and have an entire revolution completed in twenty-four hours.[41] This is the least disorder that would follow; for someone may first want to go from

39. In the sequence of subarguments in this selection, this one provides a third reason why the geostatic system is less simple and so less likely; that is, the unlikelihood of the geostatic system is increased by a third feature that makes it less simple. But this third argument is important in its own right and possesses considerable strength. For the key premise is a generalization which appeared to Galileo to be a law of nature, and which I call the law of revolution; the key point is that in the geostatic system the diurnal motion of the fixed stars violates this law, whereas in the Copernican system the diurnal rotation of the earth does not, and all revolutions around a common center are in accordance with it. This subargument may thus be called the argument from the law of revolution.

40. I call this generalization Galileo's law of revolution. It is reminiscent of, and should be compared and contrasted with, Kepler's third law. The latter states that the planets revolve around the sun in such a way that the square of the period of revolution is proportional to the cube of the mean distance from the sun (that is, the period varies as the three-halves power of the distance). Kepler advanced this law in his 1619 book *Harmonies of the World*, although Galileo was probably not sufficiently acquainted with it to have detected it there. The point is that Galileo is thinking of a more rudimentary version of Kepler's law and is doing so chiefly in the context of his own discovery of Jupiter's satellites; that is, it was already known that the periods of planetary revolutions increase as their orbits grow larger; this well-known fact could not be given the status of a law of nature as long as the planetary orbits constituted the only instance of several bodies revolving around a common center; but the methodology of the situation changed when Galileo's own discovery and detailed observations of Jupiter's satellites revealed that these four bodies follow the same relationship between size of orbit and period of revolution; he was then led to think that the pattern is true in general and represents a significant fact about motion. In this paragraph he summarizes the evidence leading him to this conclusion and uses the law to support the geokinetic hypothesis.

41. This refers to the stellar sphere or the orb of the Prime Mobile.

Saturn to the stellar sphere and make it larger than the orbit of Saturn in a proportion appropriate to its very slow motion with a period of many thousands of years;[42] but then one would have to make a much more disproportionate jump in going from the stellar sphere to an even larger one, and make the latter revolve in twenty-four hours. On the other hand, once we give motion to the earth, the order of the periods is very strictly followed, and from the very sluggish orb of Saturn we go to the fixed stars, which completely lack such motions.

The earth's rotation also enables one to escape a fourth difficulty, which must necessarily be admitted if the stellar sphere is made to move. The difficulty is the immense disparity among the motions of the stars: some would move at very great speed in very large circles, while others would move very slowly in very small circles, depending on whether they are respectively further away from or closer to the poles. This is problematic because we see those heavenly bodies whose motion is not in doubt all moving in great circles, as well as because it does not seem to be good planning that bodies which are supposed to move in circles be placed at immense distances from the center and then be made to move in very small circles.

Aside from the fact that the pattern of the magnitude of the circles and the consequent speed of the motions of these stars would be very different from the pattern of circles and motions of the others, each of these same stars would be changing its circle and speed,[43] and this is a fifth disadvantage. For there are stars which two thousand years ago were positioned on the celestial equator, and consequently described great circles with their motion; but in our time they are located away from it by several degrees, and so one must attribute to them a slower motion and make them move in smaller circles; and it may even happen that the time will come when some star which in the past always moved will become motionless by being joined to the pole, and then again (after resting [146] for some time) it will get back in motion. On the other hand, as previously stated, all the other stars that are unquestionably in motion describe the greatest circle of their orb and keep themselves constantly in it.

The unlikelihood is increased by a sixth disadvantage. Anyone with sound common sense will be unable to conceive the degree of solidity of

42. This period refers to the precession of the equinoxes, namely, the apparent westward movement of the equinoctial points on the celestial sphere. Since the time of Copernicus, it has been estimated to have a period of 26,000 years, but ancient astronomers gave the higher figure of 36,000 years.

43. Once again, this is a reference to the precession of the equinoxes.

that very large sphere in whose thickness would be embedded so many stars so firmly that they do not change their relative positions in the least, and yet they are made to revolve together with such great disparity. On the other hand, it is more reasonable to believe that the heavens are fluid,[44] so that each star wanders about in space by itself; if this belief is true, what law would regulate their motions and to what end while making sure that (when observed from the earth) they would appear as if they were produced by a single sphere? It seems to me that an easier and more manageable way of accomplishing this would be to make them motionless, rather than making them wandering, just as it is easier to keep track of the many stones cemented into the pavement of a marketplace than of the bands of children running over them.

Finally, there is a seventh objection: if we attribute the diurnal turning to the highest heaven,[45] it must have so much force and power as to carry with it innumerably many fixed stars (all very huge bodies and much larger than the earth) and also all the planetary orbs, even though both the latter and the former by nature move in the contrary direction;[46] moreover, it is necessary to admit that even the element fire and most of the air would be carried along as well, and that only the tiny terrestrial globe would be stubborn and recalcitrant vis-à-vis so much power; this seems to me to be a very problematic thing, and I would be unable to explain how the earth (as a body suspended and balanced on its center, indifferent to motion and to rest, and placed in and surrounded by a fluid environment) would not yield and be carried along the rotation. However, we do not find such obstacles in giving motion to the earth; it is an insignificant and very small body compared to the universe, and thus unable to do any violence to it.[47]

44. One of the issues in the Copernican controversy was whether the heavens are solid or fluid; that is, whether or not there exist crystalline spheres made out of impenetrable aether in which the various planets and fixed stars are embedded and whose rotation makes these heavenly bodies revolve.

45. The outermost spherical crystalline shell surrounding the earth, namely, the Prime Mobile.

46. Recall that in the Ptolemaic system, the diurnal motion is really westward, whereas the individual planetary revolutions are really eastward.

47. This is an important argument, partly because it raises the question of the connection between physics and astronomy, partly because it underscores the crucial importance of the earth-heaven dichotomy in the Ptolemaic system, partly because it questions the internal coherence of the Aristotelian world view, and partly because it seems to have considerable cogency (although this depends on some mechanical intuitions that are not completely self-evident). The point is that the Ptolemaic diurnal motion is transferred all the way from the Prime Mobile to the stellar sphere, the planets, the moon, and (according to

SAGREDO. I feel some of these concepts whirling in my mind, and indeed I am very confused after the discussions we have just had; if I want to be able to concentrate on what remains to be said, I must try to put some order in my ideas and draw some useful lesson (if possible). Proceeding by questioning will perhaps help me to explain myself better. So I first ask Simplicio whether he believes that [147] different natural motions may belong to the same simple body, or else only one is appropriate as its own natural motion.

SIMPLICIO. For a simple movable body, only one, and no more, can be the motion that naturally belongs to it; all other motions can belong to it only by accident or by participation.[48] For example, for someone walking on a ship, his own motion is that of walking, and by participation he has the motion bringing him to port; for he will never arrive there unless the ship's motion takes him there.

SAGREDO. Tell me a second thing, in regard to the motion that by participation is transferred to some moving body while the latter moves on

---

some versions of Aristotelianism) even the sphere of the element fire just below the moon and the atmosphere below that; but then it strangely stops at the terrestrial globe. On the other hand, the Copernican diurnal motion is the earth's axial rotation, and there is no reason why the rotation of such a small ordinary body should make other bodies move; so the diurnal rotation belongs just to the earth. The Aristotelians could reply, and did reply, that the circular motion originating from the Prime Mobile stops at the earth because the earth is made out of elements which do not obey the same laws as the heavenly bodies; for example, the natural place of the element earth is the center of the universe, and its natural motion is toward that center. But this would not end the discussion, for Galileo could and did counter (selection 8, [167]) by saying that Aristotle himself (*Meteorology*, I, 7, 344a11) admits that some circular motion is transferred to the elements fire and air (and so he is not completely self-consistent). Galileo could also give the following argument which he hints at here but had elaborated in his earlier works: even from an Aristotelian viewpoint, the element earth has a tendency to go toward the center of the universe and a repugnance to move away from it; hence, earth has neither a tendency nor a repugnance to move in a way that does not bring it nearer to or farther from the center; but axial rotation does not bring the earth either nearer to or farther from that center; so the earth has neither a tendency nor a repugnance to undergo axial rotation; thus, the smallest force is sufficient to start the earth rotating; in short, terrestrial rotation would be a *neutral* motion (neither natural nor violent), and so the diurnal motion of even aethereal bodies would start it spinning.

48. Cf. Aristotle, *Physics*, II, 14, 296b31–32. This was a basic principle of Aristotelian physics. Here it is applied to derive the existence of the Prime Mobile. But it could also be used to formulate a more direct and explicit objection to Copernicanism: if Copernicanism were correct, the earth would have at least three natural motions, namely, the downward fall of its parts, axial rotation, and orbital revolution of the whole around the sun; but a simple body like the earth cannot have more than one natural motion; so Copernicanism is wrong. In other parts of the *Dialogue* omitted here (Favaro 7:281–89 and 423–41; Galileo 1967, 256–64 and 397–415), Galileo countered by undermining the second premise; he did this partly by criticizing this principle of Aristotelian physics (on which that premise is based) and partly by questioning whether the earth is indeed a simple body.

its own with some motion different from the shared one; must this transferred motion belong to some subject by itself, or can it exist in nature without other support?

SIMPLICIO. Aristotle answers all these questions. He says that, just as to a given moving thing there corresponds one particular motion, to a given motion there corresponds one particular moving thing; consequently, no motion can exist or be imagined without it inhering in its subject.[49]

SAGREDO. Thirdly, I should like you to tell me whether you believe that the moon and the other planets and heavenly bodies have their own proper motions and what these motions are.

SIMPLICIO. They have them, and the motions are those whereby they run through the zodiac: the moon in one month, the sun in one year, Mars in two years, and the stellar sphere in so many thousands; these are their own proper and natural motions.

SAGREDO. But in regard to the motion whereby I see the fixed stars and all the planets proceed together from east to west and return to the east in twenty-four hours, in which way does it belong to them?

SIMPLICIO. They have it by participation.

SAGREDO. Therefore, this motion does not reside in them; now, since it does not reside in them, and since there must be some subject in which it resides, it is necessary that it should be the proper and natural motion of some other sphere.

SIMPLICIO. In this regard, astronomers and philosophers have found a very high sphere without stars to which the diurnal rotation naturally belongs; it is called the Prime Mobile and carries along with it all the lower spheres, thus transferring its motion to them and sharing it with them.

SAGREDO. However, suppose everything fits and agrees with perfect harmony without the introduction of unknown and very huge spheres, [148] without additional shared motions and transfers, by giving each sphere only its own simple motion, without mingling contrary motions, but having them all go in the same direction (as they must when they all depend on a single principle); then, why reject this proposal and accept those very strange and problematic complications?

SIMPLICIO. The point is to find an easy and handy way of accomplishing this.

SAGREDO. The way is promptly found, I think. Let the earth be the Prime Mobile, that is, let it rotate on itself every twenty-four hours in the same direction as all the other planets or stars; then they will all appear

49. Cf. Aristotle, *Physics*, VIII, 1, 251a10.

to rise and set in the usual way and exhibit all the other phenomena without that terrestrial motion being transferred to any of them.

SIMPLICIO. The important point is to be able to move the earth without a thousand inconveniences.[50]

SALVIATI. All the inconveniences will be removed as you propose them. The things said so far are only the initial and more general reasons why it seems not to be entirely improbable that the diurnal turning belongs to the earth rather than to the rest of the universe; I do not advance them as inviolable laws but as likely reasons. Now, I understand very well that a single contrary experience or conclusive demonstration suffices to shoot down these and a hundred thousand other probable arguments; thus, one must not stop here, but proceed and hear what Simplicio has to say and what better probabilities and stronger reasons he advances against them.[51]

SIMPLICIO. I will first say something in general about all these considerations taken together, and then I will come to particulars. It seems to me that in general you base yourself on the greater simplicity and facility of producing the same effects; you do this when you judge that, in regard to the fact of causing them, it is the same to move the earth alone as to move the rest of the universe without the earth, but in regard to the manner of operation, the former is much easier than the latter. To this I answer that it seems the same to me too as long as I consider my own strength, which is not only finite but very puny; but from the standpoint of the power of the Mover, which is infinite, it is no harder to move the universe than the earth or a straw. Now, if the power is infinite, why should He not exercise a greater [149] rather than a smaller part of it? Thus it seems to me that your account in general is not cogent.[52]

SALVIATI. If I had ever said that the universe does not move due to insufficient power in the Mover, I would have made a mistake and your

50. The latest exchange reminds us that the simplicity argument assumes that both hypotheses can explain the observed phenomena. Here Simplicio correctly points out that the fact that Copernicanism can explain apparent diurnal motion does not necessarily imply that it can explain other phenomena. As Salviati goes on to admit, such phenomena (especially the allegedly conflicting ones) will have to be examined one by one.

51. It is worth repeating that the main argument in this selection is alleged to be merely probable.

52. Here Simplicio questions the principle of simplicity precisely in the version used by Salviati; that version states that a simpler arrangement is more likely to be true than a less simple one. Recall that this principle is a crucial premise of the geokinetic argument advanced here because even if the earth's rotation can explain all that heavenly diurnal motion can and is simpler, that does not make it true or probable unless something is said about the connection between simplicity and truth or probability. Thus, Simplicio's objection reduces to saying that from the viewpoint of an infinite God, the more complicated arrangement is at least as likely to have been created.

correction would be appropriate; for I admit that to an infinite power it is the same to move one hundred thousand things as to move one. What I said does not regard the Mover but only the bodies moved; that is, not only their resistance, which is undoubtedly less for the earth than for the universe, but also the other particulars mentioned above. Moreover, I want to respond to your saying that an infinite power is such that it is better to exercise a greater than a smaller part of it: a part of the infinite is never greater than another, if both are finite; nor can one say that one hundred thousand is a greater part of an infinite number than two, even though the former is fifty thousand times greater than the latter; if to move the universe one needs a finite power (although very great in comparison to what would suffice to move the earth alone), one would not thereby be using a greater part of the infinite, nor would the unused part be less than infinite; thus it makes no difference to use a little more or a little less power to bring about a particular effect. It should also be mentioned that the action of such a power does not aim at the diurnal motion alone, but that there are in the world many other motions known to us and there may be many others unknown to us. So, from the standpoint of the things moved, there is no doubt that the shorter and quicker mode of operation is to move the earth rather than the universe; let us also keep in mind the many other conveniences and benefits it brings about, and let us remember the very true Aristotelian principle saying that it is useless to do with more means what can be done with fewer;[53] all these considerations render it more probable that the diurnal motion belongs only to the earth rather than to the universe except the earth.

SIMPLICIO. In mentioning this principle you left out a clause that is all important, especially in the present context; it is the phrase "equally well."[54] Therefore, one must examine whether everything can be accommodated equally well with each of the two assumptions.

SALVIATI. Whether both positions satisfy equally well is something that will be understood from the particular examinations of the phenomena which must be accommodated; so far we have discussed, and

53. Cf. Aristotle, *Physics*, I, 6, 189b17–29. This methodological principle is quoted by Galileo in italics in Latin (*Frustra fit per plura quod potest fieri per pauciora*); it may be called teleological on account of its reference to purpose, and here it is advanced as a possible justification for the principle of simplicity. Remembering that what is now in question is the principle of simplicity itself, note that it is being evaluated partly on the basis of theological considerations (about divine omnipotence) and partly on the basis of teleological considerations; if such justifications are deemed wrong, we should be prepared to suggest alternative ways of justifying or criticizing simplicity.

54. Here Galileo uses the Latin phrase *aeque bene*.

we are now discussing, hypothetically,[55] namely, by supposing that in regard to accommodating the phenomena, both [150] positions are equally satisfactory. Moreover, in regard to the phrase you say I have left out, I suspect that instead you have superfluously added it; for "equally well" is a relationship, which necessarily requires at least two terms since a thing cannot have a relation with itself (for example, one cannot say that rest is equally good as rest); furthermore, when one says "it is useless to do with more means what can be done with fewer means," one understands that what is to be done must be the same thing and not two different things; now, since the same thing cannot be said to be equally well done as itself, adding the phrase "equally well" is superfluous and exemplifies a relation with only one term.[56]

## [7. The Case against Terrestrial Rotation and the Value of Critical Reasoning]

SAGREDO. If we do not want the same thing happening as yesterday, let us please return to the subject; and let Simplicio begin to produce the difficulties which seem to him to contradict this new arrangement of the world.

SIMPLICIO. The arrangement is not new but very old. That this is true may be seen from the fact that Aristotle refuted it. His refutations are the following:[57]

"First, if the earth were in motion (either around itself while located at the center, or in a circle while placed outside the center), this motion

55. Here Galileo uses the Latin phrase *ex hypothesi*.

56. There is something right about Salviati's reply to Simplicio's last objection here, though I am not sure the former does full justice to the latter. The issue is still the proper formulation and justification of the principle of simplicity, but it has now become rather complex, and a full analysis is not easy. Interesting questions to explore are whether a more crucial qualification (and perhaps the one intended by Simplicio) is "other things being equal" (*ceteris paribus*) rather than "equally well" (*aeque bene*), and whether the "equally well" qualification should instead be applied to the claim that both hypotheses can explain the observed phenomena. There is also a historical question, for as Drake (1967, 475n) suggests, Galileo may have included this discussion to answer a point raised in Christopher Clavius, *In Sphaeram Ioannes de Sacrobosco Commentarius* (Rome, 1581), 434ff.

57. Cf. Aristotle, *On the Heavens*, II, 14, 296a24–297a8. Galileo places quotation marks around the text of the next five paragraphs, thus suggesting that he is quoting from an edition of Aristotle; but the language of these quotations is Italian rather than Latin, indicating that they are his own translations. My translation is from Galileo's Italian text. I have closed the quotation marks in the middle of the statement of the third argument because the subsequent text makes it clear that what follows is paraphrase or indirect quotation.

would have to be a violent one because it is not its own natural motion; if it were natural, it would also belong to every one of its particles, whereas each of them moves in a straight line toward the center. Being thus violent and preternatural, it could not be everlasting. But the world order is everlasting. Therefore, etc.[58]

"Secondly, except for the Prime Mobile, all the other bodies moving with circular motion seem to fall behind and to move with more than one motion. Because of this, it would be necessary for the earth to move with two motions. If this were so, there would necessarily have to be variations in the fixed stars. But this is not seen; instead, the same stars always rise at the same places and always set at the same places, without any variations.[59]

"Thirdly, the motion of the parts and of the whole is naturally toward the center of the universe; therefore, the whole stands still therein."[60] He also asks whether the motion of the parts is to go naturally to the center of the universe or to the center of the earth; he concludes that their proper instinct is to go to the center of the universe, and that their accidental instinct is to go to the center of the earth. We discussed this question at length yesterday.[61]

[151] Fourthly, he confirms the same conclusion with an argument based on our experience with heavy bodies. As these fall down from on high, they move perpendicularly to the earth's surface. Similarly, projectiles thrown perpendicularly upwards come back down perpendicularly by the same lines, even when they are thrown to an immense height. These experiences provide a necessarily conclusive argument that their

58. This first Aristotelian objection is called the violent motion argument; Galileo criticizes it explicitly in a passage immediately following the present selection but omitted here (Favaro 7:159–62; Galileo 1967, 133–36). For a detailed logical analysis, see Finocchiaro (1980, 379–80).

59. This second Aristotelian objection is called the two motions argument; Galileo criticizes it explicitly in a passage omitted from our selections (Favaro 7:162–64; Galileo 1967, 136–38). For a detailed logical analysis, see Finocchiaro (1980, 380–86); see also Santillana (1953, 138n.22).

60. As noted above, Galileo's quotation marks extend to the penultimate paragraph in this speech, through the clause "there could not be such a correspondence unless it were true"; but the indirect language of what follows makes it clear that he is not giving a direct quotation beyond this point.

61. This third Aristotelian objection corresponds to the natural motion argument, a crucial aspect of which was criticized in selection 2; other aspects are examined in Galileo's criticism of Aristotle's general theory of motion in a passage immediately preceding that selection and omitted here (Favaro 7:38–57; Galileo 1967, 14–32). For more details, see Finocchiaro (1980, 349–56).

motion is toward the center of the earth, which awaits and receives them without moving at all.[62]

Lastly, he mentions that astronomers have produced other reasons to confirm the same conclusions, namely, that the earth is at the center of the universe and motionless. He gives only one of these; that is, all phenomena seen in regard to the motions of stars correspond to the position of the earth at the center, and there could not be such a correspondence unless it were true.[63]

There are other arguments produced by Ptolemy and other astronomers.[64] I can bring them up now, if you so desire; or I can do it after you tell me what occurs to you in response to these Aristotelian ones.

SALVIATI. The arguments produced in this matter are of two kinds:[65] some regard terrestrial phenomena and have no relation to the stars;

62. This fourth objection is called the vertical fall argument; it will soon be restated and elaborated, and then analyzed and criticized at length in the next selection 8. The Aristotelian text is very cryptic about this fourth argument, which may be easily missed in the context of his statement of the others; but there is a sentence which reads: "it is clear, then, that the earth must be at the centre and immovable, not only for the reasons already given, but also because heavy bodies forcibly thrown quite straight upward return to the point from which they started, even if they are thrown to an infinite distance" (Aristotle, *Physics*, II, 14, 296b22–26).

63. Galileo paraphrases this argument so vaguely that it is not surprising that it receives no further explicit discussion. Aristotle's own statement, while cryptic, is more helpful: "This view is further supported by the contributions of mathematicians to astronomy, since the observations made as the shapes change by which the order of the stars is determined, are fully accounted for on the hypothesis that the earth lies at the centre" (*On the Heavens*, II, 14, 297a3–7). This refers to the sort of argument Galileo examines at the beginning of his "Reply to Ingoli" (Finocchiaro 1989, 166–75).

64. Cf. Ptolemy, *Almagest*, I, 4–7; and Brahe (1596, 167, 188–89; 1602, 662).

65. The anti-Copernican arguments can be classified in accordance with several distinct but overlapping criteria. Here Galileo divides them into those based on astronomical evidence and those based on terrestrial evidence; then he subdivides the latter into those found in Aristotle's works and those found in the works of subsequent authors; and he further subdivides these others into the traditional ones and the "modern" ones made possible by such new inventions as artillery. Another way of classifying the anti-Copernican arguments is reflected in the fact that the *Dialogue* is divided into four Days; this classification depends on the particular element of the whole which is at issue: earth-heaven dichotomy, diurnal motion, annual motion, and cause of the tides. Another classification would be in terms of the various disciplines whose principles are involved, which is a natural classification nowadays when disciplines are more clearly separated than in the seventeenth century: scientific, philosophical, and theological; astronomical, physical, and mechanical; methodological, epistemological, metaphysical, and teleological. Another sweeping classification involves the methodological distinction between theory and observation, giving rise to observational arguments and theoretical ones. A fifth way of classifying the arguments would be in terms of the type of reasoning or logical form embodied in a given argument: syllogisms, arguments from authority, arguments from analogy, generalizations, ad hominem, probable, demonstrative, and so on. For more details, see the introduction (3.2), the appendix, and Finocchiaro (1980).

others are taken from the appearances and observations of heavenly bodies. Aristotle's arguments are mostly taken from things near us, and he leaves the others to astronomers; thus, it is appropriate, if you agree, to examine the ones taken from terrestrial experience first, and then we will come to the other kind. Moreover, Ptolemy, Tycho, and other astronomers and philosophers produced other such arguments besides accepting, confirming, and strengthening those of Aristotle; hence, these can all be considered together in order not to have to repeat twice the same or similar replies. So, Simplicio, whether you wish to relate them, or whether you want me to release you from this burden, I am here to please you.

SIMPLICIO. It will be better for you to present them since you have studied them more, and so you will be able to present them more readily and in greater number.

SALVIATI. As the strongest reason, everyone produces the one from heavy bodies, which when falling down from on high move in a straight line perpendicular to the earth's surface. This is regarded as an unanswerable argument that the earth is motionless. For, if it were in a state of diurnal rotation and a rock were dropped from the top of a tower, then during the [152] time taken by the rock in its fall, the tower (being carried by the earth's turning) would advance many hundreds of cubits toward the east and the rock should hit the ground that distance away from the tower's base. They confirm this effect with another experiment. That is, they drop a lead ball from the top of the mast of a ship which is standing still, and they note that the spot where it hits is near the foot of the mast; but if one drops the same ball from the same place when the ship is moving forward, it will strike at a spot as far away from the first as the ship has moved forward during the time the lead was falling. This happens only because the natural motion of the ball in free fall is in a straight line toward the center of the earth.[66]

This argument is strengthened with the experiment of a projectile thrown upward to a very great height, such as a ball shot by a cannon aimed perpendicular to the horizon. The time required for it to go up and down is such that at our latitude we, together with the cannon,

66. This paragraph elaborates the fourth Aristotelian argument briefly introduced above, namely, the vertical fall objection; this elaboration introduces another argument involving the experiment of dropping objects from the top of a ship's mast. This new argument is related to the vertical fall objection, but it is useful to give it a separate label, namely, the ship experiment or the ship analogy argument; it too will be examined in the next selection 8.

would be carried by the earth many miles toward the east; thus the ball could never fall back near the gun, but rather would fall as far to the west as the earth would have moved forward.[67]

Moreover, they add a third and very effective experiment, which is the following: if one shoots a cannon aimed at a great elevation toward the east, and then another with the same charge and the same elevation toward the west, the westward shot would range much farther than the eastward one. For, since the ball goes westward and the cannon (carried by the earth) goes eastward, the ball would strike the ground at a distance from the cannon equal to the sum of the two journeys (the westward one made by itself and the eastward one of the cannon carried by the earth); by contrast, from the journey made by the ball shot toward the east, one would have to subtract the one made by the cannon while following it; for example, given that the ball's journey in itself is five miles and that at that particular latitude the earth moves forward three during the ball's flight, in the westward shot the ball would strike the ground eight miles from the cannon (namely, its own westward five plus the cannon's eastward three), whereas the eastward shot would range two miles (which is the difference between the five of the shot and the three of the cannon's motion in the same direction). However, experience shows that the ranges are equal. Therefore, [153] the cannon is motionless, and consequently so is the earth.[68]

No less than this, shooting toward the south or toward the north also confirms the earth's stability. For one would never hit the mark aimed at, but instead the shots would always be off toward the west, due to the eastward motion of the target (carried by the earth) while the ball is in midair.[69]

These shots along the meridians would not be the only ones that would hit off the mark. If one were shooting point-blank, the eastward shots would strike high and the westward ones low. For in such shooting, the ball's journey is made along the tangent, namely, along a line parallel to the horizon; moreover, if the diurnal motion should belong to the earth, the eastern horizon would always be falling and the western

67. This is called the vertical gunshot argument and is also closely related to the vertical fall argument. Besides formulating it separately, Galileo also criticizes it separately in a passage that is omitted here (Favaro 7:197–202; Galileo 1967, 171–78); cf. Finocchiaro (1980, 394–97).

68. This is called the east-west gunshot argument and is criticized in a passage that is omitted here (Favaro 7:194–97; Galileo 1967, 167–71); cf. Finocchiaro (1980, 391–94).

69. Again, the obvious label for this is the north-south gunshot argument; it, too, is explicitly criticized in a passage that is omitted here (Favaro 7:203–5; Galileo 1967, 178–80); cf. Finocchiaro (1980, 398–99).

one rising (which is why the eastern stars appear to rise and the western ones to fall); therefore, the eastern target would drop below the shot and so the shot would strike high, while the rising of the western target would make the westward shot hit low. Thus, one could never shoot straight in any direction; but, because experience shows otherwise, one is forced to say that the earth stands still.[70]

SIMPLICIO. Oh, these arguments are beautiful, and it will be impossible to find answers to them.

SALVIATI. Do they perhaps strike you as novel?

SIMPLICIO. Frankly, yes. Now I see how many beautiful observations nature has graciously provided to help us come to know the truth. Oh, how well one truth agrees with another, and all conspire to make themselves invulnerable!

SAGREDO. What a pity that there were no cannons in Aristotle's time! With them he would have indeed conquered ignorance and spoken without hesitation of the things of the world.[71]

SALVIATI. I am very glad you find these arguments novel, so that you will not remain of the opinion held by most Peripatetics; they believe that if anyone disagrees with Aristotle's doctrine, this happens because of not having heard or properly grasped his demonstrations. However, you will certainly hear other novelties, and you will hear the followers of

70. This is called the point-blank gunshot argument, and it is criticized in a passage that is omitted here (Favaro 7:205–9; Galileo 1967, 180–83). This argument should not be confused with the east-west gunshot objection, as is done even by otherwise good scholars, for example Koyré (1966, 215–38) and Drake (1967, 126 and 168); for a criticism of these confusions and an analysis of the argument, see Finocchiaro (1980, 208, 240, 399–403, and 412n.13). The point is that both arguments involve shooting alternately toward the east and toward the west and both raise the difficulty that on a rotating earth the targets would not be hit properly; but the east-west argument alleges that when a gun on the ground is aimed high in order to hit a distant target also on the ground, what would be hit instead (*if* the earth were rotating) would be a point that is either farther or nearer than that target; whereas the point-blank argument alleges that when a gun on a tower or hill is aimed horizontally in order to hit a target at the same height on another tower or hill, what would get hit instead (*if* the earth were rotating) would be a point either above or below the target.

71. Sagredo's witticism and Simplicio's exclamation in the latest exchange constitute one of many examples giving the *Dialogue* considerable rhetorical and aesthetic value. Indeed the book can be read from the viewpoint of literature, though such an appreciation depends on understanding its scientific and philosophical content. Like all great literature, the book's aesthetic dimension is hard to translate since such an accomplishment would involve essentially the creation of a new work of art. I do not claim that my translation does justice to the literary power of the original, for I have focused on intellectual and conceptual accuracy. We should neither be oblivious to such passages nor let ourselves be distracted from the main thread of the discussion. For more details, see the appendix (3) and Finocchiaro (1980, 46–66).

the new system produce against themselves observations, experiments, and reasons much stronger than those produced by Aristotle, Ptolemy, and other opponents of the same [154] conclusions; you will thus establish for yourself that it is not through ignorance or lack of observation that they are induced to follow this opinion.[72]

SAGREDO. I must take this opportunity to relate to you some things which have happened to me since I began hearing about this opinion. When I was a young man and had just completed the study of philosophy (which I then abandoned to apply myself to other business), it happened that a man from Rostock beyond the Alps (whose name I believe was Christian Wursteisen)[73] came into these parts and gave two or three lectures on this subject at an academy; he was a follower of Copernicus and had a large audience, I believe more for the novelty of the subject than anything else. However, I did not go, having acquired the distinct impression that this opinion could be nothing but solemn madness. When I asked some who had attended, they all made fun of it, except one who told me that this business was not altogether ridiculous. Since I regarded him as a very intelligent and very prudent man, I regretted not having gone. From that time on, whenever I met someone who held the Copernican opinion, I began asking whether he had always held it; although I have asked many persons, I have not found a single one who failed to tell me that for a long time he believed the contrary opinion but that he switched to this one due to the strength of the reasons supporting it; moreover, I examined each one of them to see how well he understood the reasons for the other side, and I found everyone had them at his fingertips; thus, I cannot say that they accepted this opinion out of ignorance or vanity or (as it were) to show off. On the other hand, out of curiosity I also asked many Peripatetics and Ptolemaics how well they had studied Copernicus's book, and I found very few who had seen it and none who (in my view) had un-

72. This passage introduces the topic of open-mindedness, by which I mean the willingness and capacity to know, understand, and learn from the arguments against one's own views. In this selection Galileo gives an illustration and justification of open-mindedness. The illustration is the case of the Copernicans, who are acquainted not only with the arguments for the geokinetic system but also with those for the geostatic view. The value of open-mindedness is that it strengthens the view we hold (since we can defend it from objections), and it facilitates the discovery of the truth (which emerges more easily from the clash of opposing arguments); and open-mindedness need not lead to mental confusion, if we combine it with the art of analyzing and evaluating arguments (namely, critical reasoning) and with the willingness and capacity to accept the views supported by the best arguments (which I call rational-mindedness).

73. Wursteisen (1544–1588) was a German-Swiss astronomer and one of the first and at the time very few to be favorably disposed toward Copernicanism.

derstood it; I also tried to learn from the same followers of the Peripatetic doctrine whether any of them had ever held the other opinion, and similarly I found none who had. Now, let us consider these findings: that everyone who follows Copernicus's opinion had earlier held the contrary one and is very well-informed about the reasons of Aristotle and Ptolemy; and that, on the contrary, no one who follows Aristotle and Ptolemy has in the past held Copernicus's opinion [155] and abandoned it to accept Aristotle's. Having considered these findings, I began to believe that when someone abandons an opinion imbibed with mother's milk and accepted by infinitely many persons, and he does this in order to switch to another one accepted by very few and denied by all the schools (and such that it really does seem a very great paradox), he must be necessarily moved (not to say forced) by stronger reasons. Therefore, I have become most curious to go, as it were, to the bottom of this business, and I regard myself very fortunate to have met the two of you; without any great effort I can hear from you all that has been said (and perhaps all that can be said) on this subject, and I am sure that by virtue of your arguments I will lose my doubts and acquire certainty.[74]

SIMPLICIO. But beware that your belief and hope will not be frustrated, and that you will not end up being more confused than before.[75]

74. Here Galileo advances an interesting, important, and controversial argument. It was held against him by the special commission appointed by the pope soon after the book's publication to investigate the many complaints against it; the charge was "that he [Galileo] gives as an argument for the truth the fact that Ptolemaics occasionally become Copernicans, but the reverse never happens" (Finocchiaro 1989, 222). This charge gives an essentially accurate interpretation and would be a good place to start the critical analysis of the argument. But the following refinements would have to be worked out: Is Galileo's conclusion the claim that Copernicanism is true, or that it is *probably* true, or that it is *truer* than the Ptolemaic view? And is he reaching a conclusion primarily about the probable truth of Copernicanism or about the probable superiority of the Copernican reasons? That is, is he perhaps concluding merely that the Copernican arguments are probably better than the Ptolemaic arguments? When suitably refined, it is not obvious that the argument is fallacious. This is not the only place where Galileo advances such an argument; a more elaborate version is found at the beginning of his "Considerations on the Copernican Opinion" of 1615 (Finocchiaro 1989, 71–73); that earlier version makes it clearer that the main conclusion he reaches is about the Copernican *arguments* and explicitly contains probabilistic language.

75. I call this the problem of *misology* (hatred of logic) because this label is explicitly used in a similar passage in Plato's *Phaedo* (88A–91D), where Socrates tries to dispel the confusions and discouragement that some of his interlocutors began to feel after hearing several arguments and counterarguments about the immortality of the soul. It is also amazing that Galileo's resolution of the problem is essentially identical to the Socratic one; Socrates argues that the way out of a misologic attitude is to learn what one translator renders as "the art of logic" and explains as the skill of the "critical understanding" of arguments (Tredennick 1969, 144–45).

SAGREDO. I think I am sure that this cannot happen in any way.

SIMPLICIO. Why not? I myself am a good witness that the further we go, the more confused I become.

SAGREDO. That is an indication that those reasons, which so far seemed conclusive to you and kept you certain of the truth of your opinion, are beginning to feel different in your mind and to gradually let you, if not switch, at least incline toward the contrary one. However, I, who am and have been so far undecided, am very confident to be able to reach a state of serenity and certainty; and you yourself will not deny it, if you want to listen to my reasons for this expectation.

SIMPLICIO. I will be glad to listen, and no less glad if the same effect should be produced in me.

SAGREDO. Please, then, answer my questions.[76] Tell me, first, Simplicio, whether the conclusion whose correctness we are trying to determine is not one of the following: whether one must hold, with Aristotle and Ptolemy, that the earth stands still at the center of the universe and all the heavenly bodies move; or whether the stellar sphere stands still, the sun is placed at the center, and the earth is located off the center and has those motions which appear to belong to the sun and to the fixed stars.

SIMPLICIO. These are the conclusions about which we are disputing.

SAGREDO. Are these two conclusions such that it is necessary for one of them to be true and the other false?

[156] SIMPLICIO. That is correct. We are facing a dilemma in which it is necessary that one alternative should be true and the other false.[77] For rest and motion are contradictories, and there is no third alternative such

76. What follows is an example of Galileo's frequent use of the Socratic method.

77. This is not exactly right. The two alternatives stated by Sagredo (in the speech before his last one) could both be false, and indeed they are both false, though admittedly they could not both be true; their relationship is described by what logicians call *contrariety* rather than *contradictoriness*. That is, two propositions are contraries of each other when they cannot both be true, though they can both be false; whereas two propositions are contradictories of each other when they cannot both be true and cannot both be false, but one must be true and the other false. This distinction implies the following principles of reasoning when one is deciding between two alternative positions: if the two positions are contradictories, then we can prove the truth of one by showing the falsity of the other; but if the two positions are just contraries, then from the falsity of one we cannot derive the truth of the other. Some writers (for example, Morpurgo-Tagliabue 1981) interpret Galileo as trying to prove the truth of Copernicanism by showing the falsity of the Ptolemaic view, and thus as committing the fallacy of confusing contraries and contradictories; though interesting, this interpretation is ultimately untenable due to insufficient attention to the details of what Galileo says and does. For example, in his "Considerations on the Copernican Opinion," he clearly thinks that all he can derive from the falsehood of the Ptolemaic system is that Copernicanism *may* be true; cf. remark number 7 of part III (Finocchiaro

that one might say: "The earth neither moves nor stands still, and the sun and stars neither move nor stand still."

SAGREDO. Are the earth, sun, and stars insignificant or substantial bodies in nature?

SIMPLICIO. These bodies are the most important, magnificent, huge, substantial, and integral parts of the universe.

SAGREDO. What kind of phenomena are motion and rest in nature?

SIMPLICIO. They are so pervasive and important that nature herself is defined in their terms.[78]

SAGREDO. Thus, to be eternally in motion and to be completely immobile are two very significant conditions in nature, especially when attributed to the most important bodies of the universe; as a result of those conditions one can get only very dissimilar occurrences.

SIMPLICIO. Certainly.

SAGREDO. Now, respond to another point. Do you believe that in logic,[79] rhetoric, physics, metaphysics, mathematics, and reasoning in general, there are good arguments proving false as well as true conclusions?

SIMPLICIO. No, sir! Instead I firmly believe and am sure that for the proof of a true and necessary conclusion there are in nature not just one but many very powerful demonstrations, that one can discuss and approach it from thousands of points of view without ever encountering any contradiction, and that the more a sophist[80] would want to taint it, the clearer its certainty would become. On the contrary, to make a false proposition

---

1989, 85). Similarly, although Copernicanism and geostaticism as a whole are contraries and not contradictories, the propositions that Simplicio goes on to consider in this speech are indeed contradictories and not just contraries; that is, in the next sentence he asserts correctly that "the earth moves" and "the earth stands still" are contradictories, and that "the sun moves" and "the sun stands still" are also contradictories; he does *not* attribute contradictoriness to the pair "the earth moves" and "the sun moves," which are not even contraries, let alone contradictories; nor does he attribute contradictoriness to the pair "the earth stands still" and "the sun stands still," which are mere contraries.

78. Aristotle's definition of nature reads: "*nature is a source or cause of being moved and of being at rest in that to which [one] belongs primarily*, in virtue of itself and not in virtue of a concomitant attribute" (*Physics* 192b21–24, italics in the original).

79. Galileo's word here reads *dialettica*, which literally means "dialectics" and was a common traditional term for what we nowadays call logic. For the possible significance of the difference, see Moss (1993, 282–83).

80. The term *sophist* usually has a pejorative connotation and refers to someone skillful in deception by means of reasoning, for example, by making the weaker argument appear stronger; accordingly, a sophism is a deceptively incorrect argument, such as a sophist is likely to give. Originally a sophist was one of a group of controversial thinkers in ancient Greece who were interested in public speaking and discussing political and moral questions. Plato portrayed them unfavorably, and his account gave the word its current connotation.

appear true and to persuade someone of it one can produce nothing but fallacies, sophisms, paralogisms, equivocations, and arguments that are pointless, incoherent, and full of inconsistencies and contradictions.[81]

SAGREDO. Now, if eternal motion and eternal rest are such important properties in nature and so different that their effects must be very different, especially when attributed to such huge and noteworthy bodies in the universe as the earth and the sun; if it is impossible that one of the two contradictory propositions [157] should not be true and the other false; and if to prove a false proposition one can only produce fallacies, whereas a true one is supportable by all kinds of conclusive and demonstrative arguments; if all this is true, how can it be that someone undertaking to support a true proposition would not be able to persuade me? I would have to have a stupid understanding, a perverse judgment, a dull mind and intellect, and a dim-witted common sense; and I would have to be unable to discern light from darkness, gems from coals, and truth from falsehood.

SIMPLICIO. As I have said other times, I tell you that the greatest master from whom to learn how to recognize sophisms, paralogisms, and other fallacies is Aristotle; in this regard, he can never be deceived.

SAGREDO. You again mention Aristotle, who cannot speak; and I tell you that if Aristotle were here, he would be persuaded by us or he would dissolve our reasons and persuade us with better ones. At any rate, in hearing the gunshot experiments related, did you yourself not admire them and recognize and admit them to be more conclusive than those of Aristotle? Nevertheless, I do not see that Salviati (who has produced them, has undoubtedly examined them, and has probed them most fastidiously) is admitting being persuaded by them, or even by the stronger ones which he indicates he is about to present.[82] I do not know

81. This paragraph expresses an important but controversial methodological principle that may be called naive rationalism; for such a label suggests a belief in the power of human reason to arrive at the truth, and Simplicio's principle claims that there is a simple one-to-one correspondence between truth and good reasoning and between falsity and bad reasoning. But in this selection Galileo is making a plea for rational-mindedness and critical reasoning, not naive rationalism. For more details, see the appendix (2.2).

82. Here Sagredo introduces two qualifications in Simplicio's naive rationalism mentioned in the previous note: Sagredo's talk of Aristotle's *persuasiveness* and *better* reasons presupposes that arguments must be evaluated for degrees of strength, and not merely for conclusive validity and worthless fallaciousness; and Sagredo's reference to Salviati's being unconvinced by the strong anti-Copernican arguments he presents presupposes a distinction between appearance and reality in logical strength. See the appendix (2.2) for more details. The exchange between Sagredo and Simplicio of the last several paragraphs should also be compared with Galileo's "Considerations on the Copernican Opinion," especially number 6 of part III (Finocchiaro 1989, 85).

why you would want to portray nature as having become senile and having forgotten how to produce theoretical intellects, except those who make themselves servants of Aristotle in order to understand with his brain and perceive with his senses. However, let us listen to the remaining reasons favorable to his opinion, and then go on to test them by refining them in the assayer's crucible and weighing them in his balance.

SALVIATI. Before proceeding further, I must tell Sagredo that in these discussions I act as a Copernican and play his part with a mask, as it were. However, in regard to the internal effect on me of the reasons I seem to advance in his favor, I do not want to be judged by what I say while we are involved in the [158] enactment of the play, but by what I say after I have put away the costume; for perhaps you will find me different from what you see when I am on stage.[83] Now, let us go on.

Ptolemy and his followers[84] advance another observation, similar to that of projectiles: it concerns things that are separate from the earth and remain at length in the air, such as clouds and birds in flight. Since clouds are not attached to the earth, they cannot be said to be carried by it, and so it does not seem possible that they could keep up with its speed; instead, they should all appear to us to be moving very fast toward the west. And, if we are carried by the earth and in twenty-four hours move along our parallel (which is at least sixteen thousand miles), how could birds keep up with so much drift? On the contrary, we see them fly toward the east as well as toward the west and toward any other direction, without any sensible difference.[85]

Furthermore, when we run on horseback we feel the air strike very hard against our face, and so what a wind should we constantly feel blowing from the east if we are carried with such rapid motion against the air? Yet, no such effect is felt.[86]

83. This speech is one of many disclaimers Galileo places in Salviati's mouth to avoid offending Church authorities. For more details, cf. the introduction (4), the appendix (3.2), selection 1, and Finocchiaro (1980, 6–18).

84. The next four arguments (about clouds, birds, wind, and extrusion) are important and were advanced by critics of Copernicanism in Galileo's time; but as noted at the beginning of selection 9, it is unclear whether they (in particular, the last one) can be found in Ptolemy's *Almagest*.

85. This is called the birds argument and is criticized in a passage that is omitted here (Favaro 7:209–12; Galileo 1967, 183–86). It was a distinct argument with a special strength because birds (as animals) have the power of self-movement, and it seemed that on a rotating earth they would lose such power if they were to fly as they are observed to fly; cf. Finocchiaro (1980, 403–6).

86. This is called the wind argument, and it is criticized in selection 10.

Here is another very ingenious argument, taken from the following observation; it is this: circular motion has the property of extruding, scattering, and throwing away from its center the parts of the moving body whenever the motion is not very slow or the parts are not attached together very firmly. For example, consider those huge treadmill wheels designed so that the walking of a few men on their inner surface causes them to move very great weights, such as the massive rollers of a calender press or loaded barges dragged overland to move them from one river to another; now, if we made one of these huge wheels turn very rapidly and its parts were not very firmly put together, they would all be scattered along with any rocks or other material substances however strongly tied to its external surface; nothing could resist the impetus that would throw them with great force in various directions away from the wheel, and consequently away from its center. If, then, the earth were rotating with a very much greater speed, what weight and what strength of mortar or cement would keep rocks, buildings, and entire cities from being hurled toward the sky by such a reckless turning? And think of people and animals, which are not attached to the earth at all; how would they resist so much impetus? On the contrary, we see them and other things with much less resistance (pebbles, sand, [159] leaves) rest very calmly on the earth and fall back to it even when their motion is very slow.[87]

Here, Simplicio, are the very powerful reasons taken from terrestrial things, so to speak. We are left with the other kind, namely, those that relate to heavenly phenomena. Actually, those reasons tend to demonstrate instead that the earth is at the center of the universe and consequently lacks the annual motion around it, which Copernicus attributed to the earth; since they deal with a somewhat different subject, they can be produced after we have examined the strength of the ones presented so far.

SAGREDO. What do you say, Simplicio? Does it seem that Salviati knows and can explain the Ptolemaic and Aristotelian reasons? Do you think that any Peripatetic is equally knowledgeable of the Copernican demonstrations?[88]

87. This is called the argument from the extruding power of whirling, or more simply the extrusion argument, or (in modern terminology) the centrifugal force argument. It will be elaborated, clarified, and criticized in selection 9.

88. This praise for the open-mindedness of the Copernicans is part of Galileo's methodological plea for the value of this mental attitude in general, independently of the argument that such open-mindedness is a clue to the greater probability of Copernicanism.

SIMPLICIO. If the discussions so far had not produced in me such a high opinion of Salviati's well-founded understanding and of Sagredo's sharp intelligence, I (with their permission) would be ready to leave without listening to anything else. For it seems to me impossible that one can contradict such palpable observations; moreover, I would like to keep my old opinion without having to hear anything else, because it seems to me that even if it were false, the fact that it is supported by such likely reasons would render it excusable. If these are fallacies, what true demonstrations were ever so beautiful?

SAGREDO. Still, it will be good to hear Salviati's answers. If these should be true, they must be even more beautiful and infinitely more beautiful, and those others must be ugly, indeed very ugly; this would follow if there is truth in the metaphysical proposition that truth and beauty are the same thing, as falsehood and ugliness also are. However, Salviati, let us not lose any more time.

. . .[89]

## [8. Vertical Fall, Superposition of Motions, and the Role of Experiments]

SALVIATI. . . . [164] . . . So we can now go on to the fourth argument, which should be discussed at great length since it is based on an observation from which most of the remaining arguments then derive their strength. Aristotle says[90] that a most certain argument for the earth's immobility is based on the fact that we see bodies which have been cast upwards return perpendicularly by the same line to the same place from which they were thrown, and that this happens even when the motion reaches a great height; this could not happen if the [165] earth were moving because, while the projectile moves up and down separated from the earth, the place of ejection would advance a long way toward the east due to the earth's turning, and in falling the projectile would strike the ground that much distance away from the

89. The passages omitted here correspond to Favaro 7:159.29–164.30 and Galileo 1967, 133–38. They contain a criticism of Aristotle's first two objections—the violent motion argument and the two motions argument. For an analysis, see Finocchiaro (1980, 36, 115–16, and 379–86). Aristotle's third objection, the natural motion argument, was criticized in the "First Day," in selection 2 and in passages that are omitted here (Favaro 7:38–57; Galileo 1967, 14–32). The next selection examines Aristotle's fourth argument, the vertical fall objection.

90. Cf. *Physics*, II, 14, 296b22–26.

said place. Here we may also include the argument from the cannonball shot upwards, as well as another one used by Aristotle and Ptolemy, namely, that one sees bodies falling from great heights move in a straight line perpendicular to the earth's surface.[91] Now, to begin to untie these knots, I ask Simplicio how Aristotle and Ptolemy would prove, if someone denied it, that bodies falling freely from on high move in a straight and perpendicular line, namely, in the direction of the center.

SIMPLICIO. By means of the senses: they assure us that the tower is straight and perpendicular; they show us that the falling rock grazes it without inclining so much as a hairbreadth to one side or the other; and they show that the rock lands at the foot of the tower exactly under the place from which it was dropped.[92]

SALVIATI. But if by chance the terrestrial globe were rotating and consequently were also carrying the tower along with it, and if the falling rock were still seen to graze the edge of the tower, what would its motion have to be?

SIMPLICIO. In that case one would rather have to speak of "its motions"; for there would be one that would take it from above downwards, and it would have to have another in order to follow the course of the tower.

SALVIATI. Therefore, its motion would be a compound of two, namely, one with which it grazes the edge of the tower and another one with which it follows the tower; the result of this compound would be that the rock would no longer describe a simple straight and perpendicular line, but rather an inclined, and perhaps not straight, one.[93]

91. This is a fuller restatement of the vertical fall argument; for more details, see the appendix (1.5), Feyerabend (1975), Finocchiaro (1980, 115–16, 192–200, 208–13, 277–91, 329–30, and 387–89), Goosens (1980), and Machamer (1973).

92. Salviati's latest question and Simplicio's answer start a critical analysis of the vertical fall argument that is best understood by reference to the following reconstruction of its last step: (a) if the earth rotated, then bodies would not fall vertically; but (b) bodies do fall vertically; therefore, (c) the earth does not rotate. This inference is formally valid, being an instance of a conditional argument of the form of denying the consequent; that is, the conclusion (c) is correctly deduced by combining two premises, a conditional proposition (a) and another proposition that denies its consequent ("then") clause (b). Thus, Galileo does not question the correctness of this inference; instead, he begins his critical analysis by asking how one knows the truth of the second premise (b), which below is referred to as the middle term.

93. The resulting compound path would not be straight if (as indeed is the case) the downward fall is accelerated while the horizontal motion is uniform; thus, the actual path is parabolic, as Galileo proved in *Two New Sciences*.

SIMPLICIO. I am not sure about its not being straight; but I understand well that it would have to be inclined and different from the straight perpendicular one it would describe on a motionless earth.

SALVIATI. Therefore, from just seeing the falling rock graze the tower, you cannot affirm with certainty that it describes a straight and perpendicular line unless you first assume the earth to be standing still.

SIMPLICIO. That is correct; for if the earth were moving, the rock's motion would be inclined and not perpendicular.

[166] SALVIATI. Here, then, is the paralogism of Aristotle and Ptolemy made clear and evident, and discovered by yourself; the argument is assuming as known what it is trying to prove.

SIMPLICIO. In what way? To me it seems to be a syllogism in proper form and not a fallacy of question begging.

SALVIATI. Here is how. Tell me: does not the demonstration regard the conclusion as unknown?

SIMPLICIO. Yes, unknown, for otherwise it would be superfluous to demonstrate it.

SALVIATI. But, should not the middle term be known?

SIMPLICIO. That is necessary, for otherwise it would be an attempt to prove the unknown by means of what is equally unknown.[94]

SALVIATI. Is not the conclusion to be proved, and which is unknown, the proposition that the earth stands still?

SIMPLICIO. It is.

SALVIATI. Is not the middle term, which must be already known, the straight and perpendicular fall of the rock?

SIMPLICIO. That is the middle term.

SALVIATI. But, did we not just conclude that we can have no knowledge that this fall is straight and perpendicular unless we first know that the earth is standing still? Therefore, in your syllogism the certainty of the middle term is inferred from the uncertain conclusion. So you see the type and the seriousness of the paralogism.[95]

94. Galileo writes this last phrase in italics and in Latin: *ignotum per aeque ignotum*.

95. This completes Galileo's criticism of the original version of the vertical fall argument, from (a) and (b) to (c) as reconstructed above. This criticism may in turn be reconstructed as follows: the vertical fall argument claims that bodies actually fall vertically; it supports this claim on the basis of observation, namely, the fact that bodies *appear* to our eyes to fall vertically; that is, actual vertical fall is justified by apparent vertical fall; but this justification assumes that *apparent* vertical fall implies *actual* vertical fall; and this implication does not hold unless the earth stands still, because on a rotating earth apparent vertical fall would imply actually slanted fall; therefore, to assume that apparent vertical fall implies actual vertical fall presupposes that the earth stands still; but the proposition that the

SAGREDO. On behalf of Simplicio, I should like to defend Aristotle, if possible, or at least to understand better the strength of your inference. You say: seeing the rock graze the tower is not enough to become certain that its motion is perpendicular (which is the middle term of the syllogism) unless one assumes that the earth stands still (which is the conclusion to be proved); for, if the tower were moving together with the earth and the rock grazed it, the rock's motion would be inclined and not perpendicular. However, I will answer that, if the tower were moving, it would be impossible for the falling rock to graze it; hence, from seeing the falling rock graze it one infers that the earth is motionless.

SIMPLICIO. That is correct. For, if the falling rock should graze the tower while the latter was carried along by the earth, the rock would have to have two natural motions (namely, straight toward the center and circular around the center); and this is impossible.[96]

SALVIATI. Therefore, Aristotle's defense consists in its being impossible, [167] or at least in his having regarded it as impossible, that the rock could move with a motion mixed of straight and circular; for, if he

---

earth stands still is the conclusion the argument tries to prove; therefore, the argument assumes the same thing it tries to prove. This criticism depends on the claim that apparent vertical fall does not imply actual vertical fall unless the earth stands still; but this claim is correct, and Simplicio himself admits it in the preceding discussion, as does Sagredo, who goes on to summarize this criticism. The criticism also presupposes the distinction between actual and apparent vertical fall, but it is not obvious how one could question this distinction; in fact, in the next two speeches Sagredo and Simplicio go on to reformulate the original vertical fall argument in a way immune to this criticism, and so they may be interpreted as granting the correctness of this question-begging charge. The function of this first criticism is also to clarify the original argument and force a reformulation that does not beg the question. For future reference, it is useful to call the version of the vertical fall argument criticized so far—from (a) and (b) to (c)—the argument from *actual* vertical fall; the reformulated version, to be discussed presently, is called the argument from *apparent* vertical fall. This criticism of the argument from actual vertical fall should be compared with the criticism of other question-begging arguments, such as the basic geocentric argument in selection 2 ([59–61]) and the Peripatetic proof of the axiom that a straight line is the shortest distance between two points in selection 9 ([231–32]).

96. The last two speeches by Simplicio and Sagredo represent a revision of the vertical fall argument into the following version, which does not beg the question and is called the argument from *apparent* vertical fall: (d) if the earth rotated then bodies would not *appear* to fall vertically, because (e) if the earth rotated and bodies appeared to fall vertically then they would move with a mixture of downward and horizontal natural motions, and (f) such a mixture is impossible; but (g) bodies appear to fall vertically; therefore, (h) the earth does not rotate. This new argument is a series of two instances of denying the consequent, from (e) and (f) to (d), and from (d) and (g) to (h); the whole argument is thus formally valid; proposition (e) is more or less obviously true; so the only question to raise is that of the impossibility of mixed motion, premise (f). That is the point where Galileo focuses his next criticism. For more details, see the appendix (1.5) and Finocchiaro (1980, 387–89).

had not regarded it as impossible that the rock could move simultaneously toward the center and around the center, he would have understood that it could happen that the falling rock could graze the tower when it is moving as well as when it is standing still; consequently, he would have realized that from this grazing nothing could be inferred regarding the motion or the rest of the earth. However, this does not in any way excuse Aristotle because he should have said so if he had had this thought in mind, it being such a key point in his argument; moreover, one cannot say either that this effect is impossible or that Aristotle regarded it as impossible.[97] The first cannot be said because I will soon show that it is not only possible but necessary. Nor can one say the second, for Aristotle himself grants[98] that fire goes naturally upward in a straight line and turns by participation with the diurnal motion, which is transferred by the heavens to all of the element fire and to most of the air; if, then, he did not regard it as impossible to mix straight upward motion with the circular one communicated to fire and air by the inside of the lunar orb, much less should he regard it as impossible to mix the rock's straight downward with the circular one that would be natural for the whole terrestrial globe of which the rock is a part.

SIMPLICIO. It does not look that way to me; for, if the element fire turns together with the air, it is very easy and indeed necessary that a particle of fire rising from the earth and going through the rotating air should receive the same motion since it is such a rarefied and light body and most ready to move; but it is completely incredible that a very heavy rock or cannon ball falling through the air should let itself be carried along by it or anything else. Furthermore, there is the very appropriate experiment of the rock dropped from the top of a ship's mast; that is, when the ship is standing still it falls at the foot of the mast, but when the ship is going forward it falls away from the same place at a distance equal

97. This criticism attributes three flaws to the argument from apparent vertical fall. The first is that the argument is an ad hoc reformulation of the original; even if it is correct, this is a criticism that I would regard as rhetorical rather than as logical or scientific; but this point is not pursued in the text. The second is that the argument is not consistent with the rest of Aristotelianism; this is an ad hominem criticism, ad hominem not in the pejorative sense common today, but in the seventeenth-century sense meaning that it refutes a claim on the basis of ideas accepted by the argument's proponent (but not necessarily by the critic and not necessarily acceptable in general); this charge undergoes some discussion in the text. The third alleged flaw is more substantive, claiming that the impossibility of mixed motion cannot be justified; this is the criticism stressed by Galileo.

98. Aristotle, *Meteorology*, I, 7, 344a11.

to that traversed by the ship during the rock's fall (which amounts to many cubits when the ship's course is fast).[99]

SALVIATI. There is a great disparity between the case of the ship and that of the earth, if the diurnal motion should belong to the terrestrial globe. For it is most evident that the ship's motion does not belong to it naturally, [168] just as it is an accidental property of all things in it; so it is not surprising that when the rock is let go after being held at the top of the mast, it should fall without any obligation to follow the ship's motion. However, the diurnal rotation would be attributed to the terrestrial globe (and consequently to all its parts) as their own natural motion, and it would be regarded as indelibly impressed in them by nature; hence, a primary instinct of the rock at the top of the tower would be to go around the center of the whole of which it is a part every twenty-four hours, and it would eternally exercise this natural inclination regardless of the conditions in which it might be placed. To be persuaded of this, you have only to change an old impression and say to yourself: "Up to now, I have thought it is a property of the terrestrial globe to stay motionless at the center, and so I have never felt any difficulty or repugnance in understanding that every one of its particles is also naturally in the same state of rest; similarly, if the terrestrial globe had the natural instinct to rotate in twenty-four hours, then one would have to say that every one of its parts has the intrinsic and natural inclination to follow the same course and not to stand still." Thus, without encountering any inconvenience, one may conclude that when the rock is separated from the ship, it must regain its natural state and return to exercise its pure and simple natural instinct, for the motion transmitted from the power of the oars to the ship and from the ship to all the things it contains is not natural but foreign to them.

It should be added that it is necessary that the lower part of the air below the higher mountains would be captured and carried around by the roughness of the earth's surface, or that it would naturally follow the diurnal motion insofar as it is mixed with many earthly vapors and ema-

99. The ship experiment is introduced at this point not (not yet) as a free-standing argument in its own right, but rather as evidence for the impossibility of mixed motion, and thus as part of the argument from apparent vertical fall; the subargument is that bodies in general cannot move with a mixture of downward and horizontal natural motions because experiment reveals that bodies do not do so on a ship that is advancing forward. But the immediately following criticism applies to the ship experiment whether it is construed as a step in that longer argument or as a self-contained argument; that is, the next few speeches by Salviati and Simplicio may be regarded either as part of the criticism of the argument from apparent vertical fall or as part of the criticism of the ship analogy argument.

nations; this does not happen to the air around the ship, which is propelled by the oars. Therefore, to argue from the case of the ship to the case of the earth has no inferential force.[100] For the rock falling from the top of the mast enters a medium that does not share the ship's motion; but the one released from the top of the tower finds itself in a medium that shares the same motion as the terrestrial globe, and so it can follow the general course of the earth without being hindered by the air but rather being favored by its motion.

SIMPLICIO. I do not understand how the air can impart its own motion to a very large rock or a large iron or lead ball, which, [169] for example, might exceed two hundred pounds. Perhaps it transmits its motion to feathers, snow, and other very light objects; but I see that a weight of that kind is not displaced by a single inch even when exposed to the fiercest wind. Now, think whether the air can carry it along.

SALVIATI. There is a great disparity between your experiment and our case. You have the wind come upon the rock lying at rest, whereas we expose to the already moving air a rock which is itself moving at the same speed; thus, the air does not have to impart to it some new motion, but rather must keep it in motion or (to be more exact) not hinder the motion already acquired. You want to push the rock into a motion foreign to it and against its nature; we want to conserve it in its natural motion. If you want to present a more appropriate experiment, you could say

100. This criticism reduces to questioning the analogy between a rotating earth and a moving ship; the analogy is questioned in two or three ways. If right, this criticism would not disprove the physical impossibility of mixed motion (would not prove its physical possibility), but would only show that the assertion of this impossibility is groundless; it would follow that the argument from apparent vertical motion (of which this subargument is a part) is groundless, namely, it relies on a premise which is unjustified (not false, but merely unsupported). As applied to the self-contained ship analogy argument, this criticism questions the relevance of the alleged evidence, independently of the truth of the allegation (which will also be questioned later). Here Galileo does not discuss a more theoretical argument which the Aristotelians had to support the impossibility of mixed motion, namely, the argument based on their general theory of natural motion; a relevant reason would be the principle that every body can have only one natural motion; another would be the principle that there are only two basic types of natural motions (straight and circular), corresponding to the two basic types of simple geometrical lines. But he already criticized Aristotle's general theory of natural motion at the beginning of the "First Day" in a passage that is omitted above (Favaro 7:38–57; Galileo 1967, 14–32); that criticism implies that the two natural motions being mixed in falling bodies on a rotating earth would be natural in two different senses, the straight-downwards fall in the sense of spontaneous inclination, and the horizontal-circular rotation in the sense of potentially perpetual state; for more details, see Finocchiaro (1980, 33–34, 107, 349–53, and 387–89). Finally, the issue of the impossibility of mixed motion will reemerge at the end of this selection, as underlying the dispute about the results of the ship experiment.

that one should observe (with the mind's eye, if not with the real one) what would happen when an eagle carried by the wind releases a rock from its claws; because the rock is moving like the wind at the moment of separation from the claws, and thereafter it enters a medium which is moving at the same speed, I am strongly inclined to think that we would not see it fall perpendicularly, but that it would follow the course of the wind and add to this the motion due to its own gravity, and so it would move with an inclined motion.

SIMPLICIO. One would have to be able to make such an experiment and then form a judgment depending on the result; however, so far the ship experiment seems to favor our opinion.[101]

SALVIATI. Well said, "so far"; for perhaps before long, appearances may change. In order not to keep you in suspense any longer, tell me, Simplicio, Do you think the ship experiment fits our purpose so well that it is reasonable to believe that what is seen to happen on the ship should likewise happen on the terrestrial globe?

SIMPLICIO. Up to now I think so; although you have advanced some small differences, they do not seem to me to be of such import as to make me change my mind.

SALVIATI. On the contrary, I wish you to continue believing firmly that the result on the earth should correspond to the one on the ship, as long as you do not feel like changing your mind if this were discovered to be prejudicial to your cause.[102]

101. This ends Galileo's criticism of the argument from apparent vertical fall and of the inferential *strength* of the ship analogy argument. Now he begins discussing the factual *accuracy* of the alleged experimental results and of the corresponding premises of the latter argument.

102. Here some readers might charge Galileo with the following logical inconsistency or rhetorical trickery: that he first denies and then affirms the soundness of the analogy between the moving ship and the rotating earth. But I do not think he is being inconsistent; the point is that the evaluation of arguments has two aspects, the truth of the premises and the validity of the inferential link between premises and conclusion; and these two aspects are independent of each other. Thus, Galileo has just finished questioning the inferential link between what happens on a moving ship and what would happen on a rotating earth; now he explores whether, *even if this link were valid*, what happens on the moving ship is really what the anti-Copernicans claim. This situation is analogous to what happens in his "Letter to the Grand Duchess Christina" when he criticizes the argument that the earth must be standing still because the Bible says so. He first replies that the Bible is not a scientific authority but an authority only on questions of faith and morals; so what it says about natural phenomena carries no weight; thus, the premises of the biblical objection do not lend any support to its conclusion; so this argument is invalid. Then he argues that, independently of this difficulty, even if the argument were valid, a careful examination of what the relevant passages say and imply about the two world systems reveals that the Bible does not really state or imply that the earth is motionless; instead those passages are

You say: because when the ship [170] stands still the rock falls at the foot of the mast, and when the ship is in motion it falls away from the foot, therefore, inverting, from the rock falling at the foot one infers the ship to be standing still, and from its falling away one argues to the ship being in motion; but what happens to the ship must likewise happen to the terrestrial globe; hence, from the rock falling at the foot of the tower, one necessarily infers the immobility of the terrestrial globe. Is this not your reasoning?

SIMPLICIO. Exactly. You have made it concise and very easy to understand.[103]

SALVIATI. Now, tell me, If the rock released from the top of the mast were to strike the same spot on the ship when it is going forward at great speed as when it is standing still, what use would these experiments have for ascertaining whether the vessel is standing still or going forward?

SIMPLICIO. Absolutely none. Similarly, for example, from a pulse beat we cannot learn whether someone is asleep or awake since the pulse beats in the same manner in people who are asleep and who are awake.

SALVIATI. Very well. Now, have you ever made the ship experiment?

SIMPLICIO. I have never made it, but I really believe that those authors who put it forth have diligently made the observations. Furthermore, the cause of the disparity is so well known that there is no room for doubt.

SALVIATI. It is possible that those authors put it forth without having made it; you are a good witness to this yourself, for without having

---

more in accordance with the Copernican than with the Ptolemaic system. In short, after questioning the validity of the biblical argument, he questions the crucial premise; for more details, see Finocchiaro (1986; and 1989, 87–118). There is no more contradiction or trickery at this point in the *Dialogue* than there is in the "Letter to the Grand Duchess."

103. The previous paragraph contains a clear statement of the ship analogy argument; but this statement involves hidden complexities, some of which are analyzed in Finocchiaro (1980, 389–91). For example, its structure seems to be "because (A), therefore (B); but (C); hence (D)," where these four letters correspond respectively to the first, second, third, and fourth clauses of the previous paragraph. One problem is that proposition (D) looks like an inference rule, which leads to the question as to whether it is legitimate to justify an inference rule by means of an argument from analogy. Another complication is the nature of the connection between (A) and (B): each proposition has two parts, which may be labeled respectively (A1), (A2), (B1), and (B2); but to deduce (B1) from (A1) and (B2) from (A2) would involve committing the fallacy of conversion, which is a version of the fallacy of affirming the consequent; instead (B1) can be deduced from (A2) and (B2) from (A1) by the principle of contraposition; so the "inverting" mentioned in the text must be interpreted as contraposition rather than conversion. Finally, this matter is complicated by the fact that Galileo's original text justifies this step explicitly by saying "by the principle of the converse" (*per il converso*); my translation reflects my judgment that I see no reason for interpreting Galileo's *converso* as meaning what present-day logic textbooks take it to mean.

made it you present it as certain and in good faith rely on their assertion. At any rate, it is not only possible but necessary that they too relied on their predecessors, without ever arriving at someone who made it; for whoever performs the experiment will find it to show the complete opposite of what is written; that is, it will show that the rock always falls at the same spot on the ship, whether it is standing still or moving at any speed. Hence, since the same holds for the earth as for the ship, from the rock falling always perpendicularly to the foot of the tower nothing can be inferred about the earth's motion or rest.

SIMPLICIO. If you were referring me to some means other than experiment, I really think our disagreements would not end very soon; for this seems to me an issue so remote from any human speculation that it leaves no room for considerations of credibility or probability.

[171] SALVIATI. And yet I think it does.

SIMPLICIO. So, you did not make one hundred tests, or even one, and yet you claim the result to be certain and unequivocal? I am skeptical about this, and I go back to my certainty that the experiment has been made by the principal authors who use it, and that it shows what they claim.

SALVIATI. Without experiment I am certain the result will happen as I say because it is necessary that it should happen that way;[104] I add that

104. This remark along with the latest series of interchanges between Salviati and Simplicio raises important historical and methodological issues, such as whether either side actually performed the experiment, whether either side thought it necessary to perform it, and what each side thought the actual experiment would contribute to the resolution of the controversy. This passage seems to suggest that neither side had actually performed the experiment, that the Aristotelian side thought the experiment was necessary and would settle the issue, and that Galileo had not performed the experiment and did not think it necessary to perform it. This in turn has generated a methodological controversy about whether Galileo was indeed an apriorist (as he appears to be here), and by extension, whether science in general is apriorist. I would argue that careful attention to the textual details and context of this passage shows that the methodological lesson derivable from Galileo here is that experiments are *sometimes* unnecessary and their results can *sometimes* be predicted in advance, not that experiments are always or usually unnecessary and their results always or usually predictable; the cases in question would be those, like the ship experiment, where one can construct a good argument based on known or more easily observable facts to arrive at a conclusion about what will happen in the experiment; in short, while Galileo is portraying the Aristotelians as naive empiricists, he is not advocating apriorism but rather a position that could be described as either critical empiricism or critical apriorism. Regarding the historical question, despite the impression conveyed by this passage, Galileo probably performed the actual experiment since in an earlier discussion of this problem in his "Reply to Ingoli" he stated, "I have been a better philosopher than you in two ways: for, besides asserting something which is the opposite of what actually happens, you have also added a lie by saying that it was an experimental observation; whereas I have made the experiment, and even before that, natural reason had firmly persuaded me that the effect had to happen the way it indeed does" (Finoc-

even you yourself know that it cannot happen otherwise, although you pretend (or try to pretend) not to know it. However, I am so good at picking people's brains that I will make you admit it by force.[105] Sagredo is very quiet, but I thought I saw him gesturing to say something.

SAGREDO. Truly I wanted to say something or other. But then I heard you threaten Simplicio with violence, to make him reveal the knowledge he wants to conceal from us; this made me so curious that I put away any other desire. So I beg you to make good your boast.

SALVIATI. As long as Simplicio is willing to answer my questions, I will not fail.

SIMPLICIO. I will answer what I know and am certain I will have little difficulty; for knowledge is about truths and not about falsehoods, and thus I do not think I know anything about the things I regard as false.

SALVIATI. I do not want you to say or answer anything but what you are sure you know. So, tell me, Suppose you had a plane surface very polished like a mirror and made of a hard material like steel; suppose it was not parallel to the horizon but somewhat inclined; and suppose that on it you placed a perfectly spherical ball made of a heavy and very hard material like bronze, for example; what do you think it would do when released? Do you not think (as I believe) that it would stand still?[106]

---

chiaro 1989, 184); and the experiment was apparently performed by other Copernicans, for example, earlier by Giordano Bruno, in 1639 by Gio. Battista Baliani (1582–1666), and in 1640 by Pierre Gassendi (1592–1655); but some Aristotelians also claimed to have performed it, for example, a professor at the University of Padua named Giovanni Cotunio (cf. Chiaramonti 1633, 339; Barenghi 1638, 183). These historical facts suggest that neither side adopted an apriorist attitude on this particular issue; but, since the experimental results conflicted, they also suggest another reason why the issue could not be simply resolved empirically (another methodological lesson): reasoning is necessary and preferable to simple observation not only when argumentation can be based on facts more easily ascertainable than what is in question, but also when the experiments yield conflicting results. For more details on this experiment, see Drake (1982, 580n.38), Favaro (18:102–3), Feyerabend (1975), Finocchiaro (1980, 116–17, 211, and 389–91), Koyré (1966, 225; 1965, 176), Santillana (1953, 140n.26 and 159n.41), and Strauss (1891, 521n.38); on the issue of the role of experimentation in Galileo's work, recent scholarship has provided conclusive documentation that he was an indefatigable practitioner of real experiments (cf. Drake's works, Hill 1988, MacLachlan 1973, Naylor 1976 and 1990, Segre 1980, and Settle 1961); and on the role of experiments in physical science in general, see Franklin (1986; 1990) and Hacking (1983).

105. This speech by Salviati and the ensuing passage provide an example of Galileo's frequent employment of the Socratic method and further evidence of his high regard for it; for more details, see Finocchiaro (1980, 116–17 and 145–64). They also add dramatic impact to the discussion and are amenable to being analyzed from a literary viewpoint; cf. Finocchiaro (1980, 51–52).

106. As the ensuing dialogue shows, Salviati does not really believe this; what is happening is best described in terms of comic relief or rhetorical zest.

SIMPLICIO. If that surface was inclined?

SALVIATI. Yes, for this is the supposition.

SIMPLICIO. I do not think it would stand still; rather I am sure it would spontaneously move downward along the incline.

SALVIATI. Be very careful about what you say, Simplicio; for I am sure it would stand still in any spot you had placed it.

SIMPLICIO. Salviati, when you make this sort of [172] assumption, I begin to be less surprised that you should arrive at very false conclusions.

SALVIATI. Are you thus very sure that it would spontaneously move downwards along the incline?

SIMPLICIO. What is there to doubt?

SALVIATI. And you firmly believe this not because I taught it to you (for I tried to persuade you of the opposite), but because you arrived at it on your own using your natural judgment.

SIMPLICIO. Now I understand your trick; you said what you did in order to lead me on and (as the popular expression goes) to trap me, not because you really believed that.

SALVIATI. That is correct. Now, how long would the ball's motion last, and what speed would it have? Notice that I am referring to a perfectly round ball and a fastidiously polished plane, in order to remove all external and accidental impediments; similarly, I want you to disregard the impediment offered by the air through its resistance to being parted, and any other accidental obstacles there may be.

SIMPLICIO. I understand everything very well. As for your question, I answer that the ball would continue to move ad infinitum, as far as the inclination of the plane extends; that it would move with continuously accelerated motion, for such is the nature of falling bodies, which "acquire strength as they keep going";[107] and that the greater the inclination, the greater would be the speed.

SALVIATI. However, if someone wanted to have the ball move upward along the same surface, do you think it would move that way?

SIMPLICIO. Not spontaneously; but it would if dragged along or thrown by force.

SALVIATI. So, if it were propelled by some impetus forcibly impressed on it, what would its motion be and how long would it last?

SIMPLICIO. Its motion would keep on being continuously reduced and retarded, due to its being against nature; and it would last more or

107. Galileo writes this clause in italics and in Latin (*vires acquirant eundo*); it comes from Virgil (*Aeneid*, IV, 175), who refers to rumors and how they spread.

less depending on the greater or smaller impulse and on the steeper or gentler inclination.

SALVIATI. Therefore, I think that up to now you have explained to me the following properties of a body moving along a plane in two different directions: when descending on an inclined plane, the heavy body is spontaneously and continuously accelerated, and it requires the use of force to keep it at rest; on the other hand, in an [173] ascending path a force is needed to make it move that way (as well as to keep it at rest), and the motion impressed on it is continuously diminishing, so that eventually it is annihilated. You also say that in both cases there is a difference stemming from the greater or smaller inclination of the plane; so that a greater inclination leads to a greater downward speed, but on an upward path the same body thrust by the same force moves a greater distance when the inclination is less. Now tell me what would happen to the same body on a surface that is not inclined.

SIMPLICIO. Here I must think a little before I answer. Since there is no downward slope, there cannot be a natural tendency to move; since there is no upward slope, there cannot be a resistance to being moved; thus, the body would be indifferent to motion and have neither a propensity nor a resistance to it; I think, therefore, that it should remain there naturally at rest. Sorry to have forgotten, for I now remember that not long ago Sagredo explained to me that this is what would happen.[108]

SALVIATI. I think so, if one were to place it there motionless; but if it were given an impetus in some direction, what would happen?

SIMPLICIO. It would move in that direction.

SALVIATI. But with what sort of motion? A continuously accelerated one, as on a downward slope, or a progressively retarded one, as on an upward slope?

SIMPLICIO. I see no cause for acceleration or retardation since there is neither descent nor ascent.[109]

108. This is a puzzling reference. Galileo may be referring to a passage at the beginning of the "First Day" that discusses speed, acceleration, and falling bodies but which is omitted here (Favaro 7:45–55; Galileo 1967, 21–29).

109. Simplicio's answer strikes me as plausible but does not seem to be faithful to Aristotelian physics; for its most basic principle is that all motion requires a force (internal or external) to continue and ceases (gradually) when it is no longer present; this implies that when the body is placed on the horizontal surface and is given an impetus along the horizontal, it will start to move but will be progressively retarded; thus, from an Aristotelian viewpoint there is cause for retardation, namely, the fact that the body is separated from the force which gave it the initial impetus. It is unclear what conclusion to draw from Simplicio's un-Aristotelian behavior. Some readers might say that Galileo is not doing justice to Aristotle, and so his usual judiciousness is failing him here; they would thereby be

SALVIATI. Yes. But if there is no cause for retardation, still less is there cause for rest. So, how long do you think the moving body would remain in motion?

SIMPLICIO. As long as the extension of that surface which is sloping neither upward nor downward.

SALVIATI. Therefore, if such a surface were endless, the motion on it would likewise be endless, namely, perpetual?

SIMPLICIO. I think so, as long as the moving body was made of durable material.

SALVIATI. This has already been supposed, for we have already said that all accidental and external impediments should be removed, and in this regard the body's fragility is one of the accidental impediments.[110] Now tell me, What do you think is the reason why that ball moves spontaneously on the downward path and not without force on the upward one?

---

advancing a methodological criticism. Others might take a rhetorical viewpoint and stress that Galileo has contrived to insinuate that the Aristotelian Simplicio is beginning to change his mind—is beginning to see the light, so to speak. From the (logical) viewpoint of critical reasoning, we should first reconstruct the Aristotelian argument for retardation and the Galilean argument for neither retardation nor acceleration; then we should evaluate these arguments, keeping in mind that we are not dealing with rigorous reasoning susceptible of conclusive demonstration, but yet not despairing of the possibility of some kind of comparative judgment; for example, we cannot but be impressed by the continuity and symmetry of Galileo's reasoning.

110. The agreement here reached by Salviati and Simplicio embodies an important physical principle, which is called the principle of the conservation of motion; it represents Galileo's approximation to two fundamental laws of classical physics—the law of inertia and the law of the conservation of momentum. In his own words, extracted from this discussion, the principle may be formulated as follows: on an horizontal plane a body will remain motionless or conserve its motion forever unless acted upon by some external force or disturbed by some accidental or external impediment; examples of such impediments are the body's fragility, air resistance, and friction between the body and the plane. This statement should be compared and contrasted to Newton's first law of motion: "Every body continues in its state of rest, or of uniform motion in a right line, unless it is compelled to change that state by forces impressed upon it" (Newton 1934, 13). At an abstract level of analysis, it is unclear whether the Galilean horizontal or noninclined direction is truly rectilinear or ultimately circular along the terrestrial circumference; but the comparison should also be carried out in regard to their applications; and the application of the general principle to the case of the ship experiment that Galileo later gives seems to be hardly distinguishable from that of Newton. This passage is one of several whose scientific content relates to the law of inertia (for example, selection 9 contains others). They have generated a dispute about Galileo's concept of inertial and natural motion; some scholars attribute to him a theory of "circular inertia," for example, Franklin (1976, 58–62 and 84–87), Koyré (1966, 205–90; 1978, 154–236), Shea (1972, 116–38), and Shapere (1974, 87–121); a more correct interpretation is one along the lines of Chalmers and Nicholas (1983, 328–40), Coffa (1968), Drake (1970, 240–78; 1978, 126–32), Finocchiaro (1980, 33–34, 87–92, and 349–53), McMullin (1967, 27–31), and Sosio (1970, l-li).

[174] SIMPLICIO. Because the tendency of heavy bodies is to move toward the center of the earth, and only by force do they move upward away from it; and by moving down on an inclined surface one gets closer to the center, and by moving up one gets further away.

SALVIATI. Therefore, a surface sloping neither downward nor upward would have to be equidistant from the center at all of its points. But are there any such surfaces in the world?

SIMPLICIO. There is no lack of them: one is the surface of our terrestrial globe, if it were smoothed out, and not rough and mountainous, as it is; another is the surface of the water when it is calm and tranquil.

SALVIATI. Therefore, a ship moving in a calm sea is a body going over a surface that slopes neither downward nor upward, and so it has the tendency to move endlessly and uniformly with the impulse once acquired if all accidental and external obstacles are removed.

SIMPLICIO. It seems that it must be so.

SALVIATI. Now, when the rock at the top of the mast is being carried by the ship, does it not also move along the circumference of a circle around the center, and consequently with a motion indelibly inherent in it as long as external impediments are removed? And is this motion not as fast as that of the ship?

SIMPLICIO. So far, so good; but what about the rest?

SALVIATI. You should be able to draw the last consequence yourself, if on your own you have discovered all the premises.

SIMPLICIO. By the last conclusion you mean that, since the rock is moving with a motion indelibly impressed on it, it will not leave but follow the ship, and at the end it will fall at the same spot where it falls when the ship stands still; and I, too, say that this would follow if there were no external impediments to disturb the rock's motion after being released. However, there are two such impediments: one is that the moving body is incapable of parting the air merely by means of its impetus, once it loses that of the oars' power, which it shared when it was part of the ship while still at the top of the mast; the other is the newly acquired motion of falling down, which must be an impediment to its horizontal motion.[111]

111. Simplicio's latest objection is well taken and is formulated in terms of the principle of the conservation of motion. This states that motion is conserved *unless* external impediments interfere, and hence that conservation of motion is contingent on the absence of these interferences; Simplicio now claims that the rock's downward fall is an external impediment which interferes with its horizontal motion, and thus prevents this motion from being conserved. The issue is then whether vertical fall interferes with horizontal motion, or to be more exact, how

SALVIATI. Regarding the impediment of the air, I do not deny it; and if the falling body were made of a light material like a feather or a lock [175] of wool, the retardation would be very great; but for a heavy rock it is very little. A short while ago you said yourself that the force of the strongest wind does not suffice to displace a big rock; now, think what will happen when the calm air meets the rock moving no faster than the ship as a whole. However, as I said, I grant you the small effect that may result from this impediment, just as I know you will grant me that if the air were moving at the same speed as the ship and the rock, the impediment would be absolutely nil.

Regarding the impediment of the newly acquired downward motion, first it is clear that these two motions (namely, the circular around the center and the straight toward the center) are neither contrary nor incompatible nor destructive of each other; for the moving body has no repugnance toward such motion; you yourself already granted that its repugnance is to motion which takes it farther from the center, and that its inclination is to motion which brings it closer to the center; so it follows necessarily that the moving body has neither repugnance nor propensity to motion that takes it neither farther from nor closer to the center, and consequently there is no reason for any decrease in the power impressed on it. Moreover, the cause of motion is not a single one, which might diminish on account of the new action; instead there are two distinct causes, of which gravity attends only to drawing the body toward the center, and the impressed power to leading it around the center; therefore, there is no reason for an impediment.

. . . [112]

---

exactly the two motions are to be combined. Salviati's answer will be that the newly acquired motion of fall does not destroy the old horizontal one, but that the two subsist and act independently of each other, thus yielding a resultant motion which takes the rock to the foot of the mast even on the moving ship; this answer presupposes the principle of the superposition of motion, another one of the fundamentals of classical mechanics. It is instructive to identify and evaluate the reasons advanced by Salviati as to why the second motion does not interfere with the first. This issue of whether and how to combine these two motions is a version of the issue about the alleged impossibility of mixed motion, which emerged earlier in the criticism of the argument from apparent vertical fall; thus, Salviati's reasons why the two motions can be superimposed also strengthens the previous criticism of that argument; more generally, the importance of Galileo's intuition of the principle of superposition is thereby enhanced.

112. The passages omitted here correspond to Favaro 7:175.24–214.22 and Galileo 1967, 149–88. They contain first a discussion of the cause of the motion of projectiles and an elaboration of Galileo's ideas about the conservation and superposition of motion; these strengthen his criticism of the vertical fall argument. Then he criticizes the four gunshot objections (east-west, vertical, north-south, and point-blank) and the birds argument; these critiques contain a further elaboration of the principle of the relativity of motion. For more details, see Finocchiaro (1980, 36–38, 116–19, and 391–406).

## [9. Extruding Power and the Role of Mathematics]

[214] . . . SAGREDO. . . . There remains now the objection based on the experience of seeing how a rapid turning has the power of extruding and scattering the objects attached to the device which is turning; thus, many (including Ptolemy)[113] have thought that, if the earth were turning around itself at such a speed, rocks and animals should be cast toward the stars, and even buildings could not be attached to their foundations with mortar strong enough to spare them from such mass destruction.[114]

SALVIATI. Before coming to the resolution of this objection, I cannot be silent about something I have observed a thousand times, not without amusement. This regards what comes into the mind of almost all men when they first hear about the earth's motion: they believe it to be so fixed and immobile that not only do they never doubt such rest, but they firmly believe that all other men also regard it as having been created immobile and having been kept in this state for all past centuries; being fixated upon this idea, they are then surprised upon [215] hearing that someone attributes motion to it, as if he were foolishly thinking that it first started moving when motion was first attributed to it by Pythagoras (or whoever it was who first did so). Now, I am not surprised that such a very foolish thought should find its way into the minds of ordinary men with little common sense (I mean the thought that those who attribute motion to the earth regard it to have been immobile from its creation until the time of Pythagoras, and to have been set in motion only after Pythagoras regarded it as being in motion); but I really think it is a much stranger and an inexcusable simplemindedness that even

113. It is uncertain whether Ptolemy really advanced this argument. In so stating, Galileo probably follows Copernicus, who attributed it to Ptolemy (Copernicus, *On the Revolutions*, I, 7; 1992, 15). But the relevant passage from Ptolemy (*Almagest*, I, 7) is insufficiently clear and explicit to determine whether he indeed had the extrusion objection in mind or was talking about terrestrial bodies being left behind if the whole earth were moving; cf. Drake (1967, 478n; 1982, 580n.30), Hill (1984, 110–15), Rosen (1992, 351, note to 15:17), Strauss (1891, 519n.30), and Santillana (1953, 146n.33),

114. The objection restated here was first introduced in selection 7; again, I call it the argument from the extruding power of whirling, or the extrusion argument, or (in modern terminology) the centrifugal force argument. Others, for example, Drake (1986a) and Hill (1984), call it the projection argument.

people like Aristotle and Ptolemy should have held such a childish thought.

SAGREDO. So, Salviati, you believe Ptolemy thought that in this dispute he had to maintain the earth's rest against men who grant it to have been motionless till the time of Pythagoras, and who claim it to have started moving only when Pythagoras attributed motion to it?

SALVIATI. We cannot believe otherwise if we consider carefully the way he goes about refuting their opinion. This refutation refers to the destruction of buildings and to rocks, animals, and men themselves being cast toward the heavens; but such destruction and scattering cannot happen to buildings and animals that do not already exist on the earth, nor can men be born and buildings erected on the earth unless it stands still; therefore, it is clear that Ptolemy is arguing against those who grant the earth to have been at rest for a long time (that is, while animals and masons could live on it, and palaces and cities could be built), but then suddenly set it in motion, to the ruin and destruction of buildings, animals, etc. On the other hand, if he had intended to argue against those who attribute this turning to the earth since its original creation, he would have refuted them by saying that, if the earth had always been in motion, then neither beasts nor men nor rocks could ever have been formed on it, and still less could buildings have been erected and cities founded, etc.

SIMPLICIO. I do not well understand this Aristotelian and Ptolemaic impropriety.

SALVIATI. Ptolemy argues either against those who regard the earth as being always in motion, or against those who regard it to have been [216] still for some time and then to have been set in motion. If he is arguing against the first, he should have said: "the earth has not always been in motion because terrestrial rotation would not have allowed men, animals, or buildings to exist on the earth, and so there would have never been any of them on it." However, he argues by saying: "the earth does not move because the beasts, men, and buildings already found on the earth would be cast off"; hence, he supposes the earth to have been once in such a state as to have allowed beasts and men to form and live on it; this implies the consequence that once it was motionless, namely, suitable for animal life and the construction of buildings. Do you now grasp what I am trying to say?[115]

115. This comment on the extrusion argument may be interpreted as a formal criticism or an analytical clarification. The criticism is that, as ordinarily stated, the argument alleges

SIMPLICIO. I do and I do not. But this matters little as far as the merits of the issue are concerned; nor can a little error by Ptolemy, committed inadvertently, be sufficient to move the earth if it is motionless. But let us put jokes aside, and let us come to the crux of the argument, which to me seems unanswerable.

SALVIATI. Simplicio, I want to strengthen and tighten it further by showing even more sensibly how true it is that, when heavy bodies are rapidly turned around a motionless center, they acquire an impetus to move away from that center, even if they have a propensity to go toward it naturally. Tie a small pail full of water at the end of a string, and make it go around rapidly in such a way that it describes the circumference of a circle, by firmly holding the other end of the string in your hand, and using the string and your arm as a radius and your shoulder's joint as the center; whether the circular path is parallel, perpendicular, or in any way inclined to the horizon, in every case it will happen that the water does not fall out of the vessel, and the person turning it will feel the string pull and exert a force away from the shoulder. Moreover, if a hole is made in the bottom of the little pail, the water will be seen to spurt out no less toward the sky than sideways and toward the ground; and if pebbles are used instead of water, by turning the pail in the same manner one will feel them exert the same force against the string. Finally, children are seen to sling stones to great distances by turning a piece of reed which has a stone wedged in at one end. All these provide arguments for the truth of the conclusion that turning gives a moving body an impetus toward the circumference, when the turning is rapid. Now, [217] if the earth were turning around itself, the surface motion (especially near the equator) would be incomparably

---

to be proving one conclusion (that the earth is not in motion) but instead at best proves another (that the earth did not recently begin to move); that is, the argument reaches an irrelevant conclusion, namely, a proposition that is not disputed. So interpreted, the argument as originally stated is an example of a classic fallacy called *ignoratio elenchi* (cf. Aristotle, *On Sophistical Refutations* 167a21). But the comment may also be taken as introducing an essential clarification in the statement of the extrusion argument; for the criticism just mentioned cannot even be stated without indicating the proper way to reformulate the argument, and so once the criticism is understood, we can immediately reformulate the argument; this is how the dialogue proceeds. This first comment is thus similar to the first comment on the vertical fall argument in selection 8; that comment amounts to showing why the vertical fall argument should not be stated in terms of actual vertical fall but in terms of apparent vertical fall.

faster than those mentioned, and so everything should be extruded toward the sky.[116]

SIMPLICIO. The objection seems to me to be very well established and secured, and in my opinion a great effort will be required to remove and answer it.

SALVIATI. The answer to it depends on some facts which you know and believe no less than I do; but, because you do not remember them, you do not see the answer. So, without my teaching them to you (because you already know them), but by simply reminding you of them, I will make you answer the objection yourself.

SIMPLICIO. I have thought several times about your manner of reasoning; it makes me think you are inclined to accept Plato's doctrine that "our knowledge is a form of recollection."[117] So, please resolve my doubt by telling me what you mean.

SALVIATI. I can explain by words and also by deeds how I feel about Plato's doctrine. In the arguments discussed so far, I have already explained myself more than once by deeds; I will follow the same style in the particular case at hand. You may later use it as an example to understand better my views on the acquisition of knowledge, if there is time left on another day and Sagredo is not annoyed with such a digression.

116. This "strengthening and tightening" of the extrusion argument by Salviati is an indication of Galileo's open-mindedness and part of his plea for the value of this mental attitude, which is discussed in selection 7. This feature of the discussion also adds another reason for regarding the first comment on the extrusion argument constructively as an analysis designed to clarify its formulation; for we may regard the discussion so far in this selection as a constructive elaboration of an argument soon to be refuted, consisting first of a formal clarification and second of a substantive strengthening; this strengthening is a subargument supporting the argument's basic premise with which Galileo agrees; this premise is a generalization about the existence of an extruding power in whirling. This procedure is in exact compliance with the formulation of the principle of open-mindedness that he had occasion to state (to no avail) in the tragic circumstances of the 1633 trial, namely, that "when one presents arguments for the opposite side with the intention of confuting them, they must be explained in the fairest way, and not be made out of straw to the disadvantage of the opponent" (Finocchiaro 1989, 287).

117. Galileo writes this sentence in Latin: *nostrum scire sit quoddam reminisci*; cf. Plato's *Meno* and *Phaedo* (72C-77B). Plato (c. 427–347 B.C.) was a Greek thinker whose writings focused on questions of metaphysics, epistemology, ethics, and political theory, but not natural philosophy (although he did write one work on the latter subject, the *Timaeus*); his writings laid the foundations for much subsequent Western philosophy. He was a pupil of Socrates and the teacher of Aristotle and wrote almost all of his works in dialogue form, portraying Socrates as one of the speakers.

SAGREDO. On the contrary, I will be very pleased, for I remember that when I studied logic I could never understand Aristotle's so-much-discussed concept of most powerful demonstration.[118]

SALVIATI. Let us go on then.[119] Simplicio should tell me what is the motion of the stone wedged in the notch of a sling reed while the boy moves the reed to cast the stone far away.

SIMPLICIO. As long as the stone is in the notch, its motion is circular; that is, it follows the arc of a circle whose motionless center is the shoulder's joint and whose radius is the reed together with the arm.

SALVIATI. But when the stone leaves the reed, what is its motion? Does it continue to follow the preceding circular motion, or does it follow some other line?

118. "Concept of most powerful demonstration" is my translation of Galileo's Latinized Italian phrase *dimostrazion potissima*; I adapt it from Wallace's translation of the Latin term *demonstratio potissima* in Galileo's Latin logical notebooks, manuscript MS 27 (Wallace 1992a; 1992b). Suffice it to know that Aristotle's "most powerful demonstration" is a form of demonstrative argument which he thought had a key role in science, and that Sagredo's remark is a trace of Galileo's early study of Aristotle. It is interesting to quote the definition from Galileo's notebook: "There are three kinds of demonstration, of the fact, of the reasoned fact, and most powerful. That of the fact demonstrates [a cause] from an effect; that of the reasoned fact gives the reason why a property exists in a subject; and that which is most powerful both gives the reason why a particular property exists in a subject and proves the existence of that property" (Wallace 1992a, 102); Galileo argues later in these notebooks that the most powerful demonstration is not really a distinct kind, but a special case of demonstration of the reasoned fact (Wallace 1992a, 177; 1992b, 183); thus, Sagredo's difficulty seems to reflect Galileo's own view. Finally, the passages that gave rise to this type of commentary are in Aristotle, *Posterior Analytics*, I, 2–13.

119. The exchange in the last four speeches is methodologically significant. Salviati's unwillingness to commit himself to Plato's doctrine of recollection, his emphasis on actual practice, and Sagredo's injection of the context of justification all suggest that Galileo's commitment is to the pedagogical power and epistemological value of the Socratic method rather than to the metaphysics of Platonic anamnesis. By the Socratic method I mean a method of teaching and of justification (invented and practiced by the historical Socrates) in which the teacher or the proponent engages in a dialogue with the pupil or the opponent, focuses on asking questions rather than giving answers, stresses negative criticism to awaken the interlocutor's own curiosity, and leads him gradually to arrive at the truth or work out the answer himself (as if the interlocutor were discovering it himself); the last feature is especially important and is given the special label *maieutics* (namely, "midwifery"). By *anamnesis* is meant Plato's doctrine that learning is a form of recollection from a previous life in which the soul had the knowledge which it then lost at birth when it became incarnated into its current body. Socrates' own unquestionable commitment to the Socratic method (exhibited in Plato's *Meno*) goes hand in hand with a certain amount of skepticism about anamnesis (suggested in Plato's *Phaedo*, 72C–73C and 114B–115B); Galileo's attitude is analogous to that of Socrates. For more details, see Feyerabend (1975), Finocchiaro (1980, 120 and 150–64), and Santillana (1953, 202n.66).

SIMPLICIO. It does not in the least continue to move in a circle because in that case it would not go away from the thrower's shoulder, whereas we see it go very far.

[218] SALVIATI. So with what motion does it move?

SIMPLICIO. Let me think about it a little, for I do not yet have a mental image.

SALVIATI. Listen carefully, Sagredo; here is the "form of recollection"[120] in action, as it were. You are thinking about it a long time, Simplicio!

SIMPLICIO. For me the motion acquired upon leaving the notch can only be in a straight line; indeed it is necessarily in a straight line, if we refer merely to the impressed impetus. I was somewhat bothered by seeing the stone describe a curve; but, because this curve bends constantly downward and not in any other direction, I understand that this inclination derives from the stone's gravity, which pulls it naturally downward. Without a doubt, I say that the impressed impetus is in a straight line.

SALVIATI. But which straight line? For infinitely many lines in all directions may be drawn from the reed's notch at the point where the stone separates from the reed.

SIMPLICIO. It moves in a direction that is directly in line with the motion the stone had together with the reed.

SALVIATI. You have already said that the stone's motion was circular while it was in the notch; now, it is inconsistent to be both circular and directly in line with it since no part of a circular line is straight.

SIMPLICIO. I do not mean that the projected motion is directly in line with the whole circular motion, but with the last point where the circular motion ended. I understand it in my mind, but I do not know how to explain myself properly.

SALVIATI. I, too, notice that you understand the thing but do not know the proper terms to express it. Now, I can easily teach these to you; that is, I can teach you the words, but not the truths, which are things. To make you tangibly grasp the fact that you know the thing but only lack the terms to express it, tell me, When you fire a gun, in what direction does the bullet acquire an impetus to go?

SIMPLICIO. It acquires an impetus to go along the straight line which is directly in line with the barrel, that is, which deviates neither to the right nor to the left and neither up nor down.

120. Here Galileo writes this phrase in Latin: *quoddam reminisci*. Such reminders are frequently given by Socrates in the best available example of Socratic method, Plato's *Meno*.

SALVIATI. Which, in short, is to say that it makes no angle with the line of the straight motion along the barrel.

SIMPLICIO. That is what I wanted to say.

SALVIATI. If, then, the line of motion of the projectile is to be extended without making an angle with the circular line described while it was [219] in the sling, and if it must change from this circular motion to a straight motion, which one must this straight line be?

SIMPLICIO. It must be the one that touches the circle at the point of separation because I think all others would (if extended) cut the circumference, and so would make some angle with it.

SALVIATI. You have reasoned very well and have shown yourself to be something of a geometrician. So make it a point to remember that your real thought is explained by these words, namely, that the projectile acquires an impetus to move along the tangent to the arc described by the thrower's motion at the point where the projectile separates from the thrower.[121]

SIMPLICIO. I understand very well; that is what I wanted to say.

SALVIATI. When a straight line touches a circle, which one of its points is closest to the center of the circle?

SIMPLICIO. The point of contact, undoubtedly; for it is on the circumference of the circle and the others are outside, and all points on the circumference are equidistant from the center.

SALVIATI. Therefore, when a moving body leaves the point of contact and moves along the tangent line, it continuously moves away from the point of contact and also from the center of the circle.

SIMPLICIO. That is certainly correct.

SALVIATI. Now, if you remember all the propositions you told me, put them together and tell me what one gathers from them.

SIMPLICIO. I do not believe I am so forgetful that I do not remember them. From the things said one gathers that, when a projectile separates from a thrower who has been rapidly moving it around, it retains an impetus to continue its motion in a straight line that is tangent to the circle described by the thrower's motion at the point of separation; by this motion the projectile always increases its distance from the center of the circle described by the thrower's motion.

121. This statement is reminiscent of, and should be compared and contrasted to, the law of inertia. It should also be compared to the formulation given in the discussion of the ship experiment in selection 8 ([173]); the main difference is that here the rectilinear nature of the tendency is explicit.

SALVIATI. So you now know the reason why heavy bodies are extruded from the surface of a rapidly turning wheel, extruded in the sense of being hurled beyond the circumference to increasingly greater distances from the center.

SIMPLICIO. I think I comprehend this point very well; but this [220] new knowledge rather increases than diminishes my incredulity that the earth could turn around at such a speed, without extruding toward the sky stones, animals, etc.

SALVIATI. You will learn, or rather you already know, the rest in the same manner you learned so far; by thinking about it, you will remember it on your own.[122] However, to speed up the process, I will help you to remember it. So far you discovered on your own that, when separation occurs, the circular motion of the thrower impresses on the projectile an impetus to move in a straight line tangent to the circle of motion at the point of separation, and that the projectile keeps on moving farther and farther away from the thrower as long as motion along the tangent continues; and you said that the projectile would continue moving along such a straight line if its own weight did not add a downward tendency which produces a bending in the line of motion.[123] I also think you seemed to know on your own that this bending is always in the direction of the earth's center, because all heavy bodies have a tendency to go there. Now I go one step further and ask you whether the moving body, after separation and while continuing its straight motion, keeps on uniformly increasing its distance from the center (or, if you will, from the circumference) of the circle of which the preceding motion was a part; that is, whether a projectile leaving the point of tangency and moving along the tangent increases its distance equally from both the point of contact and the circumference of the circle.

SIMPLICIO. No, sir. For near the point of contact the tangent is very little off the circumference, with which it makes an extremely small angle; but, as its distance increases, the increase always takes place in greater proportion. Thus, for example, in a circle with a diameter of ten cubits, if a point on the tangent is two palms away from the point of con-

122. This remark makes it clear that the remembering crucial for the doctrine of recollection is really a form of thinking; so this doctrine may be reduced to the claim that learning is a form of thinking. This claim is plausible enough, but it suggests another reason why it is preferable to interpret talk of Platonic anamnesis as an embellishment on the power of the Socratic method.

123. As McMullin (1967, 27–31) has argued, this sentence, together with its reiteration in Salviati's first speech on p. [221], provides one of the clearest Galilean approximations to the law of inertia of modern physics.

tact, it will be three or four times farther from the circumference of the circle than a point that is one palm away from the contact; and if a point is half a palm away, I think likewise its distance will hardly be one fourth that of the second; so that at one or two inches near the contact, one hardly notices that the tangent is separated from the circumference.

SALVIATI. So, at the beginning the projectile's distance from the circumference of the preceding circular motion is extremely small?

SIMPLICIO. Almost imperceptible.

[221] SALVIATI. Now, tell me something about this projectile that from the thrower receives an impetus to move straight along the tangent, and that would so move if its own weight did not pull it downward; how soon after separation does it begin to bend downward?

SIMPLICIO. I believe this begins immediately because it lacks support, and so its own gravity cannot fail to act.

SALVIATI. Thus, if a rock extruded from a wheel turning at high speed had a natural propensity to move toward the center of the same wheel (as it does toward the center of the earth), it would be easy for it to return to the wheel, or rather for it not to leave the wheel; for at the beginning of the separation the distance from the wheel is so very small (due to the infinite acuteness of the angle of contact) that the least tendency to go back toward the center of the wheel would suffice to keep it on the circumference.

SIMPLICIO. I have no doubt that, if we suppose what is not and cannot be the case (namely, that the tendency of heavy bodies is to go to the center of that wheel), then they would be neither extruded nor cast off.

SALVIATI. I am not really supposing, nor do I need to suppose, what is not the case because I do not want to deny that those rocks are cast off; I am only speaking hypothetically[124] so that you will tell me the rest. Imagine now that the earth is the great wheel which is turned at great speed and has to cast off the stones. You have already been able to tell me very well that the projected motion must be in a straight line tangent to the earth at the point of separation; how noticeably does this tangent recede from the surface of the terrestrial globe?

SIMPLICIO. I believe that in a thousand cubits it does not recede by an inch.

SALVIATI. And do you not say that the projectile, pulled by its own weight, drops down from the tangent toward the center of the earth?

124. This word reads *per supposizione* in Galileo's text.

SIMPLICIO. I already said that, and now I say the rest. I understand perfectly that the stone will not be separated from the earth since at the beginning its separation would be so very minimal that the stone's tendency to move toward the earth's center would be a thousand times greater;[125] and in this case this center is also the center of the wheel. Indeed one must admit that stones, animals, and the other heavy bodies cannot be extruded. However, now I see a new difficulty with very light objects, which have a very weak tendency [222] to go down toward the center; since they lack the power to return to the surface, I do not see what could prevent them from being extruded; and you know that "one counterinstance suffices to refute a generalization."[126]

SALVIATI. We will give satisfaction to this too. But first tell me what you mean by light objects; that is, whether you mean really light substances (so that they go up), or else things which are not absolutely light but heavy to such a small degree that they do indeed come down (although slowly). For, if you mean absolutely light objects, I will grant you that they are extruded more than you want.

125. This number of 1,000 is meant as a rough estimate, as a claim about the order of magnitude. At this point, Galileo could have been more quantitatively precise about the comparison. For example, he could have calculated the distance covered in one second by a body falling from rest; before his telescopic discoveries, he had done extensive research on this topic, and in another passage of the *Dialogue* (Favaro 7:250; Galileo 1967, 223), he gives some relevant approximate figures (namely, 100 cubits in five seconds). Then he could have calculated the distance separating the extrusion tangent and the terrestrial circumference in one second of time at the equator; this can be done from a knowledge of the earth's radius, the rate of terrestrial rotation, and the geometry of the situation. He could have easily arrived at the number 266 as the ratio between these two distances; that is, at the conclusion that in one second free fall takes a body 266 times farther downwards than terrestrial rotation extrudes it away from the center. In fact, a few years after the book's publication, Mersenne (while working directly from Galileo's book) used this method of calculation and arrived at essentially this figure; see MacLachlan (1977, especially 176–78) for this account and other details. Later in the century, Huygens and Newton both arrived at a more adequate solution of the problem, corresponding to the viewpoint of modern physics; that is, at the equator the centripetal acceleration due to gravity is 289 times the centrifugal acceleration due to terrestrial rotation, which is to say that a body weighs 1/289 less at the equator than it would on a motionless earth. For a more modern argument along the same lines, see Chalmers and Nicholas (1983, 322). But Galileo chose not to perform the Mersenne-type of calculation. Though his motivation is unclear, one likely reason is that he thought he could prove something much stronger than the contingent fact that the downward tendency due to weight happens to exceed the extruding tendency due to rotation; his stronger claim is that the downward tendency (however small it might be, as long as it not zero) can always overcome the extruding tendency (however large it might be). In fact, the next several pages contain several distinct mathematical arguments attempting to prove this stronger claim.

126. Galileo writes this principle in Latin: *ad destruendum sufficit unum*.

SIMPLICIO. I mean the second type, such as feathers, wool, cotton, and the like; the least force suffices to lift them, and yet they are seen lying very calmly on the earth.

SALVIATI. As long as this feather has some natural propensity to descend toward the earth's surface, however minimal it is, I say it is sufficient to prevent the feather from rising; and this is not unknown to you either. So tell me, If the feather were extruded by the earth's rotation, by what line would it move?

SIMPLICIO. By the tangent at the point of separation.

SALVIATI. And if it should come back to join the earth, by what line would it move?

SIMPLICIO. By that which goes from it to the earth's center.

SALVIATI. Thus, here we are considering two motions: the motion of projection, which begins at the point of contact and follows the tangent, and the motion of the downward tendency, which begins at the projectile and follows the secant[127] toward the center. To want the projection to continue, it is necessary that the impetus along the tangent should prevail over the tendency along the secant; is it not so?[128]

SIMPLICIO. So it seems to me.

SALVIATI. But what do you think there must be in the motion of projection for it to prevail over the motion tending downward, so that it would follow that the feather is detached and recedes from the earth?

SIMPLICIO. I do not know.

SALVIATI. How can you not know? Here we have one and the same moving body, namely, the same feather; how can the same moving body surpass itself and prevail over itself?

SIMPLICIO. I think it can prevail over itself and yield to itself in motion only by moving sometimes faster and sometimes slower.

[223] SALVIATI. So you did know it after all. If, then, the feather is to be projected and its motion along the tangent is to prevail over the motion along the secant, what must the speeds be?

SIMPLICIO. The motion along the tangent must be greater than that along the secant. Alas, poor me! Is it not one hundred thousand times greater than, not only the downward motion of the feather, but also that

127. In geometry, a secant is a straight line from a point outside a circle to its center, thus intersecting the circumference at right angles. For example, in figure 4 (p. [224]), GD is a secant.

128. This is crucial and correct: we must compare the centrifugal tendency along the tangent due to inertia and the centripetal tendency along the secant due to weight, and extrusion means that the former overcomes the latter.

of the stone? Like a real simpleton, I had let myself be persuaded that stones could not be extruded by the earth's turning! So I correct myself and say that, if the earth were rotating, then rocks, elephants, towers, and cities would necessarily fly toward the sky; because this does not happen, I conclude that the earth does not rotate.

SALVIATI. Oh, Simplicio, you get so easily aroused that I begin to worry more about you than about the feather. Calm yourself a little, and listen. If to keep the stone or feather attached to the earth's surface the descent had to be greater than or equal to the motion along the tangent, you would be right in saying that it would have to move as fast or faster downward along the secant than eastward along the tangent; but did you not tell me a while ago that along the tangent a distance of a thousand cubits from the contact produces a separation from the circumference of barely one inch? Hence, it is not enough that the motion along the tangent (which derives from the diurnal rotation) should be simply faster than the motion along the secant (which derives from the feather's downward tendency); instead, the former must be so much faster that, for example, the time sufficient to carry the feather a thousand cubits along the tangent would not be enough to move it a single inch downward along the secant. Now, I say this will never happen, even if you make the former motion as fast as you like, and the latter as slow as you like.[129]

129. This important claim has been made twice on the two preceding pages and is reiterated in the next several pages. There are four main questions to ask. First, how does Galileo support this claim? For now, let us note simply that he supports it with three lines of reasoning, not just one; that these three arguments are mathematical; but that the claim is a physical statement about nature. Second, how does he use the claim? It is clearly the crucial claim in his second criticism of the extrusion objection, in the sense that if it is true then bodies could not be extruded on a rotating earth and the objection is thereby invalidated; but, as already suggested, it is a stronger claim than he needs, for it is enough that weight and rotational extrusion happen to have values such that the former would de facto overcome the latter (a weaker claim that he has already made). Third, is the claim physically true? Here, most readers think that (according to modern physics) the claim is incorrect because it says that terrestrial bodies would never be extruded regardless of how fast the earth was rotating, which is clearly false. But to be more precise, we should distinguish two types of extrusion: (1) leaving the earth's surface and going into a geocentric orbit and (2) escaping from the earth's gravitational field. As for orbital extrusion, to see that it could happen as a result of (increased) terrestrial rotation, we may reason as follows: when a body moves in a circle, a centripetal force is required, otherwise it would move in rectilinear uniform motion (by inertia); the centripetal force is directly proportional to the square of the linear speed and inversely proportional to the radius ($F = mV^2/R$); for terrestrial bodies this force is provided by the earth's gravitational attraction, which is inversely proportional to the square of the radius, but does not depend on speed ($F = GmM/R^2$); thus, as the rate of terrestrial rotation increases, the required centripetal force also increases, but

SIMPLICIO. Why could not the motion along the tangent be so fast that the feather would not have the time to return to the earth's surface?

SALVIATI. Try to formulate the question in quantitative terms, and I will answer you. So, tell me how much you think the former motion would have to be faster than the latter.

SIMPLICIO. I will say, for example, that, if the former were one million times faster than the latter, the feather and even the stone would be extruded.

[224] SALVIATI. You say this, which is false, for lack of neither logic nor physics nor metaphysics, but rather of geometry. For, if you only understood its basic principles, you would know that from the center of a circle one can draw a straight line to the tangent such that the segment of the tangent between the contact and the secant is one, two, or three million times longer than the segment of the secant between the tangent and the circumference; further, as the secant approaches the contact, this proportion becomes greater ad infinitum; so there is no danger that the feather (or something lighter) can begin to rise, regardless of how fast is the rotation and slow the downward motion, for the downward tendency always exceeds the speed of projection.

SAGREDO. I do not completely understand this business.

SALVIATI. I will give you a demonstration that is both most general and very easy. Consider the ratio of line BA to line C, and let BA be longer than C as much you wish; and let D be the center of the circle from which one must draw a secant such that the ratio of the tangent to

---

the available gravitational attraction remains constant; so the point would be reached when the former exceeds the latter and orbital extrusion occurs; the minimum orbital speed $V_o$ is such that $V_o^2 = GM/R$; but once the body is in orbit, any increase in terrestrial rotation does not affect its velocity, and so it simply remains in orbit and does not escape. To escape, a body must be given enough speed so that it will never be pulled back to earth, namely, so that it can reach an infinite distance from the earth, so to speak; that is, the body's kinetic energy must be sufficient to do the work required to move it from the earth's surface to infinity; this work is equal to the body's change in potential energy; setting potential energy to zero at infinity, this change equals the body's *initial* potential energy; that is, the escape velocity is such that $(mV_e^2)/2 = GmM/R$; or, in terms of orbital velocity, $V_e^2 = 2V_o^2$; that is, the escape velocity is greater than the orbital velocity (by a factor of the square root of two), and so the escape velocity will never be reached by just increasing terrestrial rotation. For more details, see Chalmers and Nicholas (1983), Drake (1986a), Hill (1984), MacLachlan (1977), and Pagnini (1964, 2:383n.1 and 387n.1). Finally, independently of the physical truth or falsehood of Galileo's claim, we may ask whether his supporting reasoning is correct; clearly, if the claim is false, that does not tell us where the error in his reasoning lies; and if the claim is true, that does not mean that the supporting argument is correct; as always, the truth value of a conclusion does not by itself determine the inferential soundness of an argument.

the exsecant[130] is the same as BA to C [see fig. 4]. Now, construct the third proportional[131] IA to BA and C; then in the same proportion as BI to IA, extend the diameter FE by EG; and from G draw the tangent GH. I say this is what we needed: GH is to EG as BA is to C. For, FE is to EG as BI is to IA; therefore, by composition,[132] FG is to EG as BA to IA; now, C is the mean proportional[133] between BA and IA, and GH is the mean proportional between FG and EG;[134] therefore, as BA is to C, so is FG to GH, namely, GH to EG. This is what had to be done.[135]

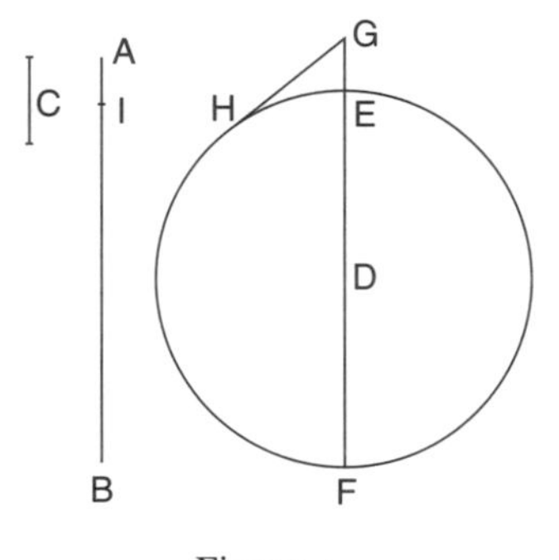


Figure 4.

130. Galileo's word is *segante* (meaning secant), but he clearly intends *exsecant*; an exsecant is defined as the part of a secant external to the circle and thus between the circumference and the tangent. Salviati's previous speech is clear in this regard, although it does not use the word exsecant.

131. The third proportional to two quantities $a$ and $b$ is a third quantity $c$ such that $a:b = b:c$; so this construction is such that BA:C = C:IA. That is, he takes a segment within BA which is as much smaller than C as C is smaller than BA.

132. "Composition" involves adding EG/EG and IA/IA to the two respective sides of the previous equation, FE/EG = BI/IA, to yield [(FE + EG)/EG] = [(BI + IA)/IA]; this in turn is equivalent to the consequence drawn in the next clause.

133. The mean proportional between $a$ and $c$ is the quantity $b$ such that $a:b = b:c$; this is not the same as the third proportional, for when a quantity $b$ is the mean proportional between two others $a$ and $c$, the last quantity $c$ is the third proportional to the other two, $a$ and $b$. The point is that in the proportion $a:b = b:c$, $b$ is called the mean proportional and $c$ the third proportional.

134. The reason why GH is the mean proportional between FG and GE involves an application of the Pythagorean theorem to the right triangle GHD. Here the English translation in Galileo (1967, 198) reads "since C is the mean proportional between BA and AI, GH is the mean between FG and GE," thus giving the wrong impression that this second mean proportionality can be derived from the first; this is one of many examples where insufficient attention to the reasoning indicators renders understanding impossible.

135. This passage contains a correct mathematical proof of a geometrical theorem—specifically of the proposition that in the neighborhood of the point of tangency, as one approaches that point, the ratio of the tangent segment to the exsecant grows without

SAGREDO. I comprehend this demonstration. However, it does not entirely remove all my scruples; instead I feel some confusion whirling in my mind; like a dense and dark fog, it prevents me from discerning the clarity and necessity of the conclusion with the lucidity that is usually appropriate in mathematical reasoning. What confuses me is this. It is true that the distances between the [225] tangent and the circumference go on diminishing ad infinitum as one approaches the contact; but, on the other hand, it is also true that the propensity of the movable body to descend becomes smaller and smaller as it finds itself closer and closer to the initial point of its descent, namely, the state of rest; this is evident from what you told us[136] when you showed that the falling body starting from rest must pass through all degrees of slowness intermediate between rest and any given degree of speed, which are smaller and smaller ad infinitum. Add to this that the speed or propensity to motion goes on diminishing ad infinitum also for another reason; this stems from the possibility of diminishing ad infinitum the gravity of the movable body.[137] Thus, there are two causes that diminish the propensity to descend and consequently favor the projection, namely, the lightness of the movable body and the proximity to the point of rest; and both can increase ad infinitum. On the other hand, they are opposed by a single cause for preventing[138] the projection; and, although it likewise can increase ad infinitum, I do not comprehend how it, being only one, should not be overcome by the union and combination of the other two, which can also increase ad infinitum.

---

limit. The proof is based on two premises which are true by construction, namely, that BA:C = C:IA and FE:EG = BI:IA. From this mathematical truth Galileo infers the physical claim that, however large the extruding tangential tendency or speed is, it will always be insufficient to overcome the downward secant tendency or speed. This inference assumes that the tangential speed may be identified with the tangent segment GH, and the downward speed with the secant segment GE. The difficulty lies in this identification, for perhaps these lines should be identified only with physical distances rather than with speeds. This critical interpretation helps to make sense of the methodological discussion that follows, which is about the problem of applying mathematics to physical reality.

136. This is discussed earlier in the *Dialogue*, in a passage that is omitted here (Favaro 7:45–47 and 51–53; Galileo 1967, 21–23 and 27–29).

137. This possibility corresponds to Aristotle's idea that lighter bodies fall slower than heavier ones, which view Galileo rejected and replaced with the principle that all bodies fall at the same rate, independently of weight. At the end of this discussion ([228]), Salviati makes it clear that he considered this possibility for argument's sake, not because he (or Galileo) believes it.

138. Here Galileo speaks of *accomplishing* the projection (*far la proiezione*); I take this to be a slip of the pen, unnoticed by previous editors.

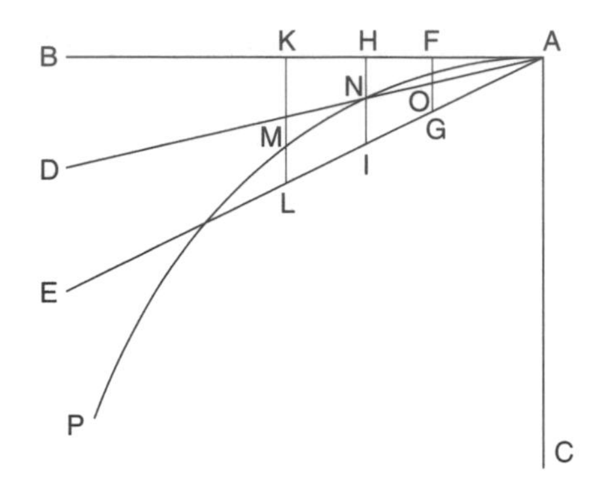


Figure 5.

SALVIATI. This is a difficulty worthy of Sagredo. Since you say you are still somewhat confused about it, in order to elucidate it, so that we can understand it more clearly, let us clarify it by drawing a diagram [fig. 5];[139] this will perhaps also help us to resolve it. Let us then draw a line, namely AC, directed perpendicularly toward the center, and let the horizontal AB meet it at right angles; the motion of projection would be along the latter and the projectile would continue to advance with uniform motion if gravity did not incline it downwards.

Now, from point A draw a straight line, namely AE, intersecting AB at an arbitrary angle; along AB let us then mark some equal distances, AF, FH, and HK; and from these let us draw the perpendiculars FG, HI, and KL as far as AE. As stated earlier, the heavy body falling from rest acquires a constantly greater degree of speed as time goes on, in accordance with the increase of the time elapsed; thus, we can imagine the lines AF, FH, and HK to represent equal times, and the perpendiculars FG, HI, and KL to represent the degrees of speed acquired in these times; and so the degree of speed acquired [226] during the entire period AK is to the degree acquired during the period AH as KL is to HI,

139. From the viewpoint of the history of science and mathematics, figure 5 embodies an important advance; this is the technique of representing the values of two magnitudes as points along two perpendicular axes and representing their functional interdependence as a line in the plane defined by these axes. The technique was partly anticipated by Nicole Oresme (c. 1320–1382), and after Galileo it became common. As we shall soon see, he commits several errors, one of which is the equivocation of letting the vertical axis represent both speeds and distances; but these errors, too, can be instructive insofar as they illustrate the difficulties involved in such advances.

and to the degree acquired during the period AF as KL is to FG; these degrees of speed KL, HI, and FG (as is evident)[140] are to each other as the times KA, HA, and FA; if other perpendiculars are drawn from other points arbitrarily chosen along line FA, we will always find degrees of speed which are smaller and smaller ad infinitum as we approach point A, which represents the first instant of time and the initial state of rest. This convergence toward A represents the first infinite diminution of the propensity to downward motion due to the movable body approaching the initial state of rest, which approach can be increased ad infinitum.

Let us now look for the other diminution of the speed, which can also go on ad infinitum and is due to the diminution of the gravity of the movable body. This can be represented by drawing other lines from point A that generate angles smaller than angle BAE, for example, AD; by cutting the parallels KL, HI, and FG at points M, N, and O, this line enables us to represent the degrees of speed FO, HN, and KM, which are acquired during the times AF, AH, and AK and which are smaller than the other degrees of speed FG, HI, and KL acquired during the same times; but the latter involve a heavier body, the former a lighter one. It is evident that by bringing line EA toward AB and narrowing angle EAB (which can be done ad infinitum, just as gravity can be diminished ad infinitum), we likewise diminish ad infinitum the speed of the falling body and consequently the cause that prevents projection.

So, it seems that, by diminishing ad infinitum both of these two causes opposing projection, the projection could not be prevented. Reducing the whole argument to brief words, we can say: by narrowing the angle EAB, we diminish the degrees of speed LK, IH, and GF; further, by pulling the parallels KL, HI, and FG toward the angle at A, we again

140. It is indeed evident from figure 5 that the instantaneous velocity (degree of speed) increases in direct proportion to the time elapsed. But if this assertion is interpreted as a physical principle about the actual behavior in nature of bodies falling freely from rest, the claim was far from self-evident; in fact, Galileo had been struggling with this problem for about twenty years before his astronomical discoveries in 1609. Some of the difficulties he had to overcome were to determine that the instantaneous velocity is not simply proportional to the *distance* fallen (but rather to its square root), that space-proportionality is *not* equivalent to time-proportionality, and that time-proportionality implies that the distance fallen from rest varies as the square of the time elapsed (the law of squares); eventually he was able to establish that falling bodies accelerate in such a way that their acquired velocity is directly proportional to the time elapsed, and this demonstration represents one of his major scientific achievements. He did not publish the full details of this until his *Two New Sciences* of 1638, but he had essentially completed the research thirty years earlier; so we find various bits of these ideas in the *Dialogue* (Favaro 7:45–54 and 244–60; Galileo 1967, 21–30 and 218–33).

diminish the same degrees of speed; therefore, the speed of downward motion can be diminished so much (being subject to a twofold diminution ad infinitum) that it is not sufficient to bring the body back onto the circumference of the wheel, and consequently to ensure that the projection is prevented or overcome.[141]

On the other hand, to make sure that projection does not happen, it is necessary that the distances through which the projectile descends to reunite with the wheel should become so short and narrow that the descent of the body would be sufficient to bring it back there, regardless of its slowness and its diminution ad infinitum; so it would be necessary to find a diminution of these distances that not only would [227] continue ad infinitum, but also such that its infinity would exceed the double infinity that characterizes the diminution of the speed of the falling body. But how can a magnitude be diminished more than another that is diminished ad infinitum in two ways? Now, let Simplicio note to what extent one can philosophize well about nature without geometry.

The degrees of speed are always determinate, although they diminish ad infinitum due to the diminution of the body's gravity as well as to the approach to the first point of motion (namely, the state of rest); they correspond proportionately to the parallels lying between the two straight lines converging at an angle, which may be the angle BAE or BAD or some other angle that is infinitely more acute but still always rectilinear. However, the diminution of the distances over which the body has to go in order to return to the circumference of the wheel is proportionate to another sort of diminution taking place between lines converging at an angle that is infinitely more narrow and acute than any rectilinear acute angle whatever; it is the following. Take any point you wish C along the perpendicular AC; using C as center, draw the arc AMP[142] at the distance CA; this arc will cut the parallels representing the degrees of speed, regardless of how small they are and regardless of how

141. Salviati's speech so far, together with Sagredo's previous one, is a statement of an objection to the previous proof of the mathematical impossibility of extrusion. Although Salviati proceeds to refute this objection, its presence is an indication that Galileo had some doubts regarding the soundness of his claim about the mathematical impossibility of extrusion. Moreover, as we shall see, that refutation provides two other mathematical reasons why extrusion is supposedly impossible; thus, despite his rhetoric to the contrary, Galileo may have been unsure about that claim, for in mathematical reasoning a single proof is sufficient and decisive (if it is really valid).

142. The label AMP suggests that the arc would necessarily intersect point M, which had been previously defined as the intersection of lines AD and KL; this suggestion is incorrect, and Galileo's own diagram corrects this suggestion by showing the arc intersecting line KL at a different point. But figure 5 is misleading in another regard, namely, inso-

narrow is the rectilinear angle within which they lie; the portions of these parallels lying between the arc and the tangent AB are the magnitudes of the distances necessary for returning to the wheel; they are always smaller than the parallels of which they are parts, and the more they approach the point of contact the more they become proportionately smaller than these same parallels.[143]

Moreover,[144] as one moves inward into the angle, the parallels lying between the two straight lines diminish always in the same proportion;

---

far as it shows the arc intersecting point N, which had been previously defined as the intersection of lines AD and HI; and in general, the arc need not intersect point N. However, both errors are minor oversights because Galileo's argument does not actually exploit these two features of the wording and diagram.

143. The best critical interpretation of this subargument has been advanced by Hill (1984, 121): "It contains an interesting and fairly well-concealed fallacy which can be characterized either as a *non sequitur* partly disguised by the vagueness of a key term or as a classical equivocation on that term. Galileo successfully argues that as we approach the point of contact, *A*, the distances which need to be covered to prevent projection necessarily vanish more quickly than the speeds of fall. But this does not imply that centripetal tendencies must overwhelm centrifugal tendencies. To prove this Galileo would have to show that the distances which *need* to be covered to prevent projection necessarily vanish more quickly than the distances a falling body would *actually cover* (as the point of contact is approached). This, however, cannot be established. These two distances vanish at the same rate, both being as the square of the speeds (and times). This may also be construed as an equivocation because Galileo uses the vague term 'downward tendency' or 'tendency toward downward motion' in such a way as to fuzz the distinction between speed of fall and distances covered in fall." I believe that Hill is essentially correct, although at this point his account does not provide a sufficiently detailed analysis and an adequate textual documentation; but my hunch is that these could be provided. Moreover, although Hill's account of this argument is penetrating, it is somewhat one-sided insofar as he does not pay equal attention to the other parts of Galileo's argument about extruding power. Finally, Hill uses his critical interpretation to motivate a series of considerations that lead him to a rhetorical interpretation of this passage, and while this is interesting from the viewpoint of rhetoric, I do not find it convincing. At any rate, Hill (1984, 121n.35) makes another insightful remark when he further characterizes the error by saying that "as I see it, Galileo has *misapplied* his geometrical apparatus, contrasting the diminishing exsecants with the *wrong set of lines*"; this statement is methodologically important because it paves the way for appreciating that he is involved in an analogous misapplication of geometry in the preceding subargument, and that in the following major portion of this passage he shows some awareness of these difficulties when he engages in a long and explicit methodological reflection about the relationship between mathematical truths and physical reality. See also Chalmers and Nicholas (1983, 321) and Gaukroger (1978, 189–93).

144. Here Galileo gives two distinct reasons for his conclusion, and these reasons involve two quantities which presumably become infinitely small as we approach the point of tangency. As just asserted by Salviati, the first of these quantities is the ratio of a parallel segment lying above the arc to the whole parallel lying between the two straight lines; this reason is mathematically correct, although Galileo's physical interpretation is erroneous. The second quantity, as he goes on to argue, is the ratio between the next parallel segment above the arc and the previous parallel segment; as explained in the next note, this ratio does not diminish without limit but tends to the limit of one-fourth.

for example, having divided AH in the middle at point F, the parallel HI will be double the parallel FG; if we subdivide FA in half, the parallel drawn from the dividing point will be half of FG; and continuing the subdivision ad infinitum, the subsequent parallels will always be half of the immediately preceding ones. However, this is not what happens with the lines lying between the tangent and the circumference of the circle; for, making the same subdivision of FA, if for example the parallel originating from point H were double the one originating from F, the latter will then be more than double the following one; and, continuing, as we come toward the contact A, we will find that the preceding lines can contain the following ones three, four, ten, one hundred, one thousand, one hundred thousand, and one hundred million times, and more ad infinitum.[145] Thus, the shortness of these lines is reduced to such an extent that it greatly exceeds the need [228] to make sure that the projectile (however light it might be) returns to the circumference, or rather, remains there.

SAGREDO. I understand very well the whole argument and the strength that ties it together. However, I think that if one wanted to belabor it further, one could advance a difficulty. That is, of the two causes that render the body's descent slower and slower ad infinitum, it is evident that the one which depends on the proximity to the initial point of descent grows always in the same proportion, just as the aforementioned parallels always maintain the same proportion among themselves; but it is not equally evident that the diminution of the same speed which depends on the diminution of the body's gravity (which was the second cause) takes place also in the same proportion. Who assures us that it does not take place in accordance with the proportion of the lines interposed between the tangent and the circumference, or even with a greater proportion?

145. This mathematical assertion is incorrect, as MacLachlan (1977, 176) and Hill (1984, 120) have pointed out. For, near the point of tangency, if the distance along the tangent is halved, the distance between tangent and circumference is quartered. That is, if near the point of tangency we consider the series of parallel segments between tangent and circumference such that the next segment is half as far from the point of tangency (measuring along the tangent) as the previous segment, then the ratios of one segment to the next converge to the limit of four; this ratio does not become infinite. Again, these segments do get smaller and smaller, and are infinitely small at the point of tangency; and they do get smaller at a faster rate than their distance from that point; but their rate of decrease has a limit, namely, twice the rate at which the distance from that point decreases. Finally, we may also say that these parallels are to each other as the square of their respective distances from the point of tangency. Here Galileo seems to have committed a purely mathematical error involving a failure of his mathematical intuition. The effect of this error is to render groundless the subargument he advances for the conclusion he draws in the next sentence.

SALVIATI. I had taken it as true that the speed of naturally falling bodies is proportional to their gravity for the sake of Simplicio and of Aristotle, who in several passages affirms it as an evident proposition.[146] You, playing the devil's advocate, cast doubt on this and assert that maybe the speed grows by a greater proportion than gravity, perhaps greater ad infinitum; hence, the whole earlier argument would fall to the ground. To support it, it remains for me to say that the speed grows by a much smaller proportion than gravity, thus not only upholding but strengthening what I said; as proof for this I refer to experience, which will show that a body thirty or forty times heavier than another (as, for example, a lead ball would be in relation to one of cork) will hardly move even twice as fast.[147] Now, if projection would not happen even when the speed of the falling body diminishes in proportion to gravity, still less will it happen when the speed decreases little as the result of the weight having been greatly reduced.

But let us assume that the speed were to diminish by a much greater proportion than gravity and that this proportion were the same as that by which are diminished the parallels between the tangent and the circumference; even so, I do not grasp any necessary reason which would persuade me that the projection must happen for bodies which are extremely

146. Aristotle, *Physics*, IV, 8, 216a12–16; cf. the Introduction (2.2).

147. Here Galileo makes a relatively modest claim about the connection between the speed of fall and the weight of a body. Later in the *Dialogue*, in a passage that is omitted here (Favaro 7:249–50; Galileo 1967, 222–23), he makes the stronger claim that weight is completely irrelevant to the speed of fall. Although this stronger claim is true only in the ideal case when all air resistance is absent, it is the theoretically more important claim, being for example a consequence of the basic principle that the speed of a falling body increases in direct proportion to the time of fall; such theoretical details were later published by Galileo in the *Two New Sciences*. In this passage he appeals to experience; this appeal is unobjectionable so far as it goes, although the issue could not be decided by mere empiricism. On the other hand, in that later work (Galileo 1974, 66–67) he also gave an elegant theoretical refutation of Aristotle's law of fall; this can easily be taken as an indication of apriorism if considered in isolation, but it is best taken as a classic instance of critical reasoning. Finally, the reference to Aristotle's law and the explicit mention of weight in this speech make it clear that all the Galilean talk of gravity in the rest of this discussion refers to the hypothetical possibility that the downward tendency decreases when we consider terrestrial bodies that are lighter than others; it would be a mistake to read into his talk of gravity a groping toward Newtonian gravitation, which is less for bodies with less mass and for the same masses at greater distance from each other. That is, from a Newtonian viewpoint, if the earth were less massive or its radius were bigger, the centripetal tendency of terrestrial bodies would be less; so gravitational force may be properly regarded as a variable quantity for the purposes of comparing the centripetal and centrifugal tendencies, of computing whether one could overcome the other, and of determining the exact point when the two exactly balance each other.

light or as light as you wish; [229] but we must understand that we are referring to bodies which are not light strictly speaking (namely, which are devoid of any weight and which by their nature go up), but to bodies which descend extremely slowly and have very little weight. Instead I claim that the projection will not happen. What moves me to believe that is this: the diminution of gravity in proportion to the parallels between the tangent and the circumference has zero weight as the final and smallest term, just as the last term in the diminution of those parallels is the contact, which is an indivisible point; but gravity never diminishes to the final term because then the body would have no weight; on the other hand, the distance necessary for the projectile's return to the circumference does reduce to the final smallness, which occurs when the body lies on the circumference at the point of tangency, so that to return there it does not need to descend a finite distance; thus, given a propensity to downward motion as small as you wish, it is always more than sufficient to return the body to the circumference from which it is separated by the minimum distance, namely, nought.[148]

148. This argument completes Galileo's second criticism of the extrusion objection. What does this criticism amount to? Perhaps his charge is that this objection is *quantitatively* invalid, namely, that its conclusion does not follow from its premises for reasons involving quantitative considerations. For, he agrees with the basic premise about centrifugal force, namely, with the existence of an extruding power of whirling in general; and he agrees that this generalization implies that on a rotating earth objects would experience *some* centrifugal tendency; but he objects that it does not follow that these objects will move off the surface or be extruded or be projected into outer space because the amount of this centrifugal tendency is insufficiently large to overcome the downward tendency due to weight; he supports this last claim in part by suggesting a factual contingent comparison of these two tendencies, yielding a negative and reassuring result; and he also thinks he can prove this claim mathematically, so that extrusion on a rotating earth would be a mathematical impossibility. This mathematical impossibility he tries to prove on the basis of the geometry of the situation in the neighborhood of the point of contact between a circle and a tangent, involving the following considerations as one approaches the point of tangency: (1) the ratio of an exsecant to the corresponding tangent segment tends toward zero, (2) the ratio of an exsecant to the corresponding speed of fall tends toward zero, and (3) the ratio of an exsecant to the previous one (at twice its distance from the point of tangency) tends toward zero; that is, the exsecants get smaller and smaller in relation to (1) the corresponding tangent segments, (2) the speeds of fall, and (3) each other; thus, the distances of fall required to prevent extrusion become infinitely smaller than the distances required to achieve extrusion, and they decrease more rapidly than the speeds of fall. By way of evaluation, I would say first that Galileo's criticism is valuable in some ways insofar as it introduces quantitative comparisons into the discussion; insofar as the physical quantities being compared are the right ones (centrifugal and centripetal tendencies); insofar as he gets their directions right (along the tangent and along the secant); and insofar as it is physically true that on a rotating earth the centrifugal tendency is insufficient to overcome the centripetal gravitational one. But as mentioned in previous notes, his criticism goes astray in

SAGREDO. The reasoning is really very subtle, but equally conclusive; one must admit that to want to treat physical questions without geometry is to attempt to do the impossible.

SALVIATI. However, Simplicio will not say so,[149] although I do not think he is one of those Peripatetics who dissuade their pupils from studying mathematics on the grounds that it degrades the intellect and renders it less fit for theorizing.

SIMPLICIO. I would not do such an injustice to Plato, but I would agree with Aristotle that he became excessively enamored of geometry and excessively absorbed with it.[150] For, Salviati, ultimately these mathematical subtleties are true in the abstract, but when applied, they do not correspond to sensible physical matter; for example, mathematicians may well demonstrate with their principles that a sphere touches a plane at a single point,[151] which is a proposition like the present one; but, when we deal with matter, things proceed otherwise;[152] and so I want to say that all those contact angles and proportions come to nothing when we deal with sensible material things.

SALVIATI. So you do not believe that a tangent touches the surface of the terrestrial globe at a point?

SIMPLICIO. Not at a single point; instead I believe that [230] a straight line touches the surface of the ocean (let alone that of the land) for many and many tens and perhaps hundreds of cubits before separation.

SALVIATI. But do you not notice that, if I grant you this, it is much worse for your cause? For, supposing that the tangent is separated from

---

several ways: he makes at least one important mathematical error (point 3 here); he misapplies a number of mathematical truths by assuming improper correspondences between physical quantities and geometrical lines; and he tries to give a purely mathematical proof of a factual claim, and this must be wrong-headed.

149. The discussion that begins here and continues for the next several pages may look like a digression from the critical analysis of the extrusion objection. Even if this were the case, we would have a discussion which is important in its own right, for methodological reasons; for Sagredo's previous speech expresses an important methodological principle about the relationship between mathematics and physical reality. But the digression may be more apparent than real, and it is instructive to ask whether there is a connection between this discussion and Galileo's second criticism (just concluded) and his third criticism (which will follow this discussion at the end of this selection). I believe there is a connection, partly because both criticisms amount to charging the extrusion argument with quantitative invalidity, and partly because he may be showing his awareness that the mathematical argument that is a key part of the second criticism is problematic and not as relevant or physically binding as some of his own language makes it seem.

150. Cf. Aristotle, *On Generation and Corruption*, I, 2, 316a5–14.

151. Galileo writes this clause in Latin: *sphaera tangit planum in puncto*.

152. Cf. Aristotle, *Metaphysics*, III, 2, 997b34–998a7.

the earth's surface at all points but one, still it was demonstrated that its angle of convergence (if you can call it an angle) is so small that the projectile would not move off; so, there would be much less reason for it to move off if that angle vanishes and the surface and tangent proceed united. Do you not see that in this manner the projection would take place along the earth's surface, which is the same as saying that it would not take place? So you see how great is the power of truth: while you try to bring it down, your own assaults raise and strengthen it.[153]

However, since I have freed you from this error, I would not want to leave you committing that other one, namely, thinking that a material sphere does not touch a plane at a single point. I hope that your conversation (even for just a few hours) with persons knowledgeable in geometry will make you appear a little more intelligent among those who know nothing about it. Now, let me show you how great is the error of those who say that, for example, a bronze sphere does not touch a steel plane at a single point; so tell me what you would think of someone who should say and constantly maintain that a sphere is not really a sphere.

SIMPLICIO. I would consider him to be out of his mind.

SALVIATI. Such is the state of someone who says that a material sphere does not touch a material plane at a single point because this is the same as saying that a sphere is not a sphere. To show that this is true, tell me what you think constitutes the essence of a sphere, namely, what is the property which distinguishes a sphere from all other solid bodies.

SIMPLICIO. I think that being a sphere consists of the property that all the straight lines drawn from its center to the circumference are equal.

SALVIATI. Thus, if these lines were not equal, that solid would in no way be a sphere.

SIMPLICIO. No, sir.

SALVIATI. Tell me next whether you think that, of the many lines which can be drawn between two points, there can be more than one which is straight.

SIMPLICIO. No, sir.

[231] SALVIATI. But you also understand that this unique straight line will necessarily be the shortest of all.

153. This paragraph is a reminder that the present discussion about the applicability of mathematics is not a mere digression but connects in various ways with the critical examination of the extrusion objection.

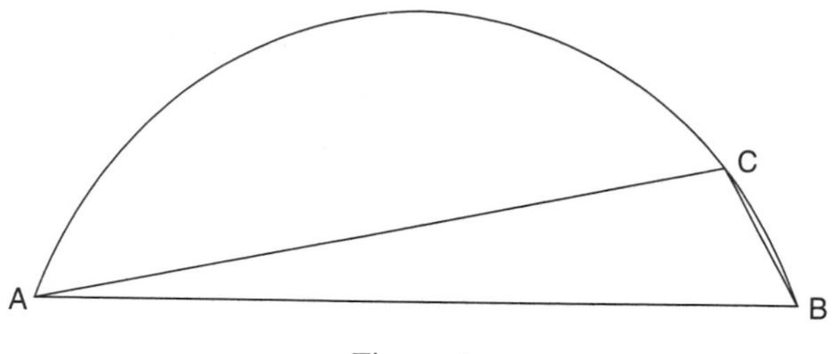


Figure 6.

SIMPLICIO. I understand it, and I even have a clear proof for it, which has been advanced by a great Peripatetic philosopher;[154] I think, if I remember correctly, that in the process he criticizes Archimedes, who takes it as self-evident even though it can be demonstrated.

SALVIATI. He must have been a great mathematician if he was able to demonstrate something that was beyond Archimedes' know-how and ability. If you recall the demonstration, I would like to hear it; for I remember very well that in his books on the sphere and cylinder, Archimedes places this proposition among the postulates, and I firmly believe he regarded it as indemonstrable.

SIMPLICIO. I believe I will be able to remember it because it is very easy and brief.

SALVIATI. That much greater will then be the shame of Archimedes and the glory of this philosopher.

SIMPLICIO. I will draw a diagram [fig. 6]. Between points A and B draw the straight line AB and the curve ACB; one wants to prove that the straight line is shorter. The proof is this. Take a point on the curve (for example, C), and draw two other straight lines, AC and CB; these two together are longer than AB alone, as Euclid[155] demonstrates; but

154. Drake (1967, 479n) identifies this philosopher as Francesco Buonamici and refers to Barenghi (1638, 11); Buonamici (1533–1603) taught at the University of Pisa from 1565 to 1603 and was one of Galileo's teachers there. Here begins a digression within a digression, dealing with the status of the claim that a straight line is the shortest distance between two points; the question is whether this claim is provable or must be assumed as an unproved axiom.

155. Euclid of Alexandria was a Greek mathematician and physicist who lived around 300 B.C.; he is the author of the *Elements* of geometry, one of the most influential books ever written. This book is not only a compendium of geometrical knowledge that is theoretically fruitful and practically useful, but it also provides the classical illustration of the axiomatic organization of knowledge; that is, one starts with a few simple definitions and a few unproved but self-evident principles, and then one derives all other truths by rigorous demonstration from them or from other previously proved propositions. The Euclidean theorem used here is the proposition that in a triangle any two sides together are longer than the third.

the curve ACB is longer than the two straight lines, AC and CB; therefore, we can be even more certain that the curve ACB is much longer than the straight line AC; and this is what had to be demonstrated.

SALVIATI. I do not think that, if one were to search through all the paralogisms in the world, one could find anything more appropriate than this to give an example of the most solemn fallacy among all fallacies, namely, to prove the unknown by means of what is more unknown.[156]

SIMPLICIO. How so?

SALVIATI. What do you mean, how so? Is not the unknown conclusion (which you want to prove) the proposition that the curve ACB is longer than the straight line AB? Is not the middle term (which is regarded as known) the proposition that the curve ACB is longer than both AC and CB together (which are known to be longer than AB)? But, if it is unknown to you that the curve is longer than the straight line AB alone, will it not be much more unknown that the curve is longer than the two straight lines AC and CB (which are indeed known to be longer than AB alone)? And yet you regard that as known.

SIMPLICIO. I do not understand too well wherein lies the fallacy.

[232] SALVIATI. Since the two straight lines are longer than AB (as we know from Euclid), if the curve were longer than the two straight lines A-C-B, would it not be much longer than the straight line AB alone?

SIMPLICIO. Yes, sir.

SALVIATI. The conclusion is that the curve ACB is longer than the straight line AB; this is better known than the middle term, which is the proposition that the same curve is longer than the two straight lines AC and CB; now, when the middle term is less well known than the conclusion, one is committing the fallacy of proving the unknown by means of what is more unknown.[157]

156. Galileo writes this last phrase in Latin: *ignotum per ignotius*.

157. This proof is an even better example of the fallacy of begging the question, although the two fallacies are obviously related to each other, as they are also related to the fallacy of proving the unknown by means of what is *equally* unknown; thus, it is instructive to compare Galileo's criticism of this proof with his criticism of the argument from actual vertical fall in selection 8 ([166]). The proof is essentially the following argument: (1) curve ACB is longer than line AB because (2) curve ACB is longer than lines AC and CB, and (3) lines AC and CB are longer than line AB; the second premise (3) is agreed to have been demonstrated by Euclid; the first premise (2) is less well known than the conclusion (1) because to know (2) we would have to know that (4) curve ACB consists of curve AC and curve CB, that (5) curve AC is longer than line AC, and that (6) curve CB is longer than line CB; but each of these last two claims is as unknown as the conclusion (1) because all three are arbitrary instances of the axiom in question (that a straight line is the shortest distance between two points); so there is more that is unknown in premise (2) than in the

Now, let us return to our subject. It is enough that you understand that a straight line is the shortest among all the lines that can be drawn between two points. As regards our main conclusion, you say that a material sphere does not touch a plane at only one point; how then will it touch the plane?

SIMPLICIO. With a part of its surface.

SALVIATI. And regarding a similar contact by another sphere equal to the first, will it also touch the plane with a similar part of its surface?

SIMPLICIO. There is no reason why it should not be so.

SALVIATI. Therefore, if the two spheres touch, they will touch with the same two parts of their surfaces; for, as each adjusts to the same plane, it is necessary that they should also adjust to one another. Imagine now two spheres, having centers A and B, and touching one another; connect their centers by means of the straight line AB, which passes through the area of contact at point C [fig. 7]. Take another point, D, in the area of contact, and draw the two straight lines AD and BD, so as to form the triangle ADB; its two sides AD and DB together are equal to the other side ACB alone since the former as well as the latter comprise two radii, which are

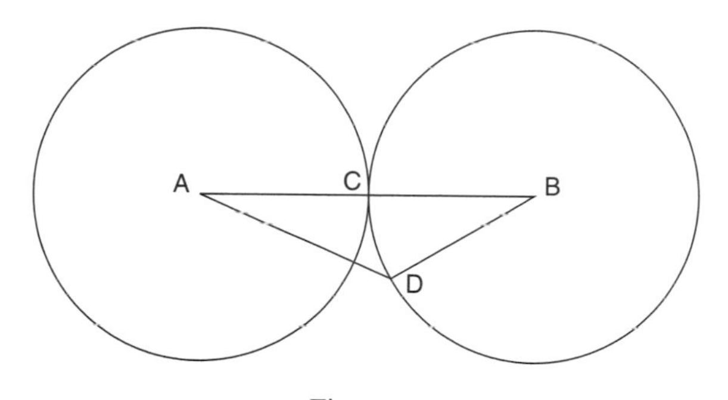


Figure 7.

conclusion (1); and this is the connection with the fallacy of proving the unknown by means of what is *more* unknown. The same considerations show that the proof begs the question and that it attempts to prove the unknown by the *equally* unknown; to see this we should focus, respectively, on the fact that propositions (5) and (6) presuppose the axiom and the fact that they are as questionable as this axiom. Note that Galileo calls the first premise, (2), "the middle term," which is a deviant usage; for by the standard definition, the middle term is the phrase "longer than lines AC and CB," whereas my modified definition (see glossary) does not apply; perhaps he meant "minor premise" rather than middle term, for that would fit; or perhaps he had in mind a more complicated argument that would add another premise in the above reconstruction, namely, the proposition that (7) if $x$ is longer than $y$ and $y$ is longer than $z$ then $x$ is longer than $z$; in this case *both* the first and second premises, (2) and (3), would be middle terms in the modified sense of being bridges between this more general premise (7) and the conclusion (1).

all equal by the definition of a sphere; thus, the straight line AB drawn between the two centers A and B is not the shortest of all since the line ADB is equal to it; and this is absurd by your own acknowledgment.[158]

SIMPLICIO. This demonstration is conclusive for abstract spheres but not for material ones.[159]

SALVIATI. Find then wherein lies the fallacy in my argument, given that it is not conclusive for material spheres, but only for immaterial and abstract ones.

[233] SIMPLICIO. Material spheres are subject to many contingencies to which immaterial ones are not subjected. Why can it not happen that when a metallic sphere is placed upon a plane, its weight should press down in such a way that the plane yields a little, or that the sphere itself should be indented at the point of contact? Moreover, the plane can hardly be perfect, if for no other reason than that matter is porous; and perhaps it will be no less difficult to find a sphere so perfect that all its lines from the center to the surface are exactly equal.

SALVIATI. Oh, I easily grant you all these things, but they are quite irrelevant. For you want to show me that a material sphere does not touch a material plane at only one point; but you use a sphere that is not a sphere and a plane that is not a plane since (as you say) these things are not found in the world, or if they are found then they get damaged when used to produce the effect. Therefore, it would have been better for you to concede the conclusion, but to do so conditionally;[160] that is, to concede that if there should be any material spheres and planes and if they should be and remain perfect, then they would touch at a single point; you could have then denied that the antecedent condition could happen.

158. This proof is an instance of *reductio ad absurdum*: it wants to prove that a sphere touches a plane at a single point; to do this, it assumes that this proposition is false; that is, it assumes that a sphere can touch a plane at more than one point; from this assumption it derives an absurd consequence; the absurdity is that a straight line is not the shortest distance between two points; so it concludes that the assumption must be false, namely, that it is false that a sphere can touch a plane at more than one point; but this is to say that a sphere touches a plane at a single point, which is the proposition to be proved. Like all cases of *reductio ad absurdum*, this proof assumes that the absurdity is really an absurdity; in this case, that it is indeed absurd to deny that a straight line is the shortest distance between two points.

159. This objection is the same one that Simplicio made to Salviati's proof that on a rotating earth the downward tendency along the secant would always suffice to overcome the centrifugal tangential tendency. The proposition about spheres and planes was introduced as an easier example than the various mathematical claims about what happens to exsecants as we approach the point of tangency. But the purpose is the same: to illustrate the methodological problem of the relationship between mathematics and physical reality.

160. Galileo's word is *condizionatamente*, a very important word for understanding his methodological claim here.

SIMPLICIO. I believe the proposition of the philosophers should be understood in this sense, for there is no doubt that the imperfection of matter is such that things taken in the concrete do not correspond to things considered in the abstract.

SALVIATI. What do you mean, "do not correspond"? On the contrary, what you are saying yourself at the moment proves that they correspond exactly.

SIMPLICIO. In what way?

SALVIATI. Are you not saying that, due to the imperfection of matter, that body which should be perfectly spherical and that plane which should be perfectly plane do not after all turn out to be such in the concrete, as others imagine them to be in the abstract?

SIMPLICIO. That is what I am saying.

SALVIATI. Therefore, whenever in the concrete you bring together a material sphere and a material plane, you bring together an imperfect sphere and an imperfect plane; and you say that these do not touch at only one point. However, I tell you that even in the abstract an immaterial sphere which is not perfectly spherical can touch an immaterial plane which is not perfectly plane with a part of its surface, rather than at a single point; thus, so far, what happens in the concrete happens in the same manner in the abstract. It would indeed be news if the computations and accounts [234] made with abstract numbers would later not correspond to the gold and silver coins and to the goods in the concrete. But do you know, Simplicio, what happens? It is like this: an accountant must make allowances for containers, straps, and other packing items if he wants his calculations to come out right in regard to sugar, silk, and wool; in the same manner, if a geometrical philosopher wants to find in the concrete the effects demonstrated in the abstract, he must deduct the impediments of matter; if he knows how to do this, I assure you things will agree no less exactly than with arithmetical computations. Therefore, errors derive neither from the abstract nor from the concrete, neither from geometry nor from physics, but from the calculator who does not know how to do the accounting correctly.[161] Hence, if you had a perfect sphere

161. This is another implicit admission by Galileo that there may be a problem with his mathematical model of the relationship between rotational extrusion and downward fall in terms of the geometry of the situation in the neighborhood of the point of tangency between a circle and a line; he may not have done the accounting correctly, to use his own words. But as mentioned before, aside from this possible connection with the previous criticism, this methodological discussion is valuable in its own right and it is not surprising that it has become a subject of great controversy; cf. Clavelin (1974, 404–21), Koyré (1966; 1978), Shea (1972), Wisan (1978), and the appendix (2.5).

and a perfect plane, even if they were material, have no doubt that they would touch at a single point; and if it was and is impossible to have that, then it was quite irrelevant to say that "a bronze sphere does not touch a plane at only one point."[162]

However, I want to ask you one more thing, Simplicio: granted that there can be neither a perfect material sphere nor a perfect material plane, do you think there can be two material bodies whose surface is curved in various parts and ways and extremely irregular?

SIMPLICIO. I think there is no lack of them.

SALVIATI. Given that there are, they will also touch at a single point, for touching at only one point is not at all a particular privilege of perfect spheres and perfect planes. Indeed, if one were to reflect more keenly on this business, one would find that it is much more difficult to have two bodies touching with parts of their surfaces than at only one point; for, to want two surfaces to mesh well together, it is necessary either that both should be exactly plane, or that if one is convex then the other should be concave (but with a curvature corresponding exactly to the convexity of the other); but these conditions are much more difficult to find, due to their excessively fine specifications, whereas the others are infinite in number on account of their wide range.

SIMPLICIO. So you think that two stones or pieces of iron taken at random and brought together would touch at a single point most of the time?

SALVIATI. In casual encounters, I think not, both because [235] on their surface there would probably be some yielding dirt and for the following reason: in bringing them together one would not be careful to avoid a collision, and any minor one would suffice to make one surface yield a little to the other, so that they would mutually take each other's shape, at least to some extent. However, if their surfaces were very clean, if both were placed on a table so that one would not weigh over the other, and if they were pushed very slowly toward each other, I have no doubt that they could be made to come into contact at only one point.

SAGREDO. With your permission, I must bring up a difficulty I have; it stems from hearing Simplicio allege the impossibility of finding a solid material body having a perfectly spherical shape and from seeing Salviati agree with him to some extent and not contradict him. So, I would like to know whether there is the same difficulty in forming a solid with some other shape; that is, to explain myself better, whether it is more

162. Galileo writes this in Latin: *sphaera aenea non tangit in puncto*.

difficult to fashion a piece of marble into the shape of a perfect sphere than into a perfect pyramid or perfect horse or perfect locust.

SALVIATI. I will give the first answer to this. Let me begin by apologizing for my apparent agreement with Simplicio, which was only temporary; for I had had in mind to say, before moving on to another subject, something that is perhaps identical or very similar to your idea. To respond to your first question, I say that a spherical shape is the easiest one of all to give to a solid; it is also the simplest and holds the same place among solid figures which a circle holds among plane ones; because the description of a circle is easier than all others, it alone has been judged by mathematicians worthy of being placed among the postulates regarding the descriptions of all other figures. How easy the construction of a sphere is, may be seen from the following: if one bores a circular hole in a flat plate of hard metal, and over it one turns at random a very crudely rounded solid (as long as this solid is not smaller than the sphere just passing through the hole), then without any other artifice the solid will acquire a spherical shape, which is as perfect as possible; what is even more worthy of note [236] is that with the same hole one can make spheres of different sizes. As regards what is required to fashion a horse or (as you say) a locust, I leave that judgment to you, who know that there are very few sculptors in the world capable of doing it; and I think Simplicio will not disagree with me on this particular.

SIMPLICIO. I do not know whether I disagree with you in any way. My opinion is that none of the mentioned figures can be obtained perfectly; but I think that to approach perfection as much as possible it is incomparably easier to fashion a solid into a spherical shape than into the form of a horse or locust.

SAGREDO. What do you think this greater difficulty depends on?

SIMPLICIO. Just as the great ease of forming a sphere derives from its absolute simplicity and uniformity, so the extreme irregularity renders very difficult the construction of the other figures.

SAGREDO. So, since irregularity causes difficulty, even the shape of a rock casually broken with a hammer will be difficult to produce, for it is perhaps more irregular than that of a horse.

SIMPLICIO. It must be so.

SAGREDO. But tell me, Whatever shape that rock has, does it have that shape perfectly or not?

SIMPLICIO. It has the shape it has so perfectly that no other one fits it so exactly.

SAGREDO. So, if one can most perfectly obtain infinitely many figures that are irregular and hence difficult to produce, what is the reason for saying that it is impossible to obtain the simplest and hence easiest one of all?

SALVIATI. Gentlemen, with your permission, I think we got involved in a dispute that is not much more relevant than a nit-picking quibble; whereas our arguments should continue to be about serious and relevant matters, we are wasting time on frivolous quarrels of no significance.[163] Please, let us remember that to inquire into the constitution of the universe is one of the greatest and noblest problems in nature; this is all the more significant insofar as it is directed toward the solution of another one, namely, the cause of tides,[164] which has been investigated by all great men who ever lived and [237] perhaps found by no one. So, if there remains nothing else to discuss in regard to the definite resolution of the objection taken from the effects of the earth's rotation (which was the last argument presented in favor of its immobility about its own center), we could go on to the examination of the evidence for and against the annual motion.

SAGREDO. Salviati, I would not want you to measure our intellects in terms of yours. You are accustomed to being engaged in very lofty speculations, and you regard as frivolous and lowly some of those that seem to us proper food for our thought; but sometimes, for our satisfaction, do not disdain lowering yourself to grant something to our curiosity. Then, regarding the resolution of the last objection, taken from the extrusion of the diurnal rotation, I would have been satisfied with much less than what was produced; nevertheless, I found the things which

163. We can sympathize with Galileo for wanting to draw to a close this type of methodological analysis, but Salviati's remark is too harsh and negative; for Sagredo's "nit-picking" has established the point that material bodies are bound to instantiate or approximate some abstract entity even though we cannot be certain we have identified it correctly; this is a crucial point for the methodological understanding of the role of mathematics in physical inquiry.

164. See selections 1 and 15. Galileo had originally wanted to entitle this book *Dialogue on the Tides*; this intention may be interpreted in either of two ways. A common interpretation claims that such a title would call attention to his tidal argument, which he considered to be a conclusive demonstration, and that it was vetoed by Church authorities whose superior judgment made them realize that the tidal argument is not conclusive. I believe a more plausible interpretation is that he did not want to advertise in the title that he was dealing with the dangerous topic of the earth's motion, which would have been introduced indirectly and hypothetically as a way of explaining a phenomenon not explicable in any other way. Cf. Drake (1986b), Finocchiaro (1980, 3–26), MacLachlan (1990), McMullin (1967, 24–42), and Shea (1972, 173–74).

have been superabundantly discussed so curious that not only did they not tire my imagination, but their novelties filled it with such a delight that I would not know how to ask for a greater one. Still, if there remains any other thought for you to add, by all means put it forth because I for my part am very willing to listen to it.

SALVIATI. I have always derived very great delight from my discoveries; making them is the greatest, but next to that I take great pleasure in discussing them with a friend who understands them and shows he enjoys them. Now, since you are one of these, I will somewhat indulge my ambition, which is gratified when I show myself to be more perceptive than someone else esteemed for his sharp insight.[165] So, to top off the previous discussion, I will explain another fallacy committed by the followers of Ptolemy and Aristotle in regard to the argument already presented.[166]

SAGREDO. I am here eager to listen.

SALVIATI. So far we have overlooked and granted Ptolemy something as if it were an indubitable fact; that is, when the stone is thrown off by the speed of the wheel turning around its own center, the cause of this extrusion grows as much as the turning speed grows; from this one inferred that, since the speed of the terrestrial rotation is immensely greater than that of any machine which we can artificially cause to turn, consequently the extrusion of stones, animals, etc., should be extremely violent. Now, I note that there is a very great fallacy in this argument, when [238] we compare the two speeds indiscriminately and absolutely. It is true that, if I compare speeds for the same wheel or for two equal wheels, the one which is more rapidly turned throws off the rocks with a greater impetus, and as the speed grows the cause of the projection also grows in the same proportion. However, let us consider the case when

165. During the trial of 1633, in the second deposition Galileo (while denying any malicious intention) agreed to plead guilty to a lesser charge, in exchange for which the Inquisition would not press more serious charges. Thus he declared, in part: "As an excuse for myself, within myself, for having fallen into an error so foreign to my intention, I was not completely satisfied with saying that when one presents arguments for the opposite side with the intention of confuting them, they must be explained in the fairest way, and not be made out of straw to the disadvantage of the opponent, especially when one is writing in dialogue form. Being dissatisfied with this excuse, as I said, I resorted to that of the natural gratification everyone feels for his own subtleties and for showing himself to be cleverer than the average man, by finding ingenious and apparent considerations of probability even in favor of false propositions. . . . My error then was, and I confess it, one of vain ambition, pure ignorance, and inadvertence" (Finocchiaro 1989, 278).

166. Here begins Galileo's third criticism of the extrusion argument, though, as already mentioned, it too will involve quantitative considerations.

the speed becomes greater not by increasing the speed of the same wheel (which would happen by making it perform a larger number of turns in equal times), but by increasing the diameter and making the wheel larger (so that, while keeping constant the time for a revolution of the small as well as of the large wheel, the speed for the large one is greater only on account of its larger circumference); here, no one should believe that the cause of the extrusion in the large wheel grows in proportion to the ratio of the speed at its circumference to the speed at the circumference of the smaller wheel; for this is most false, as for now a very quick observation can show us in a rough way. We could sling a stone farther with a reed one cubit long than with a reed six cubits long, even if the motion of the stone wedged at the end of the long reed were twice as fast as the motion at the end of the shorter reed; this would happen if the speeds were such that in the time for one complete revolution of the longer reed the shorter one would perform three of them.[167]

SAGREDO. I already understand that what you are telling me, Salviati, must necessarily be so. However, I do not readily recall the reason why equal speeds should not act equally in extruding projectiles and the speed of the smaller wheel should act much more than that of the larger wheel. So I beg you to explain to me how this business goes.

SIMPLICIO. This time, Sagredo, you appear to be unequal to yourself; for you usually grasp everything in an instant, but now you are missing a fallacy in the observation about the sling reeds which I have been able to

167. The linear speed of a body lying on the earth's equator, which is about 24,000 miles long, is about 1,000 miles per hour. The extrusion objection seems to assume that the cause of the extrusion varies as the linear speed, so that the speed of 1,000 miles per hour should give rise to an extruding tendency that would tear the earth apart. Galileo's key critical claim here is that this assumption is not exactly right, that it is incomplete; it fails to take into account another crucial variable, namely, the radius of rotation; this is a serious flaw because the radius affects the extrusion in an inverse manner; that is, the extruding power decreases as the radius increases. From the viewpoint of modern physics, this claim is essentially right, although his derivation is problematic in several ways; but the situation is even more complicated than he realized; that is, his critical claim (even disregarding the derivation) is itself incomplete because he failed to realize that although the extruding power does grow with the linear speed, the growth is in accordance with the square of this speed; in short, centrifugal force is directly proportional to the square of the linear speed and inversely proportional to the radius; thus, the extrusion is a function not simply of the angular speed (number of turns per unit time), but of the radius and the square of the angular speed. Even so, this paragraph and the following discussion contain evidence of Galileo's intuition that the extruding power depends *more* on angular speed than on the radius, which is one way of stating one consequence of the correct law of centrifugal force ($F = mw^2R$). Before the end of the seventeenth century, this law was formulated in its modern form by the Dutch physicist Christiaan Huygens (1673).

penetrate. This stems from the different manner of operation in causing the projection with the short reed and with the long one. For, to make the stone escape from the notch, what is required is not to continue its motion uniformly, but rather to stop the arm and suppress the reed's speed even if it is very high; thus, the stone, which is already moving very fast, escapes and [239] moves with some impetus; but this restraint cannot be accomplished for the longer reed, which, because of its length and flexibility, does not completely obey the arm's braking action; instead such a reed continues to accompany the stone for some distance, gently holding it back and keeping it attached, without letting it escape (as would happen if there were a collision with a hard obstacle). On the other hand, if both reeds struck an obstacle which stopped them, I think the stone would escape equally from both, as long as their motions were equally fast.

SAGREDO. With Salviati's permission, I will reply to Simplicio since he addressed himself to me. I say that there is good and bad in his account; good insofar as it is almost all true, bad insofar as it is not all relevant to our purpose. It is most true that, if whatever carries the stones struck an immovable obstacle, they would fly forward with some impetus; this is like the effect easily seen on a boat which while advancing rapidly runs aground or strikes some obstacle, namely, that all aboard are caught off guard and lose their balance and fall in the direction the ship was advancing; and if the terrestrial globe encountered an obstacle that effectively resisted its rotation and stopped it, then I really think that not only beasts, buildings, and cities, but also mountains, lakes, and seas would be toppled, and even the terrestrial globe would disintegrate. However, none of this is relevant to our purpose, for we are talking about what would result from the motion of the earth turning uniformly and calmly around itself, albeit at great speed. Similarly, what you say about the reeds is partly true; however, it was not brought up by Salviati as something that applies exactly to the subject we are treating, but rather as an example that in a rough way could stimulate our mind; it was meant to make us reflect more accurately on whether, when the speed grows in any manner, the cause of projection increases in the same proportion; whether, for example, if a wheel with a diameter of ten cubits were moving in such a way that a point on its circumference was passing through one hundred cubits per minute, and if it thus had the impetus to throw a stone, then this impetus would increase one hundred thousand times in a wheel with a diameter of one million cubits. Salviati denies this, and I am inclined to believe the same thing; but, since I did not know the reason, I asked him and am eagerly waiting for it.

[240] SALVIATI. I am ready to give you all the answers that I am capable of giving; although at first you may think I am searching for things unrelated to our purpose, nevertheless I believe that as our reasoning proceeds we will find them not to be such.[168] So, Sagredo, please tell me what things you have observed offer resistance to the motion of bodies.

SAGREDO. For now I do not see that a moving body has any internal resistance to being moved except its natural inclination and propensity toward contrary motion; for example, in heavy bodies, which have a propensity for downward motion, there is a resistance to upward motion. I said *internal resistance* because I think you mean this rather than external, of which there are many accidental instances.

SALVIATI. This is what I meant, and your perspicacity has prevailed over my cleverness. However, if I was careless in asking the question, I wonder whether your answer, Sagredo, is completely adequate, and whether a moving body has another intrinsic and natural property that makes it resist motion, aside from the natural inclination in a contrary direction. So tell me again, Do you not think that, for example, the inclination of heavy bodies to move downward is equal to their resistance to being pushed upward?[169]

SAGREDO. I think these are exactly equal; thus, in a balance I see two equal weights remain stationary in equilibrium, as the gravity of one resists being raised by the gravity of the other which is trying to raise it by pushing down.

SALVIATI. Very well; thus, to make one raise the other, we must increase the weight of the one or decrease that of the other. However, if the only resistance to upward motion is gravity, how does it happen that in a balance with unequal arms (namely, a steelyard) a weight of one hundred pounds with its downward tendency is not enough to raise a counterbalancing weight of four pounds, and how can the lowering of the latter raise the hundred-pound weight? For this is the effect of the sliding counterpoise on the heavy weight which one may want to weigh. If gravity is the only resistance to motion, how can the sliding counterpoise with its weight of only four pounds resist the

168. What follows is Galileo's justification of his third critical claim on the extrusion objection. The claim is that linear speed is not the only variable on which extruding power depends (as the objection assumes), but that the radius also affects it and does so in an inverse manner.

169. This remark is reminiscent of, and should be compared and contrasted to, Newton's third law of motion—the fundamental principle of classical physics that for every action there is an equal and opposite reaction.

weight of a bale of wool or silk that may weigh eight hundred or one thousand pounds; indeed, how can the counterpoise with its moment overcome the bale and raise it? We must, Sagredo, say that here we are dealing with some other resistance and some other force[170] than simple gravity.

[241] SAGREDO. That is indeed necessary. But tell me what this second power[171] is.

SALVIATI. It is what was missing in the balance with equal arms. Consider what is different in the steelyard, and therein must lie the cause of the different effect.[172]

SAGREDO. I think that your prodding has made me remember something. In both instruments, we are dealing with weight and motion; in the balance, one weight must exceed the other in order to move it because the motions are equal; in the steelyard, the lesser weight does not raise the greater unless the latter moves little (being hung on the shorter arm) and the former moves a great distance (being hung on the longer arm); therefore, we must say that the lesser weight overcomes the resistance of the greater because the lesser moves a lot while the other moves a little.

SALVIATI. That is to say, the speed of the lighter body compensates for the weight of the heavier and slower body.

SAGREDO. But do you think that the speed compensates exactly for the weight? That is, for example, that the moment or power of a moving

170. In the original: *forza*.

171. In the original: *virtù*.

172. Here Galileo uses and in part explicitly formulates a methodological principle of causal investigation, now called Mill's method of difference after the British philosopher John Stuart Mill (1806–1873), who systematized the methods of causal inquiry (*System of Logic*, London, 1843). Galileo suggests that sometimes the unknown cause of a known effect can be discovered by the following procedure: contrast this effect to a second effect whose cause is known; determine how the second effect differs from the first; and use this difference as a clue to how the unknown cause of the first effect differs from the known cause of the second effect. The example here is the problem of why with a steelyard, a lighter weight can balance a heavier one; contrast the steelyard with the equal-armed balance; with the latter, equal weights balance each other because they can move equally; in the steelyard, *unequal* weights balance each other; so we infer that this must happen because they can *move unequally*; that is, the cause why in a steelyard lighter weights can balance heavier ones is that a lighter weight can move more than a heavier one. Cf. Machamer (1978), Mertz (1980), Skyrms (1975), and Wallace (1983). Both the principle of the equal-armed balance and the principle of the steelyard as stated here are related to Archimedes' law of the lever; this states that a lever is in equilibrium when the applied force and the resisting force are inversely proportional to their respective distances from the fulcrum.

body weighing four pounds is equal to that of a body weighing one hundred pounds whenever the former has one hundred units of speed and the latter has only four?[173]

SALVIATI. Certainly, as I could show you with many experiments. However, for now let this confirmation with the steelyard alone suffice; in it you see the relatively light counterpoise sustain and balance a very heavy bale whenever the distance of the counterpoise from the fulcrum on which the steelyard is supported and pivots is as much greater than the shorter distance at which the bale hangs as the absolute weight of the bale is greater than that of the counterpoise. The cause of the fact that the heavy bale cannot raise the much lighter counterpoise is nothing but the disparity of the respective motions of the two bodies; for example, if the bale should drop by only one inch, that would make the counterpoise rise one hundred inches, given that the bale weighs as much as one hundred counterpoises and that the distance of the counterpoise from the fulcrum of the steelyard is one hundred times the distance between the same fulcrum and the suspension point of the bale; now, to say that the counterpoise moves a distance of one hundred inches while the bale moves only one inch is the same as saying that the speed of motion of the [242] counterpoise is one hundred times greater than the speed of motion of the bale. So, fix firmly in your mind, as a true and known principle, that the resistance deriving from the speed of motion compensates that which depends on the weight of a moving body; thus, consequently, a one-pound body moving at one hundred units of speed resists being stopped as much as another body weighing one hundred pounds whose speed is just one unit; moreover, two bodies of equal weight resist equally being set in motion at equal speeds; but, if one is to be moved faster than the other, it will offer a

173. This is physically correct if we interpret Galileo's "moment or power" as the momentum of modern physics, which is defined as the product of the mass and the velocity; but it is incorrect if the "moment or power" is interpreted as kinetic energy, which is defined as one-half the product of the mass and the velocity squared. He did not make this distinction clear, and there was a controversy at the time about which of the two quantities was the true measure of the "moment or power" of a moving body; it took physicists several more decades before the issue was resolved. I have preferred to translate Galileo's *il momento e la forza di un mobile* as "the moment or power of a moving body" because later in the sentence he uses the singular *quella di un di cento* to refer to the "power" of another body, when he speaks of "that of a body weighing one hundred pounds"; this is preferable even though it involves translating *e* (literally "and") as "or" (literally *o*). For some possible scientific ramifications of this apparently pedantic issue, see Strauss (1891, 532n.87) and Drake (1982, 582n.87).

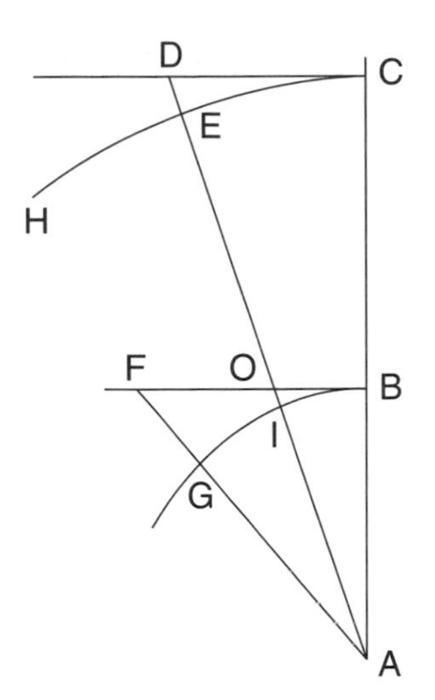


Figure 8.

greater resistance corresponding to the greater speed it is to be given.[174]

Having established these things, let us come to the explanation of our problem. To facilitate comprehension, let us draw a diagram [fig. 8]. Let A be the center of two unequal wheels, BG the circumference of the smaller, and CEH the circumference of the larger; let the radius ABC be erected perpendicular to the horizon, and draw the tangents BF and CD at points B and C; and let the arcs BG and CE be equal. Imagine then that the two wheels are turned around their center at equal speeds, so that if two bodies (for example, two stones) are placed at points B and C, they are carried along the circumferences BG and CE at equal speeds; that is, while stone B runs through the arc BG, stone C goes through the arc CE. Now, I say that the rotation of the smaller wheel is much more effective in bringing about the projection of stone B than the rotation of the larger wheel is in doing the same for stone C. For, as already explained, the projection must take place along the tangent; hence, if the stones B and C should separate from their wheels and begin the motion

174. With the formulation of this principle, Galileo is groping toward both the recognition of the importance of the concept of momentum (defined as the product of mass and velocity) and of the second law of motion (that force is proportional to acceleration). For more details, see Pagnini (1964, 2:413n).

of projection from points B and C, the impetus acquired from the rotation would hurl them along the tangents BF and CD; therefore, the two stones have equal impetuses to run along the tangents BF and CD, and they would do that if some other force did not deflect them. Is that not so, Sagredo?

SAGREDO. I think that is the way this business goes.

SALVIATI. But what force do you think can deflect the stones from moving along the tangents, which is the direction in which the impetus of the rotation really thrusts them?

[243] SAGREDO. Either their own gravity, or some glue which keeps them lying on or attached to the wheels.

SALVIATI. But to deflect a body from the motion for which it has an impetus, do we not need a force which is greater or smaller depending on whether the deflection is greater or smaller?[175] That is, depending on whether in the deflection the body must cover a greater or smaller distance in the same time?

SAGREDO. Yes, because it was concluded earlier that, in causing a body to move, when the speed at which it has to move is greater, the moving force[176] must be proportionately greater.

SALVIATI. Now, consider that to deflect the stone on the smaller wheel from the motion of projection along the tangent BF and to keep it attached to the wheel, its own gravity must pull it down by an amount equal to the length of the secant FG or of the perpendicular drawn from the point G to the line BF; on the other hand, in the larger wheel the downward pull need not be any greater than the secant DE or the perpendicular drawn from the point E to the tangent DC; but the latter secant is much shorter than FG and would become shorter and shorter as the wheel became larger and larger. Now, these two deflections must occur in equal times, namely, while the two equal arcs BG and CE are traversed; hence, the deflection of stone B, namely, the drop FG, must be faster than that along DE; so the force required to keep stone B joined to its small wheel will be much greater than that required to keep rock C on its large wheel; this is the same as saying that there will be some small force which will prevent the extrusion from the large wheel but will not prevent it from the small. It is clear, therefore, that the more

175. This remark is reminiscent of, and should be compared and contrasted to, Newton's second law of motion; that is, the fundamental principle of classical physics that acceleration is proportional to the force causing it and the change of velocity is in the direction of the force causing it.

176. In the original: *virtù*.

the wheel is enlarged, the more the cause of the projection is diminished.[177]

SAGREDO. From what I now understand, thanks to your lengthy detailed breakdown, I think I can satisfy my intellect with a very brief account as follows. The equal speeds of the two wheels impress on both stones equal impetuses along the tangents; hence, since the large circumference deviates little from the tangent, we see it favor in a sense and restrain gently the stone's appetite (so to speak) for separating from the circumference; thus, any small restraint, due to natural inclination or to some adhesive, suffices to keep it joined there. This cannot happen with the small wheel, which favors little the direction of the tangent and tries too greedily [244] to keep the stone onto itself; moreover, the restraint or adhesive is not stronger than that which kept the other stone joined to the larger wheel; so the stone breaks loose and runs along the tangent. This makes me comprehend something else besides the error of those who believe that the cause of the projection increases in the same manner that the speed of the rotation increases; that is, I also think that since the projection diminishes as the wheel becomes larger when the same speed is maintained in both wheels, perhaps it could be true that to make the large wheel extrude like the small one, its speed should increase as much as its diameter; this would happen if entire revolutions should be completed in equal times. Thus one could judge that the earth's rotation would be no more capable of extruding stones than any small wheel turning so slowly as to make just one rotation in twenty-four hours.[178]

SALVIATI. For now I do not wish to search much further; it is enough to have abundantly shown (if I am not mistaken) the ineffectiveness of an argument which at first sight seemed very conclusive and which was so judged by very great men. Moreover, I think our time and words will have been very well spent if I have gained some credence with Simplicio; I do not expect to have convinced him that the earth moves, but at least that the opinion of those who believe this is

177. It is obvious from this discussion and the diagram that this claim is restricted to situations of constant linear speed; thus, the claim is physically correct since centrifugal force does indeed vary directly as the square of the linear speed and inversely as the radius. It is equally obvious that the derivation, though interesting and suggestive, is problematic from the point of view of physics; cf., for example, Chalmers and Nicholas (1983, 323–28), Gaukroger (1978, 193–95), and Pagnini (1964, 2:415n).

178. As mentioned above, this is not exactly right since centrifugal force is proportional to *both* the square of the angular speed *and* the radius.

not as ridiculous and foolish as entire teams of ordinary philosophers take it to be.[179]

. . .[180]

## [10. The Deception of the Senses and the Relativity of Motion]

[273] . . . SIMPLICIO. . . . The first argument begins like this: "First, if Copernicus's opinion is accepted, the criterion of natural philosophy seems to be severely undermined, if not completely destroyed."[181] According to the view of all philosophical schools, this criterion requires that the senses and experience be our guides in philosophizing. However, from the Copernican perspective our senses are greatly deceived when our eyes perceive very heavy bodies fall perpendicularly in a straight line without deviating a hairbreadth from it, and they do so in close proximity and in a very clear medium; for, according to Copernicus, that motion is not really straight, but a mixture of straight and circular, and so our vision is deceived in such a clear-cut situation.[182]

179. This is an important difference; it is the distinction between refuting an argument and proving a contrary conclusion; cf. the appendix (1.3). In this selection, Galileo has shown that the extruding power of whirling does not imply that the earth stands still; he has not argued that such extruding power implies that the earth rotates.

180. The passages omitted here correspond to Favaro 7:244.25–273.6 and Galileo 1967, 218–48. They discuss two anti-Copernican arguments found in a book written by one of Christopher Scheiner's students, Locher (1614). The first is a mechanical objection involving the thought experiment of a body falling from the moon to the earth and the problem of what path it would follow and what time it would require; Galileo's criticism discusses some of his discoveries about falling bodies, such as the law that distance fallen varies as the square of the time elapsed. The second is a methodological objection based on the premise that the earth's motion would be inexplicable; his criticism discusses several methodological issues involving the nature and interrelationship of the concepts of explanation, truth, intelligibility, and comprehensibility. For more details, see Finocchiaro (1980, 38 and 122–24).

181. Galileo quotes this from Chiaramonti (1628, 472); the other quotations in this selection are taken from this book. For the rest of the "Second Day" and up to selection 11 in the "Third Day," the *Dialogue* contains a critical analysis of many anti-Copernican arguments in the same book; in selection 3, Galileo criticizes other anti-Copernican ideas found in Chiaramonti's earlier *Anti-Tycho* (1621).

182. This paragraph states what is called the argument from the deception of the senses. This argument is the most basic objection to Copernicanism, for it points out that the earth's motion contradicts direct observation; since observation may be regarded as the essential instrument through which to acquire knowledge about the world, the argument is a methodological objection pointing out that Copernicanism contradicts a fundamental principle of methodology. Note also that the earth's motion (whether axial rotation or or-

SALVIATI. This is the first[183] argument advanced by Aristotle, Ptolemy, and all their followers. We have abundantly replied to it and shown the paralogism when we explained very clearly how motion shared by us and other moving bodies is as if it did not exist.[184] However, true conclusions have countless favorable instances confirming them, and so for the sake of this philosopher[185] I want to add something else. You, Simplicio, should take his side and answer my questions. First, tell me, What effect is produced in you by a stone which falls from the top of a tower and causes you to perceive this motion? For if its fall produced in you nothing more or nothing new as compared to what its rest at the top of the tower produced in you, you certainly would not perceive its descent nor distinguish its motion from its standing still.

SIMPLICIO. I notice its descent in relation to the tower, for I now see it next to this mark on the tower, then a little below that, and so on until I perceive it land on the ground.

SALVIATI. Therefore, if the stone had fallen from the claws of a flying eagle and descended simply through the invisible air, and you had no other visible and stable object with which to keep track of it, then you could not perceive its motion?

---

bital revolution) cannot be directly observed even today, even by astronauts in outer space. In this paragraph the example of alleged deception is not the earth's motion per se but free fall, which appears straight and vertical, but which is really slanted if the earth rotates (cf. selection 8). But this argument should not be confused with the vertical fall objection: the argument from the deception of the senses argues that the earth does not rotate because our senses cannot be deceived, and so on; whereas the vertical fall objection argues that the earth does not rotate because bodies fall vertically, and so on; for more clarifications, see Finocchiaro (1980, 38–39, 124–25, 169–70, 195–96, 215–16, and 394–97). Though *distinct*, these two arguments are *related*; for example, the essential premise of the argument from the deception of the senses can be used to justify why apparent vertical fall implies actual vertical fall.

183. The first argument, I suppose, in order of importance rather that in the list quoted from Aristotle in selection 7. As we have seen, the first objection in that list is the forced motion argument.

184. This last clause states the principle of the relativity of motion, which is first explicitly formulated and applied to diurnal rotation in selection 6, and then implicitly used in the discussion of vertical fall in selection 8; the explicit application of the relativity of motion to vertical fall comes in a passage that is omitted here and that discusses the vertical gunshot objection (Favaro 7:197–202; Galileo 1967, 171–78). The "abundant reply" mentioned by Salviati refers to those discussions. In the present context the example of vertical fall is just one of many particular alleged deceptions implied by Copernicanism; before this selection, there has been no statement of (let alone reply to) the argument from the deception of the senses considered in its own right as a methodological objection to the earth's motion in general.

185. The philosopher referred to is Scipione Chiaramonti.

SIMPLICIO. I would still perceive it because to see it when it was [274] very high I would have to raise my head, and as its descent continued I would have to lower it; in short, I would have constantly to move either my head or my eyes to follow its motion.

SALVIATI. Now you have given the true answer. You know the rock is at rest when you always see it in front of you without moving your eyes at all, and you know it is in motion when you must move your organ of vision (namely, your eyes) in order not to lose sight of it. Therefore, if without ever moving your eyes you saw an object constantly looking the same, you would always judge it to be motionless.

SIMPLICIO. I think it must necessarily be so.

SALVIATI. Imagine, now, being on a ship and fixing your eyes on the tip of the sail yard; do you think that, even if the ship were moving very fast, you would have to move your eyes in order to keep looking always at the tip of the sail yard and follow its motion?

SIMPLICIO. I am sure there would be no need of any changes, and the effect would be similar for something else besides vision; that is, if I should aim a gun at the tip of the sail yard, to keep it aimed there I would never have to move it a hairbreadth regardless of any motion by the ship.

SALVIATI. This happens because the motion transmitted by the ship to the sail yard is also transmitted to you and your eyes, so that you need not move them at all in order to keep aiming at the tip of the sail yard; consequently, it appears motionless to you. Now, apply this account to the earth's rotation and the rock at the top of the tower; you could not perceive the rock move because you and the rock together would receive from the earth the motion needed to follow the rock's motion, and so you would not need to move your eyes; when it then acquires the downward motion, which is its own particular motion but not yours and is mixed with the circular, the circular part shared by the rock and your eyes would continue to be imperceptible; only the straight motion would become perceptible [275] because to follow it you would have to move your eyes downward.

To free this philosopher from his error, I would like to tell him that if he happens to take a ride on a boat, he should have a deep vase full of water and make a ball of wax or of some other substance which sinks very slowly (for example, barely a cubit in one minute); then he should gently immerse the said ball in the water and let it descend freely, and he should diligently observe its motion. He would see it go straight toward the point at the bottom of the vase where it would go if the boat were standing still, and to his eyes and in relation to the vase this motion would appear very straight and perpendicular; and yet one must say that it would be a mixture of straight downward and circular around the element water. Now, these things happen with motions that are not natural and with substances in regard to which we can make experiments when they are in a state of rest and when they are in a state of motion; yet, as far as appearances go, no difference is observed here, and it seems the senses are being deceived; so what can we expect to be able to distinguish in regard to the earth, which has perpetually been in the same state of either motion or rest? And how can we expect to be able to test whether we can detect any difference between these motion phenomena in its different states of motion and of rest, if it is eternally kept in only one of these states?

. . .[186]

[278] . . . SIMPLICIO. The author continues to show how in the Copernican doctrine one must deny the senses and the most basic sensations; for example, although we can feel the breeze of the lightest wind, we do not feel the impetus of a perpetual wind striking us at a speed of more than 2,529 miles per hour; this is the distance which would be covered in one hour by the earth's center in its annual motion along the circumference of the ecliptic, as he diligently calculates.[187] In fact, he states that according to Copernicus, "the surrounding air moves together with the earth; nevertheless, we would not feel its motion, despite its being faster and more rapid than the strongest wind; instead, we would consider it to be extremely calm, unless some other motion were added. If this is not a deception of the senses, what is?"[188]

SALVIATI. This philosopher must be thinking that the earth which Copernicus sets in motion, together with the surrounding air and along the [279] circumference of the ecliptic, is not the one we inhabit but a

186. The passage omitted here corresponds to Favaro 7:275.24–278.23 and Galileo 1967, 250–53. It contains a digression on the problem of measuring longitude at sea by means of telescopic observations on a ship; this was an application of the telescope dear to Galileo's heart, and he was involved in negotiations with the king of Spain about its feasibility.

187. As Pagnini (1964, 2:464n.1) explains, this estimate of the speed of the earth's annual motion is only about 1/25 of the true value; the latter is actually about 105,000 km/hr, whereas the figure in the text corresponds to about 4,183 km/hr, taking the Tuscan "mile" to equal 1,654 meters. The low estimate was common at the time and was due to a similar underestimation of the distance between the earth and the sun (about 1,150 earth radii rather than about 23,439).

188. Cf. Chiaramonti (1628, 472–73).

different one; for ours carries us with it at the same speed it and the surrounding air have. What kind of wound can we feel when we are running away at the same speed as someone who wants to stab us? This gentleman forgot that we too, no less than the earth and the air, are carried around; that consequently we are always touched by the same part of the air; and that, hence, it does not strike us.

SIMPLICIO. No, he did not forget. Here are the words which immediately follow: "Moreover, we too are carried around by the circulation of the earth, etc."[189]

SALVIATI. Now I can no longer help or excuse him; excuse him and help him yourself, Simplicio.

SIMPLICIO. At the moment, I cannot readily think of a satisfactory excuse.

SALVIATI. Oh, well; you will think about it tonight and then defend him tomorrow. In the meantime, let us hear the other difficulties.[190]

189. Cf. Chiaramonti (1628, 473).

190. The exchange in the last several paragraphs constitutes Galileo's second criticism of the argument from the deception of the senses. Here he objects to the example of the lack of a perpetual wind as an alleged deception; his objection is that there would be no wind deception on a moving earth because there would be no wind for us to perceive; there would be no wind because wind is defined as air moving relative to the observer, and on a moving earth the air as well as we the observers would be carried along. At this point Simplicio could question whether the air would indeed be carried along, but doing so would raise issues related to the conservation of motion, which were discussed earlier; the stress now is on the problem of the deception of the senses, and from this viewpoint Galileo's answer seems decisive. But the introduction of annual motion into the discussion is puzzling. Given that the "Second Day" is supposed to focus on diurnal motion, the reference to annual motion is out of place; it might be less confusing to think of diurnal rather than of annual motion, and of the perpetual wind that should result from terrestrial rotation; in fact, one of the many arguments against terrestrial rotation stated in selection 7 ([158]) is the one based on the lack of a perpetual wind. If we think in terms of annual motion, the following should be noted: if (as the Aristotelians believe) the air could not follow the earth's motion, then on the assumption that the earth has the *diurnal* motion, a perpetual westward wind would follow; but on the assumption that the earth has the *annual* motion, what would follow is that the air should have been left behind long ago and that there should not be any air on the earth's surface now; so the wind objection should be reformulated to read "the earth cannot have the annual motion because if it did then it would now have no atmosphere; since it is obvious that the earth has an atmosphere, the earth cannot move annually around the sun"; that is, the "wind objection" should be restated as the "air objection." This point is in the spirit of Galileo, being similar to the first critical clarification he makes to the extrusion objection, in selection 9; but since he does not offer such a critique for the wind objection, he could be criticized for having failed to do so. In short, either he should not have injected the annual motion into this discussion at all (not even by way of quoting from Chiaramonti's book), or having done so, he should have made the above mentioned critical clarification.

SIMPLICIO. He continues with the same objection, showing that according to Copernicus one must deny one's own sensations. For, this principle by which we go around with the earth is either intrinsically ours or external to us (namely, a case of being forcibly carried by the earth); if it is the latter, (since we do not feel being forcibly carried) we will have to say that our sense of touch does not feel the very object being touched and does not receive its impression in the sensorium;[191] but if the principle is intrinsic, then we will not be feeling a motion deriving from within us, and we will not be perceiving a propensity perpetually inherent in us.[192]

SALVIATI. Thus, the objection of this philosopher reduces to this: whether the principle by which we move with the earth is external or internal, we would have to feel it in any case; since we do not feel it, it is neither the one nor the other; hence, we do not move, and consequently neither does the earth. But I say that it may be either way, without our feeling it.

In regard to the possibility that it may be external, our experience on a boat removes every difficulty more than sufficiently. I say more than sufficiently because we can make it move and can make it stand still any time, and so we can very accurately observe any difference perceivable with the sense of touch from which we can learn how to determine whether or not it moves; but, since we have not yet acquired such knowledge, why should we be surprised [280] that we lack the information about the same conditions for the earth (which may have perpetually carried us without our being able ever to experience its rest)? Simplicio, I believe you too have countless times been on the boats between here and Padua; if you want to confess the truth, you have never felt inside yourself your taking part in that motion, unless the boat ran aground or struck some obstacle and stopped; in this case, taken by surprise, you and the other passengers stumbled dangerously. The terrestrial globe would have to meet some obstacle that stopped it, and I assure you that you would then perceive the impetus inherent in you, as you would be hurled toward the stars. It is indeed true that you can perceive the boat's motion by means of another sense (though accompanied by reason); that is, by means of vision, as you look at trees and buildings

191. The sensorium is the part of the brain or mind where sensory impressions are received and recorded.

192. Here Simplicio refers to our internal kinesthetic sense, so that another (a third) alleged deception would be our feeling that the earth is at rest, while it really is in motion (according to the Copernicans).

in the countryside; being separate from the boat, these things appear to move in the opposite direction. But if you wanted to convince yourself of the earth's motion by means of such an observation, I would say you should look at the stars which, because of that, appear to move in the opposite direction.

If the principle should be internal to us, being surprised that we do not feel it is even less reasonable. For, if we do not feel it when it is external to us and frequently leaves us, for what reason should we feel it when it resides within us constantly and continually? Now, is there anything else in this first argument?[193]

SIMPLICIO. There is this brief interjection: "According to this view, we must distrust our senses, as being inherently deceptive or stupid in judging sensible things, even when they are very close; so, what truth can we hope to derive from such a deceptive faculty?"[194]

SALVIATI. Oh, I would deduce from it more useful and certain precepts, such as the lesson that we ought to be more careful and less confident in regard to what at first sight is presented to us by our senses, and that they can easily deceive us.[195] Moreover, I wish this author would not be so anxious to convince us by means of the senses that the motion of falling bodies is simply straight and no other, and that he would not get angry and boisterous because such a patently clear and obvious thing is questioned. For this attitude is an indication of being under the impression that those who say this motion is not really straight but rather somewhat circular sensibly see the rock move along an arc; indeed, he refers more to their senses than to their reason to clarify [281] this effect. But this impression is not true, Simplicio, as I, who am neutral in regard to these opinions and only as an actor impersonate Copernicus in this play of ours,[196] can attest; I have never seen, and I have never thought I saw, a

193. This speech by Salviati criticizes the third example of alleged deception of the senses (if the earth is moving): there is no deception of our kinesthetic sense because our experience with navigation shows that we can feel only changes of motion and not uniform motion, and so the earth's constant rotation is not something susceptible of being felt.

194. Cf. Chiaramonti (1628, 473).

195. This lesson embodies a methodological principle that may be called critical empiricism and may be contrasted with the one advanced by the argument's proponents. They make the acquisition of knowledge so dependent on sensory experience that if the senses are not reliable, then there is no reliable guide in the search for truth and knowledge is impossible; whereas Galileo says that if the senses are not always reliable, then we should learn to distinguish situations in which they are reliable from situations in which they are not.

196. Like the disclaimers in selections 5 ([133]) and 7 ([157–58]), this one is meant to show Galileo's deference to the restrictions imposed upon him by the Church. For more details, see the introduction (4), the appendix (3.2), selection 1, and Finocchiaro (1980, 6–18).

rock fall otherwise than perpendicularly, just as I think the same appears to everyone else's eyes. Therefore, it is better to set aside the appearance, on which we all agree, and to make an effort to use our reason either to confirm the reality underlying the appearance or to discover the fallacy.[197]

SAGREDO. If I could ever meet this philosopher, who seems to tower much above many other followers of the same doctrines, as a sign of affection I would like to remind him of a phenomenon he has certainly seen countless times; with great similarity to what we are now discussing, it makes us understand how easily one may be deceived by mere appearances, namely, by the representations of the senses. The phenomenon is that, to those who are walking on a street at night, it looks as if they are being followed by the moon at a pace equal to theirs; they see it graze the eaves of the roofs over which it appears, just like a cat which follows them by actually walking on the tiles. This is an appearance that very obviously would deceive the sense of sight, if reason did not intervene.[198]

197. This paragraph contains a distinct criticism of the argument from the deception of the senses. All previous criticisms in this selection were designed to show that the alleged deceptions of the senses are not really deceptions. The present criticism says, in effect, that if and to the extent that there would be sensory deception on a moving earth, that would be no reason to conclude that knowledge and the earth's motion are impossible; the more correct conclusions would be that knowledge is difficult, that it cannot rely solely on the senses, and that reason plays an equally crucial role. Another way to distinguish the two criticisms is this: the previous critiques showed that the alleged deceptions of the senses are not deceptions on the basis of considerations dealing with the nature of the phenomena in question and with the nature of sensation, the key point being that there is no deception in not perceiving something which should not be perceivable; the present critique amounts to giving a different reason why the alleged deceptions are not deceptions of the senses, namely, that they are deceptions of reason (the reason of those who from the fact that certain things appear in a certain way conclude incorrectly that they are really that way).

198. One important issue Galileo does not discuss in this selection (but cf. Finocchiaro 1989, 86) is that the deception of not perceiving the earth's motion (if Copernicus is right) is impossible because it would be a very special deception, unlike any others such as the moon illusion just mentioned. One reason is that it would be a giant, radical deception insofar as it would involve being mistaken about something whose fundamental importance would overshadow any other small or occasional error; the deception would *not* be just a local deception relative to terrestrial location. Another reason is embodied in the language of the first quotation in this selection, namely, that the alleged deception would be occurring "in close proximity and in a very clear medium . . . in such a clear-cut situation"; that is, the difficulty is that the senses would be "inherently" deceptive, as the last quotation suggests. A third reason is that perceptual illusions are normally corrigible, for normally we can discover by means of the senses themselves that they are illusions; but the deception of the earth's motion would be perceptually incorrigible; perhaps, as Galileo argues in this selection, reason would enable us to make the correction, but that is different. A fourth reason would involve theological considerations about whether it is plausible that God

. . . [199]

---

would create the world and the human faculties in such a way that mankind would be permanently subject to such a perceptual illusion. A fifth reason stems from what Santillana (1953, 270n.119) calls the Aristotelian doctrine of the *sensibile proprium*, which he finds elaborated in Colombe (1611); this is the view that each human sense has a proper domain in which it is the only infallible judge, such as the eye is for judging color and touch is for judging shape; but its relevance is questionable since it is not clear what sense the Aristotelians could propose as the proper and final judge of motion. These reasons suffice to show that there is something special about the deception of the earth's motion. Is such a radical deception possible? It is not easy to give reasons why it is or is not possible for human beings to be so radically mistaken. This is a difficult issue that relates also to whether human beings can be radically mistaken in regard to other cognitive processes, such as reasoning; the question is really whether man is a rational animal. One attempt to answer this question was made soon after Galileo by Descartes in his *Discourse on Method* (1637) and *Meditations* (1641); for some present-day discussions, see L. J. Cohen (1981), Finocchiaro (1980, 256–72 and 332–42), and Stich (1985).

199. The passages omitted here correspond to Favaro 7:281.22–298 and Galileo 1967, 256–75. They contain a critical analysis of three arguments in Chiaramonti (1628): that the earth cannot move (1) because it is impossible for a simple body to have more than one natural motion; (2) because if the earth moved, it would follow that dissimilar bodies would have similar motions; and (3) because motion causes tiring and so sooner or later the earth would get tired and stop. For more details, see Finocchiaro (1980, 39 and 125–27).

# Third Day

. . .[1]

## [11. Heliocentrism and the Role of the Telescope]

[346] . . . SALVIATI. . . . We ought to leave this question and go back to our main subject; here the next point to consider is the annual motion, which is commonly attributed to the sun, but which was taken away from the sun and given to the earth first by Aristarchus of Samos and later by Copernicus. Against this position I see Simplicio comes well equipped, in particular with the sword and shield of the booklet of mathematical conclusions or disquisitions;[2] it would be good to begin by proposing its attacks.

SIMPLICIO. If you do not mind, I would like to leave them to the end, as they are the last to have been discovered.

1. The passages omitted here correspond to Favaro 7:299.1–346.24 and Galileo 1967, 276–318. They contain a critical analysis of Chiaramonti's (1628) arguments that the "new stars" were sublunary phenomena; the problem was that the observations were inconsistent, and so some had to be discounted and the rest had to be harmonized; the main issue was how to make these corrections. Thus, they suggest instructive lessons for the analysis of observational errors, the interpretation of data, and the evaluation of alternative interpretations. For more details, see Finocchiaro (1980, 40 and 127–28).

2. Locher (1614). This book is also discussed in selections 12 and 13 and in earlier passages omitted here (Favaro 7:117–24 and 244.25–273.6; Galileo 1967, 91–98 and 218–48).

SALVIATI. Then, in accordance with the procedure followed until now, you must [347] advance in a systematic manner the contrary reasons—those of Aristotle as well as those of the other ancients; I will also contribute to this, so that nothing is left out without being carefully considered and examined; likewise, Sagredo will bring forth the thoughts that his lively intellect will awaken in him.

SAGREDO. I will do it with my usual frankness; you will be obliged to excuse it since you made this request.

SALVIATI. Our obligation will be to thank you for the favor, not to excuse you. However, let Simplicio begin to advance those difficulties that prevent him from being able to believe that the earth, like the other planets, can move in an orbit around a fixed center.

SIMPLICIO. The first and greatest difficulty is the repugnance and incompatibility between being at the center and being away from it; for if the terrestrial globe should move in the course of a year along the circumference of a circle, namely, along the zodiac, it would be impossible for it to be simultaneously at the center of the zodiac; but Aristotle, Ptolemy, and others have proved in many ways that the earth is at this center.

SALVIATI. You speak very well; there is no doubt that whoever wants the earth to move along the circumference of a circle must first prove that it is not at the center of this circle. Therefore, it follows that we should determine whether or not the earth is at this center, around which I say it turns, and at which you say it is fixed; and before doing this, we must also see whether or not you and I have the same conception of this center. So tell me what and where you understand this center to be.

SIMPLICIO. By this center I understand the center of the universe, of the world, of the stellar sphere, of the heavens.

SALVIATI. I could very reasonably dispute with you whether there is such a center in nature since neither you nor others have ever proved whether the world is finite and bounded or infinite and boundless; however, granting[3] for now that it is finite and bounded by a spherical figure,

3. Galileo may have thought that the universe is infinite, for this belief would explain why he was agnostic about whether it has a center. His agnosticism is apparent in this selection, in selection 2 ([61]), in his "Reply to Ingoli" (Finocchiaro 1989, 179), and in his letter to Fortunio Liceti of 24 September 1939 (Favaro 18:106). But it is clear from this selection that he has no difficulty with the concept of a center of planetary revolutions and no doubt about the sun's location there. However, he also felt that the idea of an infinite universe was problematic, partly for purely intellectual reasons involving various paradoxes about the concept of infinity, and partly for religious reasons; the idea was regarded as dangerous by religious authorities and had been a factor in why Giordano Bruno was tried, condemned, and burned at the stake in 1600.

and hence that it has a center, we must decide how credible it is that the earth, rather than some other body, is located at this center.

SIMPLICIO. That the world is finite, bounded, and spherical is proved by Aristotle with many demonstrations.[4]

SALVIATI. All these, however, reduce to one and this single one to [348] nothing; for he proves that the universe is finite and bounded only if it is in motion, and so all his demonstrations fall to pieces if I deny his assumption that the universe is in motion. However, in order not to multiply the disputes, let us concede for now that the world is finite and spherical and has a center. Since such a shape and center have been proved on the basis of its mobility, it will be very reasonable to proceed to the particular investigation of the exact location of such a center on the basis of the same circular motions of the heavenly bodies; indeed Aristotle himself reasoned and proceeded in the same manner, making the center of the universe that point around which all the celestial spheres turn and at which he saw fit to place the terrestrial globe. Now, tell me, Simplicio: suppose Aristotle were forced by the clearest observations to change in part his arrangement and structure of the universe, and to admit he was wrong in regard to one of the two following propositions (that is, either in placing the earth at the center or in saying that the celestial spheres move around such a center); which of the two alternatives do you think he would choose?[5]

SIMPLICIO. I believe that in this case the Peripatetics . . . [6]

4. Cf. Aristotle, *On the Heavens*, I, 5–7.

5. Salviati is asking Simplicio to determine which is more fundamental: (1) the claim that the center of the universe is the point around which the heavenly orbs revolve, or (2) the claim that the earth is the center of the universe. Salviati suggests that the first claim is more fundamental because Aristotle's reasoning for arriving at the second one (the earth is the center of the universe because it is the center of the planetary revolutions) presupposes the first. And he and Simplicio will soon agree that (on the assumption that the universe is finite and spherical) there is considerable inherent plausibility in asserting the first claim, which implies that the first is more inherently plausible than the second one. Salviati also suggests that acceptance of the first claim may lead to rejection of the second, and in fact this is the direction of the ensuing argument. This type of critical analysis is analogous to the one Galileo employs in selection 3 ([80]) in regard to the choice between the thesis of heavenly unchangeability and Aristotle's implicit manner of reasoning and the choice between the same thesis and his explicit fundamental principle of empiricism; what is involved is partly to distinguish between the procedure followed and the result arrived at and to attach more importance to the procedure, and partly to be sensitive to which ones of an author's assertions are more and which ones are less fundamental.

6. This ellipsis is in Galileo's original text, for he wants to portray an interruption in Simplicio's speech caused by Salviati's next interjection.

SALVIATI. I am not asking about the Peripatetics, but about Aristotle himself; for I know very well what they would answer. As the most submissive and slavish servants of Aristotle, they would deny all experience and all observation in the world and even refuse to use their senses in order not to have to make the confession; they would say the world is as Aristotle said and not as nature wants;[7] for if they lose the support of this authority, with what would you want them to appear in the field? So tell me what Aristotle himself would do.

SIMPLICIO. Really, I could not decide which of the two inconveniences he would regard as the lesser one.

SALVIATI. Please do not use this term; do not call inconvenient what could turn out to be necessarily so. To want to place the earth at the center of the heavenly revolutions was indeed inconvenient. However, since you do not know which way he would be inclined to go, and since I regard him as a man of great intellect, let us examine which of the two choices is more reasonable, and let us take that one to be what Aristotle would choose.[8] Let us resume our earlier discussion, then, [349] and let us assume (with Aristotle) that the universe has a spherical shape and moves circularly, so that it necessarily has a center in regard to both its shape and its motion; although we have no observational information about the size of the universe other than that deriving from the fixed stars, we are certain that inside the stellar sphere there are many orbs, one inside the other, each with its own heavenly body, and that they also move circularly; we are inquiring about which is more reasonable to believe and say, either that these nested orbs move around the same center of the universe, or that they move around some other center very far

7. Cf. the example of the anatomical dissection to determine whether the nerves originate in the heart or in the brain, in selection 5. Citing Favaro (1883, 394–95), Strauss (1891, 551n.32) states that Cesare Cremonini (1550–1631) at the University of Padua and Giulio Libri (c. 1550–1610) at the University of Pisa had refused to look through the telescope when Galileo first announced his discoveries.

8. This sentence expresses an important principle of textual interpretation and critical evaluation; I call it the principle of charity since it reduces to the precept to be charitable. Salviati is saying that, when in doubt or when more direct evidence is not available, and when the given author or opponent is sufficiently worthy, we should attribute to him the most reasonable possible view. This principle is not very precise and is silent about what would count as "more direct evidence" and "worthy opponent," but this is not to say that it is uninformative; it conveys a definite idea and recommends a definite procedure. It would make an instructive exercise to determine whether in this book Galileo himself always abides by this principle. For more details, see Finocchiaro (1980, 338–46).

from there. Now, Simplicio, tell us your opinion about this particular detail.

SIMPLICIO. If we could limit ourselves only to this issue and be sure not to encounter any other difficulty, I would say it is much more reasonable to claim that the container and the contained parts all move around a common center than around several.

SALVIATI. Now, if it is true that the center of the universe is the same as that around which the orbs of the heavenly bodies (namely, of the planets) move, then it is most certain that the sun rather than the earth is found placed at the center of the universe; thus, as regards this first simple and general point, the place in the middle belongs to the sun, and the earth is as far away from the center as from the sun itself.[9]

SIMPLICIO. But what is the basis of your argument that the sun rather than the earth is at the center of the revolutions of the planets?

SALVIATI. I conclude this from observations that are very evident and hence necessarily binding.[10] The most palpable of these observations that exclude the earth from this center and place the sun there is the fact that all planets are found to be sometimes closer to the earth and sometimes farther; these differences are so large that, for example, when Venus is farthest it is six times farther from us than when it is closest, and Mars recedes almost eight times more in one position than in the

9. This speech concludes the first step in this selection's discussion; this step may be reconstructed as an argument aiming to show that *if* it can be proved that the sun is the center of the revolutions of the heavenly orbs, *then* it will have been proved that it is the center of the universe; this conditional conclusion is justified by Aristotle's definition of the center of the universe as the center of the revolutions of the heavenly orbs. The second step will prove (from observational evidence) that the sun is the center of the revolutions of the heavenly orbs. A third step will be to draw from this conclusion and several plausibility considerations the further consequence that the earth is more likely to have the annual motion than the sun. The final step will be to criticize some relevant evidence from traditional observational astronomy that seemed to refute the earth's annual motion.

10. Here Galileo's attitude is categorical, but note that it applies not to the whole of the Copernican system but to a specific part; I call this part the heliocentrism of planetary motions and its supporting argument the basic heliocentric argument. Note also that the planets in question are five particular bodies (Mercury, Venus, Mars, Jupiter, and Saturn); the earth is not included because to do so would mean that it too revolves around the sun and in this context this is a separate issue requiring additional evidence. That is, Galileo is certain that the planets Mercury, Venus, Mars, Jupiter, and Saturn revolve around the sun; that the evidence supporting this claim is conclusive; and that (as he claims later in this selection) such heliocentrism of planetary motions provides a plausible argument to support the earth's revolution around the sun; but this is not to say that he thinks he has conclusive proof that the earth revolves around the sun. As is well known, the system of Tycho Brahe accepted the heliocentrism of the five traditional planets, but claimed that the sun revolves around the motionless earth.

other.[11] So you can see whether Aristotle was wrong by a small amount in thinking that they are always equally distant from us.

SIMPLICIO. What, then, are the indications that their motions are around the sun?

SALVIATI. For the three superior planets (Mars, Jupiter, and Saturn) this is inferred from their being always found closest to the earth when they are in opposition to the sun and farthest when they are near conjunction;[12] [350] this variation in distance is so significant that when Mars is closest it appears sixty[13] times greater than when it is farthest. Then, in regard to Venus and Mercury, we are certain of their revolving around the sun from their never receding much from it and from our seeing them sometimes beyond it and sometimes in between; the latter is conclusively proved by the changes in the apparent shape of Venus.[14] For the case of the moon, it is indeed true that it cannot be separated from the earth, for reasons which will be given more clearly as we proceed.

11. I am not sure how he arrives at the figure of "almost eight" for Mars, given that the Copernican estimate of the mean distance between Mars and the sun is 1.52 times the mean distance between the earth and the sun (Dreyer 1953, 339), which yields a ratio of about 4.85 to 1. If we add the eccentricity of Mars's orbit, which I estimate at about 0.1 (Dreyer 1953, 336), we would still have a ratio of 5.9 to 1. Adding the eccentricity of the earth's orbit would further increase this, but not by the required amount to make it correspond to 8.

12. Conjunction is a configuration in the apparent position of two heavenly bodies when they appear to be on the same side of the earth, namely, close to each other or separated by only a few degrees on the celestial sphere; opposition is the configuration when the two bodies appear to be on opposite sides from the earth, namely, about 180 degrees apart on the celestial sphere; for example, a new or thinly crescent moon occurs when it and the sun are in conjunction, and a full moon occurs when it and the sun are in opposition.

13. This factor of 60 refers to the area of the apparent disk of Mars. That is, given that (as mentioned in Salviati's previous speech) its distance changes by a factor of "almost" 8, its apparent diameter would also change by a factor of "almost" 8; but areas vary as the square of linear dimensions, and so when the apparent diameter changes from 1 to "almost 8," the area changes from 1 to "almost 64" (since the square of 8 is 64). Galileo rounds off "almost 64" to 60. It is also important to recall (as will emerge later in this selection) that this change in the apparent size of Mars became observable only with the telescope, and that before this instrument was invented the *lack* of such a change was evidence against the earth's annual motion.

14. This paragraph contains a good summary of the basic heliocentric argument; the details of this conclusive proof will soon be explained in the text. Later in this selection Galileo discusses that the observation of these changes in Venus's appearance became possible only with the telescope and that the absence of such changes constituted another traditional astronomical objection to the earth's annual motion. In fact, these changes represent one of his major discoveries and are called the phases of Venus; as this label suggests, Venus periodically changes its apparent shape as does the moon, except that the period is longer than one month and the telescope is required to see them.

SAGREDO. I expect to hear more marvelous things that depend on this annual motion of the earth than was the case for those that depend on the diurnal rotation.

SALVIATI. You are absolutely right. For, the action of the diurnal motion on the heavenly bodies was and could be nothing but to make the universe appear to us to be hastily running in the opposite direction; but this annual motion, by mixing with the particular motions of all the planets, produces very many oddities that so far have made all the greatest men in the world lose their bearings. Now, returning to the first general considerations, I repeat that it is the sun which is the center of the heavenly revolutions of the five planets (Saturn, Jupiter, Mars, Venus, and Mercury); and it will also be the center of the earth's motion if we can manage to place it in the heavens.[15] Then, as regards the moon, it has a circular motion around the earth, from which (as I said) it cannot be separated in any way; but this does not mean that it fails to go around the sun together with the earth in the annual motion.

SIMPLICIO. I still do not comprehend this arrangement too well; perhaps by drawing a diagram we will understand it better and be able to discuss it more easily.

SALVIATI. So be it. Indeed, for your greater satisfaction and amazement, I want you to draw it yourself and see that you understand it very well, even though you think you do not grasp it; by merely answering my questions, you will draw it to the last detail.[16] So, take a sheet of paper and a compass, and let this white paper be the immense expanse of the universe where you have to locate and arrange its parts in accordance with the dictates of reason. First, without my teaching it to you, you firmly believe the earth to be located in this universe; so, take a point of your own choosing around which you understand it to be located, and mark it with some symbol.

[351] SIMPLICIO. Let this, which is marked A, be the location of the terrestrial globe. [See fig. 9.]

SALVIATI. Very well. Second, I know you know very well that the earth is neither located inside the solar body nor contiguous to it, but is

15. Note the clear distinction between these two theses both in regard to their information content and their epistemological status. The first refers to the five mentioned planets, whereas the second refers to the earth; and the first is more firmly established than the second.

16. This sentence provides another indication of Galileo's belief in the power of the Socratic method. It also indicates that he thinks the construction of diagrams is very important in the process of proof.

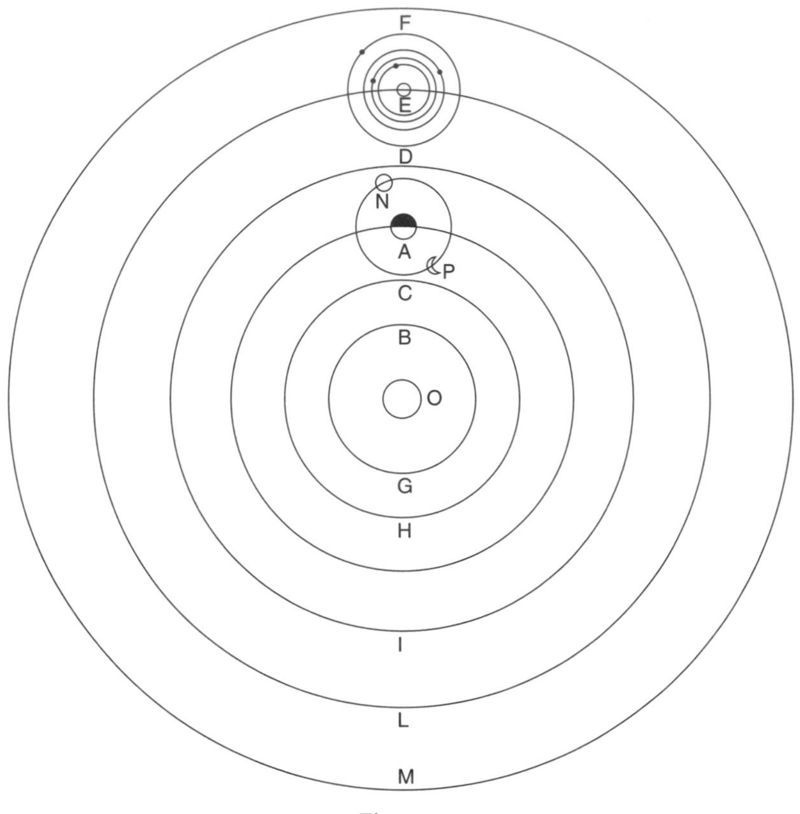


Figure 9.

separated from it by a certain distance; so, assign to the sun some other place of your choice, as far from the earth as you wish, and mark this too.

SIMPLICIO. It is done; let the location of the solar body be this, marked O.

SALVIATI. Having fixed these two, I want us to think about accommodating the body of Venus in such a way that its state and motion can satisfy what sensory appearances show us about them. So, recall what, from previous discussions or your own observations, you understand to occur in regard to this star; then assign to it the position you deem appropriate.

SIMPLICIO. Let us suppose the truth of the appearances which you related and which I also read in the booklet of conclusions: that is, that this star never recedes from the sun more than a determinate interval of little more than forty degrees, so that it not only never reaches opposition to the sun, but not even quadrature,[17] nor so much as the sextile configuration;[18] further, that it appears sometimes forty times[19] larger than at other times, namely, largest when it is in retrograde motion and approaches evening conjunction with the sun, and smallest when it is in direct motion[20] and approaches morning conjunction; moreover, that it is true that when it appears largest it shows a horned shape,[21] and when it appears smallest it is seen perfectly round. Given that all these appearances are true, I do not see how we can escape the conclusion that this star revolves in a circle around the sun; for this circle cannot in any way be said to enclose or to contain within it the earth, nor to be below the sun (namely, between it and the earth), nor to be above the sun. This circle cannot enclose the earth because then Venus would sometimes come into opposition to the sun; it cannot be below the sun because then Venus would appear sickle-shaped at both conjunctions with the sun; and it cannot be above the sun because [352] then it would appear always round and never horned. So, for its position I will mark the circle CH around the sun, without making it enclose the earth.[22]

SALVIATI. Having accommodated Venus, you should think about Mercury; as you know, the latter stays always near the sun and recedes from it much less than Venus does.

17. Quadrature is a configuration of the apparent position of two heavenly bodies when they form a right angle with the earth, namely, when they appear to be 90 degrees apart as seen from the earth; for example, a half moon is seen when the sun and moon are "in quadrature."

18. Sextile configuration is a configuration of the apparent position of two heavenly bodies when they are 60 degrees apart as seen from the earth.

19. Again, as in the aforementioned case of Mars, Galileo is thinking of the variation in the area of the planet's apparent disk: since the distance of Venus changes by a factor of 6, so does the apparent diameter; but the area changes as the square of the diameter; so the area of the apparent disk changes by a factor of 36 (the square of 6); Galileo rounds off 36 to 40.

20. Direct motion is the apparent motion which planets exhibit most of the time in their journeys among the fixed stars; its direction is eastward, namely, it follows the order of the constellations of the zodiac. The term is used primarily in contexts where we want to contrast direct motion to retrograde motion, whose direction is opposite (namely, westward).

21. A crescent shape.

22. The clarity of the exposition of the argument in this paragraph cannot be excelled, although its understanding depends, of course, on understanding the technical terms used: opposition, quadrature, sextile configuration, retrograde motion, direct motion, and conjunction. Any rephrasing of the argument would be redundant, but several points are worth repeating: it is a good example of a conclusive argument; its main conclusion states only that Venus revolves around the sun in an orbit which does not encompass the earth, but says nothing about the annual and diurnal motions; though some premises are reports of naked-eye observations, other crucial ones (involving the phases of Venus) are reports of telescopic observations.

SIMPLICIO. There is no doubt that, since it imitates Venus, a very appropriate place for it will be a smaller circle inside that of Venus and also around the sun; a very conclusive argument or reason for this, especially for its vicinity to the sun, is the vividness of its shining, which is greater than that of Venus and the other planets. So, on this basis we can draw its circle, marking it with the letters BG.[23]

SALVIATI. Where, then, shall we place Mars?

SIMPLICIO. Because Mars reaches opposition to the sun, it is necessary that its circle enclose the earth. But I see that it must necessarily enclose the sun as well; for when this planet reaches conjunction with the sun it would appear horned (like Venus and the moon) if it were not beyond the sun but rather in between; however, it always appears round. Therefore, its circle must enclose both the earth and the sun.

Moreover, I remember your having said that when it is in opposition to the sun it appears sixty times larger than when it is near conjunction; so, I think these appearances will agree very well with a circle around the center of the sun and enclosing the earth, which I am now drawing and marking DI. Here, at the point D, Mars is closest to the earth and in opposition to the sun; but, when it is at the point I, it is in conjunction with the sun and farthest from the earth.[24]

Finally, the same appearances are observed in regard to Jupiter and Saturn, although with much less variation for Jupiter than for Mars, and

23. The statement of this argument is not as precise as one might wish since the extent of Mercury's "imitation" of Venus is not specified; if this similarity were to include phases, then (like the previous argument about Venus) this one would also be conclusive; but if the similarity is restricted to the fact that Mercury too always appears at a small angular distance from the sun, then the essential step in this argument would be analogical reasoning. Later in this selection ([367]), Galileo indicates that no phases are visible for Mercury; thus, although he gives a plausible explanation of this fact, in the context of the present argument the evidence independently available remains essentially analogical. Note that here Galileo gives an additional piece of evidence, namely, that Mercury shines more vividly than Venus; if this were true, it would strengthen the point that Mercury is closer to the sun than Venus, since this shorter distance would be the only reasonable explanation of the greater vividness; but it is not true that Mercury shines more vividly than Venus, and it is unclear why Galileo puts this assertion in Simplicio's mouth.

24. This argument seems to be conclusive, but it relies on the new telescopic evidence about the periodic changes in the apparent size of Mars. Moreover, we must be careful about exactly what its conclusion claims: it says nothing about the annual and diurnal motions, and it does not deny that Mars revolves around the earth. The issue is really whether the center of Mars's orbit is the sun or the earth, given that both bodies are inside that orbit; the argument shows that this center is the sun and not the earth, but even this conclusion must be qualified as referring to the approximate and not the exact geometrical center. Similar clarifications apply with even greater force to the two other superior planets, Jupiter and Saturn, which are considered next.

still less with Saturn than with Jupiter; so, I think I understand that these two planets will also be very adequately accommodated by means of two circles also around the sun. The first one is for Jupiter and is marked EL; the other larger one is for Saturn and is labeled FM.

SALVIATI. So far you have conducted yourself splendidly. Now, as you can see, the variation in distance for the three superior planets is measured by an amount twice the distance between the earth and the sun; hence, [353] the variation is greater for Mars than for Jupiter since Mars's circle DI is smaller than Jupiter's circle EL; similarly, because EL is smaller than Saturn's circle FM, the variation is even less for Saturn than for Jupiter; this corresponds exactly to observation. What remains for you now is to think about the place to assign to the moon.

SIMPLICIO. Let us use the same argument, which seems to me to be very conclusive. Because we see the moon reach both conjunction and opposition with the sun, it is necessary to say that its circle encloses the earth; but we must not say that it encloses the sun because then near conjunction it would not appear horned but always round and full of light; furthermore, it could never produce, as it often does, an eclipse[25] of the sun by coming between it and us. Therefore, it is necessary to assign to it a circle around the earth, such as this marked NP; thus, when positioned at P, from the earth A it appears in conjunction with the sun and so can eclipse it sometimes; and when located at N it is seen in opposition to the sun, and in this configuration it can come into the earth's shadow and eclipse itself.[26]

SALVIATI. What shall we do now with the fixed stars, Simplicio? Do we want to spread them in the immense space of the universe, at different distances from any determinate point? Or do we want to place them on a surface extending spherically around its center, such that each of them is equidistant from the same center?

25. An eclipse occurs when a heavenly body becomes partially or completely invisible due to its relative position vis-à-vis the earth and the sun. In a lunar eclipse, the moon is eclipsed; that is, the earth is directly between the sun and the moon, and the moon is in the earth's cone-shaped shadow. In a solar eclipse, the sun is eclipsed; that is, the moon is directly between the sun and the earth, and the earth is in the moon's cone-shaped shadow. The explanation of eclipses was not in question in the Copernican controversy; the dispute involved which of these three bodies move and how, not the fact that eclipses result whenever they reach a straight-line configuration.

26. This argument about the moon is conclusive, but it is important to note that it depends on no new evidence (telescopic or otherwise); both sides of the Copernican controversy accepted it. Indeed, both agreed on the explanation of the monthly motion associated with the moon, of its phases, and of lunar and solar eclipses, and both accepted the geocentric character of its orbit; these things were simply not part of the controversy.

SIMPLICIO. I would rather follow an intermediate path. I would assign them an orb constructed around a determinate center and contained between two spherical surfaces, namely, a very high concave one and another convex one below it; and I would place the countless multitude of stars between them, but at different heights. This could be called the sphere of the universe, and it would contain inside it the orbs of the planets we have already drawn.

SALVIATI. So far, then, Simplicio, the heavenly bodies have been arranged just as in the Copernican system, and you have done this yourself. Moreover, you have assigned individual motions to all except the sun, the earth, and the stellar sphere; to Mercury and Venus you have attributed a circular motion around the sun, without enclosing the earth; you make the three superior planets (Mars, Jupiter, and Saturn) move around the same sun, encompassing the earth inside their circles; then the moon can move in no other [354] way but around the earth, without enclosing the sun; and in regard to these motions you again agree with Copernicus.[27]

Three things now remain to be assigned to the sun, earth, and stellar sphere: that is, rest, which appears to belong to the earth; the annual motion along the zodiac, which appears to belong to the sun; and the diurnal motion, which appears to belong to the stellar sphere and to be shared by all the rest of the universe except the earth. Since it is true that all the orbs of the planets (namely, Mercury, Venus, Mars, Jupiter, and Saturn) move around the sun as their center, it seems much more reasonable that rest belongs to the sun than to the earth, inasmuch as it is more reasonable that the center of moving spheres rather than any other point away from this center is motionless; therefore, leaving the state of rest for the sun, it is very appropriate to attribute the annual motion to the earth, which is located in the middle of moving parts; that is, between Venus and Mars, the first of which completes its revolution in nine months, and the second in two years. If this is so, then it follows as a necessary consequence that the diurnal motion also belongs to the earth; for if the sun were standing still and the earth did not rotate upon itself but only had the annual motion around the sun, then the cycle of night and day would be exactly one year long; that is, we would have six months of daylight and six months of night, as we have stated other times. So you see how appropriately the extremely rapid motion of

27. This completes the second step of this selection's main argument, a step I call the basic heliocentric argument and regard as conclusive (unlike the first and third steps). The next step uses the results already obtained to reach additional (and less certain) conclusions about the earth's motion.

twenty-four hours is taken away from the universe, and how the fixed stars (which are so many suns) enjoy perpetual rest like our sun. Notice also how elegant this first sketch is for the purpose of explaining why such significant phenomena appear in the heavenly bodies.[28]

SAGREDO. I see it very well. However, just as from this simplicity you infer a high probability[29] for the truth of this system, others, on the other hand, might perhaps draw contrary conclusions; because such an arrangement is the very ancient one of the Pythagoreans and agrees so well with the observations, one might wonder (not without reason) how it could have had so few followers in the course of thousands of years, how it could have been rejected by Aristotle himself, and how even after Copernicus it could continue to suffer the same fate.

SALVIATI. Sagredo, if you had ever happened to hear (as I have very many times) what kinds of stupidities suffice to make the common people stubbornly unwilling [355] to listen to (let alone accept) these novelties, I think you would wonder much less about the fact that there have been so few followers of this view. However, in my opinion, we should pay little attention to such brains; to confirm the earth's immobility and to remain unmoved in this belief, they regard as a very conclusive proof the fact that they cannot eat in Constantinople in the morning and have supper in Japan in the evening; and they are certain that the earth, being very heavy, cannot go up above the sun only to come back crashing down. We need not take into account these people, whose number is infinite, nor keep track of their stupidities; we need not try to gain the support of men whose definition contains only the genus but lacks the difference,[30] in order to have them as companions in very subtle and delicate discussions. Moreover, what gain would you think you could ever make

28. This plausible but inconclusive argument advances three main reasons for the earth's motion: it is more fitting to have the center (the sun) rather than a point off center (the earth) be motionless; the earth is positioned between two other bodies (Venus and Mars) which perform orbital revolutions; and the period of the earth's orbital revolution (one year) is intermediate between the periods of Venus and Mars, just as the size of its orbit is intermediate between the sizes of theirs. Salviati's talk of "elegance" (*agevolezza*) in his last sentence and Sagredo's talk of "simplicity" (*simplicità*) next are clues that this is, and is meant to be, a simplicity argument. This argument and this terminology also suggest that an aesthetic value may be involved, perhaps as an aspect of the concept of simplicity, or as a distinct principle.

29. The connection between simplicity and probability suggested here reiterates a theme discussed at greater length and more explicitly in selection 6.

30. The meaning of this witty remark should not be missed. In traditional logic, definitions were given by identifying the genus and the species (or specific difference) to which the thing defined belongs; the genus is a broader category of classification, and the species is a subdivision within the genus. The branch of modern biology called taxonomy follows this procedure to some extent; for example, it defines man as *homo sapiens*, namely, as

with all the demonstrations in the world when dealing with brains so dull that they are incapable of recognizing their extreme follies?

My wonderment, Sagredo, is much different from yours. You are surprised that there are so few followers of the Pythagorean opinion, whereas I am amazed at how there could ever have been anyone who accepted and followed it; nor can I ever sufficiently admire the eminence of mind of those who have accepted and regarded it as true, and who with the liveliness of their intellect have done violence to their own senses, so much so that they have been able to prefer what their theorizing told them over what their sensory experiences showed them very clearly to the contrary. We have already seen that the reasons against the earth's diurnal rotation, which have been examined, appear to be very good; the fact that they have been regarded as most conclusive by the Ptolemaics, Aristotelians, and all their followers, is a very good argument for their effectiveness. However, the observations that clearly contradict its annual motion appear to be even more powerful, so much so that (I repeat it) there is no end to my admiration of how in Aristarchus and Copernicus their reason could have done so much violence to their senses as to become, in opposition to the latter, mistress of their belief.[31]

---

belonging to the genus *homo* and the species *sapiens*. Aristotle defined man as *rational animal*, namely, as belonging to the genus *animal* and the species *rational*; if from this definition of man, the species is removed, we are left with *animal*; so a person whose definition (*rational animal*) contains only the genus (*animal*) but lacks the (specific) difference (*rational*) is an alleged rational animal who is not really rational but only a mere animal. In short, Galileo calls some of his opponents simply animals (or perhaps irrational animals). This rhetorical excess was one of many instances that won him many new enemies when the book was published; it must have given him pause a few years later, for he decided to insert an addendum on the next page to tone down the arrogance displayed here. After semantically understanding this remark, we should be able to appreciate it aesthetically and question it rhetorically; unfortunately, it is found in garbled form in the other English translation currently available (Drake 1967, 327). See also the appendix (3.4).

31. This is a controversial passage deserving serious reflection to derive the proper methodological lesson. Some take it as evidence of Galileo's apriorism by reading the word "admiration" literally and deriving the lesson that it is admirable and proper to invent, pursue, and accept theories contradicting observed facts; this "lesson" is a good statement of methodological apriorism; cf. Feyerabend (1975), Koyré (1978), and Shea (1972). I think Galileo's attitude is one of amazement and awe rather than endorsement; he is amazed that Aristarchus and Copernicus could be so apriorist-minded, and his amazement stems from his belief that, though such apriorism is normally undesirable, they were lucky and turned out to be right. My interpretation fits very well with the reference to the telescope in Salviati's next speech, with the explicit introduction of the empirical astronomical objections after that, and with Galileo's critical empiricism, critical apriorism, and methodological judiciousness in general; for more details, see the appendix (2.4), Drake (1978; 1990), and Finocchiaro (1988a; 1988b).

SAGREDO. Are we, then, also going to hear other powerful objections against this annual motion?

SALVIATI. We are. These are so clearly based on our sense experience that, if a higher and better sense than the common and natural ones had not joined with [356] reason, I suspect that I too would have been much more recalcitrant against the Copernican system than I have been since a lamp clearer than usual has shed light on my path.[32]

SAGREDO. Now then, Salviati, let us join the fray, for any word uttered for any other purpose seems to me to be wasted.

SALVIATI. I am ready to serve you.[33] I have already explained to you the structure of the Copernican system. [357] Against its truth the first extremely fierce assault comes[34] from Mars itself: if it were true that its

32. This is another controversial passage. Some (Drake 1967, 486n.) take the "higher and better sense than the common and natural ones" and the "clearer than usual lamp" to be the authority of the Catholic Church; thus, they interpret this remark as another one of Salviati's many instances of lip service to the Church. This interpretation strikes me as preposterous, for it seems obvious from the context that Galileo is referring to the telescope; as the ensuing discussion shows, it was this instrument that enabled him to answer most of the (otherwise conclusive) observational astronomical arguments against the earth's annual motion. Here he confesses that he is sufficiently empirically minded that he would have been unable to overcome the empirical evidence against Copernicanism if it had not been for the telescope; but this should not be equated with empiricism in the naive or dogmatic sense.

33. In the original 1632 edition of the *Dialogue*, there is no indication of a gap or an addendum at this point in the text. But in his own copy of the book, at this point Galileo made a handwritten note stating that it would be a good idea to insert some text here, and at the end of the Third Day he inserted several sheets of paper on which he wrote the text of the addendum. Favaro (7:356–62) prints this addendum in a note to the main text; Drake (1967, 328–33) inserts this additional text at this point, placing it in square brackets. I omitted the addendum from my translation because the additional text contains a discussion of two simpleminded objections to the earth's diurnal motion that does not raise any significant issues from a scientific, methodological, or logical point of view. But rhetorically speaking, it should be noted that the addendum is full of qualifications and mellow language and displays a charitable and understanding attitude toward common people and their difficulties in conceiving that the earth could move; Galileo is apparently trying to tone down the barbed and offensive remark of the previous page in which he called some of his opponents "animals" by unprosaically describing them as "men whose definition contains only the genus but lacks the difference." Finally, note that in the rest of this paragraph, the pagination of the critical edition shows a rapid succession of pages; this is because on each of these pages the addendum printed as a footnote takes up most of the space and there are only a few lines of the main text.

34. Galileo uses the present tense, although the past tense would be more appropriate; the point is that, besides explaining the structure of the Copernican system, earlier in this selection he gave a supporting argument which uses the new telescopic evidence, and so the objections he is about to state are the traditional ones that had been unanswerable until the telescope. This may be simply a slip of the pen. It may also be that Galileo insufficiently thought out the order of exposition of this whole selection, as Santillana (1953, 344n.32)

distance from the earth varies such that the farthest minus the closest distance [358] equals twice the distance from the earth to the sun, it would be necessary that when it is closest to us its disk should appear more than sixty times larger than when it is farthest; [359] however, this variation of apparent size is not perceived; instead, at opposition to the sun, when it is close to the earth, it appears barely four or five times larger than near [360] conjunction, when it is hidden behind the sun's rays.[35] Another and greater difficulty is due to Venus:[36] if (as Copernicus claims) it should turn around the sun and be sometimes beyond the sun and sometimes in between, and if it should recede from us [361] and approach us by a difference equal to the diameter of the circle it describes, then when it is positioned between us and the sun and is closest to us its disk would appear almost forty times larger [362] than when it positioned beyond the sun and is near its other conjunction; however, the difference is almost imperceptible. To this we should add another difficulty: it seems reasonable that the body of Venus is inherently dark and shines only because of the sun's illumination,[37] like the moon; if this is

---

suggests; one might then argue that these objections should have been presented before the exposition and justification of the Copernican system. However, the order of exposition Galileo actually uses makes these objections more of an integral part of the methodological discussion about the indispensability of the telescope, as I suggested above.

35. This was one of the traditional objections to the earth's annual motion; it is called the argument from the appearance of Mars.

36. This is called the anti-Copernican argument from the appearance of Venus; note that it contains two distinct but interrelated parts: changes in apparent diameter and changes in apparent shape.

37. This is a crucial assumption of the objection; without it Copernicanism would *not* imply phases for Venus, and this objection could not stand. From a Copernican viewpoint this assumption was "reasonable" because Venus was a heavenly body similar to the earth and the moon, and the latter two bodies could be seen to be opaque and dark; but the assumption was not necessary, for it was possible to be a Copernican, regard Venus as transparent or translucent, and not be bothered by its apparent lack of phases (as the example of Kepler shows). On the other hand, the assumption was made in some versions of the geostatic view, and then an analogous objection could be formulated against any arrangement that made Venus revolve around the earth in an orbit smaller than the annual motion of the sun; in those versions, the objection was answered by locating Venus beyond the sun. But most geocentrists rejected the assumption, thus keeping the orbit of Venus inside that of the sun; the rejection was relatively easy in the Ptolemaic system, either as an ad hoc maneuver or in accordance with the principle that all heavenly bodies were luminous insofar as they were composed of aether. Galileo (together with his pupil Castelli) was one of the few who felt that the lack of phases of Venus was an objection to Copernicanism. The problem of Venus's phases is further complicated by the fact that it is closely related to the question of its variation in apparent diameter, which was another effect of the Copernican arrangement. For more details, see Ariew (1987), Chalmers (1985), Drake (1984), and Thomason (1994b; forthcoming).

so, then when positioned between us and the sun it should appear sickle-shaped, as the moon does when it is likewise near the sun; but this phenomenon is not observed in Venus. Thus, Copernicus declared that either it is inherently luminous, or its substance is such as to be capable of absorbing sunlight and transmitting it through its interior, so that it appears to be always shining; this is how Copernicus excused Venus for not changing its apparent shape;[38] as regards the small variation in its apparent size, he said nothing. About Mars he said much less than was necessary; I believe the reason is that he was unable to account to his own satisfaction for a phenomenon so incompatible with his position; and yet, persuaded by many other confirmations, he stuck to it and regarded it as true. Furthermore, there is a feature that alters the order in such a way as to render it unlikely and false:[39] all planets together with the earth move around the sun, which is at the center of their revolutions; only the moon perturbs this order, by performing its proper motion around the earth; and then it, the earth, and the whole elemental sphere all together move around the sun in one year.

These are the difficulties that make me marvel at Aristarchus and Copernicus; they must have known about those difficulties but were unable to solve them; and yet, because of other wonderful confirmations, they trusted what reason told them so much that they confidently asserted that the [363] structure of the universe can have no other configuration but the one constructed by them. There are then other very serious and very beautiful difficulties which are not easily solved by mediocre intellects, but which were understood and explained by Copernicus; we will discuss them below,[40] after answering other objections which seem to undermine this position. Now, coming to the clarifications and solutions of the three very serious objections advanced above, I say that the first two not only do not contradict the Copernican system, but favor it considerably and absolutely; for both Mars and Venus do vary in apparent size in accordance with the required proportions, and Venus does appear sickle-shaped when

38. Cf. Copernicus, *On the Revolutions*, I, 10 (1992, 18–22).

39. This is called the lunar orbit argument against the earth's annual motion. As stated here, the argument speaks only of "perturbation of order," thus making it sound like a simplicity argument. But this argument can be strengthened by arguing that the difficulty with the lunar orbit in the Copernican system is that the moon revolves around the earth while the earth revolves around the sun; thus the moon has to keep up with the earth; but there was no known force or mechanism that would enable a body to behave in that manner.

40. This is a reference to the objections from stellar size, interstellar distances, and stellar parallax, discussed in selections 13 and 14.

between us and the sun and in general changes in apparent shape exactly like the moon.

SAGREDO. But how could this be hidden from Copernicus and revealed to you?

SALVIATI. These things can be understood only with the sense of vision, which nature has not granted us in such a perfect state as to be able to discern such differences; indeed the very instrument for seeing contains impediments within itself.[41] However, in our time God saw fit to allow the human mind to make a marvelous invention, which renders our vision more perfect by increasing its power by 4, 6, 10, 20, 30, and 40 times; as a result, countless objects that were invisible to us because of their distance or extremely small size are now rendered highly visible by means of the telescope.

SAGREDO. But Venus and Mars are not objects that are invisible because of their distance or small size; indeed we perceive them with our simple natural vision. So why do we not distinguish the variations in their size and shape?

SALVIATI. Here the major impediment stems from our eyes themselves, as I just mentioned. Objects that are shining and distant are not represented by our eyes as simple and sharp; instead, they are presented to us adorned with adventitious and extraneous rays, so long and thick that their bare little body appears to us enlarged 10, 20, 100, and 1,000 times more than it would be presented to us without the radiant head of hair which is not part of it.

SAGREDO. I now remember reading something on this subject, perhaps in the *Sunspot Letters* or in *The Assayer*[42] published by our common friend.[43] You ought to explain more clearly how this matter stands, both to refresh my memory [364] and for the understanding of Simplicio, who may not have seen these writings; I think this information is essential in order to comprehend what we are dealing with.

SIMPLICIO. Frankly, everything that Salviati is now advancing is new to me, for, to tell you the truth, I have not had the curiosity to read those books. Nor have I so far placed much trust in the newly introduced spyglass; on the contrary, following in the footsteps of my fellow Peripatetic

41. Two features of vision are mentioned here: one is that our eyes have limited power and so cannot see things which are too small or too far; the other is that our eyes are instruments whose operation is subject to disturbances.

42. Cf. Favaro (5:196–97; 6:273 ff.), respectively. Since the telescope, Galileo constantly reflected on this feature of human vision; he discussed it also in *The Sidereal Messenger* (Van Helden 1989) and the *Discourse on the Comets* (Favaro 6:37–108).

43. "Common friend" refers to Galileo.

philosophers, I have regarded as fallacies and deceptions of the lenses what others have admired as stupendous achievements. However, if I have been in error so far, I should like to be freed from it; enticed by the other novelties I heard from you, I will more carefully listen to the rest.

SALVIATI. The confidence these men have in their own cleverness is as unjustified as the little regard they have for the judgment of others; it is very revealing that they should consider themselves to be better qualified to judge this instrument, without having ever experimented with it, than those who have made thousands of experiments with it and continue to make them every day. However, please let us forget about such stubborn persons, who cannot even be criticized without doing them more honor than they deserve.[44]

Returning to our purpose, I say that shining objects appear to our eyes surrounded by additional rays, either because their light is refracted by the fluids covering the pupils, or because it is reflected by the edges of the eyelids (thus scattering the reflected rays onto the same pupils), or for some other reason; hence, these objects appear much larger than if their bodies were represented without such irradiation. This enlargement becomes proportionately greater and greater as such brilliant objects are smaller and smaller; for example, if we assume that the increase due to the shining hair is four inches, and that this addition is made around a circle with a diameter of four inches, then its apparent size is increased nine times, but . . . [45]

SIMPLICIO. I suspect you meant to say "three times"; for by adding four inches on one side and four on the other to a circle with a diameter of four inches, you are tripling its dimensions, not increasing them nine times.

SALVIATI. A little geometry is needed, Simplicio. It is true that the diameter [365] increases threefold, but the surface (which is what we are talking about) increases ninefold; for, Simplicio, the surfaces of circles are to each other as the squares of their diameters, and so a circle with a

44. Here Galileo does not discuss in a serious and systematic manner the controversy about the telescope and the objections to it; doing so would have been relevant because the status of the new evidence depended on the acceptance of the telescope. Ultimately he turned out to be right, and the telescope became not only an indispensable instrument in itself, but also the forerunner of an instrumentation revolution in scientific inquiry. But when the telescope was first invented, there were serious and legitimate concerns about its methodological admissibility, its theoretical explanation, its empirical reliability, and its practical operation and use. For a statement of the main objections, see the introduction (5.3) and cf. selection 3; for more details, see Crombie (1967), Feyerabend (1975), King (1955), Ronchi (1958), Rosen (1947), and Van Helden (1984; 1994).

45. This ellipsis is in the original; Galileo wants to represent an interruption of Salviati's speech by Simplicio.

diameter of four inches is to another of twelve as the square of four is to the square of twelve, namely, as 16 is to 144; hence, the latter will be nine times larger, not three. So, please be careful, Simplicio.

Now, let us go forward. If we were to add the same head of hair four inches wide to a circle with a diameter of only two inches, the diameter of the whole wreath would be ten inches, and its whole surface compared to the area of the naked little body would be as one hundred to four (for these are the squares of ten and two); therefore, the enlargement would be twenty-five times. Finally, the four inches of hair added to a small circle with a diameter of one inch would enlarge it eighty-one times. Thus, the enlargements constantly take place in greater and greater proportions as the real objects being enlarged are smaller and smaller.

SAGREDO. The difficulty that troubled Simplicio did not really trouble me; but there are some things which I want to understand better. In particular, I should like to know on what basis you claim that this enlargement is always equal for all visible objects.

SALVIATI. I already explained myself in part when I said that only brilliant objects are enlarged, not dark ones; now I shall add the rest. Brilliant objects that shine with a brighter light produce a greater and stronger reflection on our pupils, and so they appear to be enlarged much more than those which are less bright. In order not to dwell on this particular any longer, let us see what our true mentor teaches us. Tonight, when it is very dark, let us look at the planet Jupiter; we will see it appear very bright and very large. Let us look at it through a tube, or through a small hole made with a fine needle in a piece of paper, or even through the small slit we can create by closing our hand and leaving some space between our palm and fingers; we will then see the disk of the same Jupiter stripped of its rays and so small that we will easily judge it smaller than one-sixtieth the size it appears when its great torch is observed with the naked eye. [366] Let us then look at the Dog Star,[46] which is very beautiful and larger than any other fixed star, and which appears to the naked eye not much smaller than Jupiter; when we remove its head of hair in the manner indicated, its disk will be seen to be so small that it will be judged one-twentieth that of Jupiter; indeed, whoever lacks perfect vision will have great difficulty perceiving it; from this we may reasonably conclude that, insofar as the light of this star is much brighter than that of Jupiter, it produces a greater irradiation than Jupiter does. Furthermore,

46. The Dog Star (also called Sirius) is the brightest star in the sky; it is located in the southern constellation of Canis Major near the celestial equator.

the irradiations of the sun and moon are almost nothing, due to the fact that their size by itself takes up so much space in our eye as to leave no room for the adventitious rays; thus, their disks are seen shaved and clear-cut. We can ascertain the same truth by means of another experiment, which I have made several times; I am referring to ascertaining that bodies shining with a brighter light are surrounded by rays much more than those whose light is dimmer. I have observed Jupiter and Venus together several times when they were 25 or 30 degrees away from the sun and the sky was very dark; when I observed them with the naked eye, Venus appeared at least eight and perhaps even ten times larger than Jupiter; but when they were observed with a telescope, the disk of Jupiter was seen to be at least four times larger than that of Venus, and the brightness of Venus's shine was incomparably greater than the extremely dim light of Jupiter; this derived only from the fact that Jupiter was extremely far from the sun and from us, and Venus was close to us and the sun.[47]

Having explained these things, it will not be hard to understand how it can happen that, when Mars is in opposition to the sun and hence more than seven[48] times closer to the earth than when it is near conjunction, it appears to us four or five times larger in the former configuration than in the latter, although we should see it more than fifty[49] times

47. Salviati's account of vision in his last three speeches is important because it helps to answer the traditional anti-Copernican arguments based on the appearances of the planets Mars and Venus. The basic structure of these arguments is that if Copernicanism is correct, then we should be able to observe certain features in regard to these planets, which in fact we do not observe; it is not enough for Galileo to reply that with the telescope we can now observe those phenomena; he must also explain why they were not observed with the naked eye, and this account of vision provides the explanation. This account is also interesting from a more general, methodological point of view; for this is one of many passages where he treats the human eye as an instrument, and this attitude toward the eye and this type of interpretation of the process of vision are important methodologically because they started a development that became an integral part of modern scientific inquiry. Cf. selections 13 and 14, Brown (1985), and Crombie (1967).

48. Earlier in this selection ([349]) Galileo claimed that the distance between Mars and the earth changes by a factor of "almost eight." Here the factor of "more than seven" applies when Mars is "near" conjunction (as the text goes on to state), whereas when he mentioned the number eight earlier he was referring to the planet's position exactly at conjunction. At (exact) conjunction Mars is slightly farther than "near" conjunction, but sunlight renders it invisible when exactly at conjunction, and so the more relevant consideration is the one referring to its position "near" conjunction. (I thank Alan Rhoda for having suggested this resolution of the apparent discrepancy.)

49. Recall that this refers to the area of the apparent disk; the area varies as the square of the diameter, and so when the diameter changes by a factor of "more than seven," the area changes by a factor of "more than 49" (since the square of 7 is 49); Galileo is rounding off 49 to 50.

larger. The cause of this is simply the irradiation; for if we strip it of the adventitious rays, we will find it enlarged exactly by the required proportion. To strip it of its head of hair, the only excellent means is the telescope, which enlarges its disk by nine hundred or a thousand times;[50] thus, we see it bare and clear-cut like that of the moon, and different in size in the two positions exactly in accordance with the required proportion.

Then, as regards Venus, it should appear almost forty times larger at its evening conjunction below the sun than at its other morning conjunction;[51] and yet it is seen as not even doubled. [367] Here, besides the irradiation effect, what is happening is that it is sickle-shaped and its horns are not only very thin but also are receiving the sunlight obliquely; hence, this light is very dim in intensity and little in amount, and consequently its irradiation is less than when the planet's hemisphere appears entirely illuminated. On the other hand, the telescope clearly shows us its horns as clear-cut and distinct as those of the moon; and they are seen as part of a very large circle, which is almost forty times larger than its same disk when it is beyond the sun at the end of its appearance as a morning star.[52]

SAGREDO. Oh, Nicolaus Copernicus, how pleased you would have been to see this part of your system confirmed by such clear observations!

SALVIATI. Indeed; but how much less would have been his reputation among the experts for preeminence of intellect! For, as I said before, he constantly continued to claim what was in accordance with arguments even though it was contrary to sensory experiences; and I cannot stop marveling at the fact that he should have persisted in saying that Venus turns around the sun and is sometimes more than six times farther from us than at other times, although it always appears equal to itself, even when it should appear forty times larger.[53]

50. Here Galileo is thinking of a telescope with a magnification of about 30 times; this instrument would enlarge the apparent linear dimensions of an object 30 times, and its apparent surface area by the square of 30, which is 900.

51. That is, not at the morning conjunction that immediately follows the evening conjunction while Venus is still between the earth and the sun, but rather at the morning conjunction that occurs when the planet is beyond the sun, after it has appeared as a morning star for a while, and just before it turns again into an evening star. Here the value of 40 is derived as it was earlier.

52. A morning star is a bright heavenly body visible in the eastern sky just before sunrise during twilight when it is too light to see other stars.

53. This speech raises the same issues as the analogous remarks above, namely, whether Galileo is expressing admiring approval or unsettling amazement.

SAGREDO. In regard to Jupiter, Saturn, and Mercury, I think we should also see differences in their apparent size corresponding exactly to their different distances.

SALVIATI. In the case of the two superior planets, I have exactly observed these differences almost every year for the past twenty-two years.[54]

In the case of Mercury, no observation of any consequence is possible because it becomes visible only at its maximum elongations[55] from the sun (where its distances from the earth are insignificantly different), and hence these differences are imperceptible. It is similar with its changes of shape, which must occur absolutely as in Venus; that is, when we see Mercury, it should appear in the shape of a semicircle, as Venus also does at its maximum elongation; but Mercury's disk is so small and its light so bright (due to its being so close to the sun) that the power of the telescope is not enough to shave its hair and make it appear completely shorn.

There remains what seemed to be a great difficulty with the earth's motion; that is, unlike all the other planets that revolve around the sun, [368] it alone does so (in one year) accompanied by the moon together with the whole elemental sphere, while the same moon moves every month around the earth. Here we must, once again, proclaim and exalt the admirable perspicacity of Copernicus and at the same time pity his misfortune; for he does not live in our time when, to remove the apparent absurdity of the shared motion of the earth and moon, we can see that Jupiter (being almost another earth) goes around the sun in twelve years accompanied not by one moon but by four moons, together with all that may be contained within the orbs of the four Medicean stars.

SAGREDO. For what reason do you call the four planets surrounding Jupiter moons?

54. This observer is not Salviati himself but rather the Academician, namely, Galileo. As Santillana (1953, 347n.33) points out, Salviati died in 1614 when he was thirty-one and so, if this remark is taken literally, he would have been no more than nine years old and would have begun such observations before the telescope was invented; but if we take it to refer to Galileo, then the period in question is from 1609 (when he first built the telescope) to 1631 (when the *Dialogue* went to press).

55. Elongation is the angular distance (as seen from the earth) of one heavenly body from another. For the planets, elongation normally refers to their angular distance from the sun. Mercury and Venus never exhibit a large elongation; the maximum elongation of Mercury is 28 degrees, and that of Venus 47 degrees. The other planets can exhibit the greatest possible elongation, which is 180 degrees and is called opposition (to the sun).

SALVIATI. They would appear such to someone who looked at them while standing on Jupiter. For they are inherently dark and receive light from the sun, which is evident from their being eclipsed when they enter inside the cone of Jupiter's shadow; moreover, because the only part of them that is illuminated is the hemisphere facing the sun, they appear always entirely lit to us who are outside their orbits and closer to the sun; but to someone on Jupiter they would appear entirely lit when they were in the parts of their orbits away from the sun, whereas when in the inner parts (namely, between Jupiter and the sun), from Jupiter they would be seen as sickle-shaped; in short, to Jupiter's inhabitants they would show the same changes of shape which the moon shows to us terrestrials.[56]

Now you see how wonderfully in tune with the Copernican system are these first three strings that at first seemed so out of tune.[57] Furthermore, from this Simplicio will be able to see the degree of probability with which one may conclude that the sun rather than the earth is the center of the revolutions of the planets. Finally,[58] the earth is placed between heavenly bodies that undoubtedly move around the sun, namely, above Mercury and Venus and below Saturn, Jupiter, and Mars; therefore, likewise will it not be highly probable and perhaps necessary to grant that it too goes around the sun?[59]

. . . [60]

56. The existence of Jupiter's satellites helps to answer the lunar orbit objection to Copernicanism; the details of how this is accomplished are left as an exercise for the reader.

57. These strings are the observational evidence about the apparent size of Mars, the apparent size and shape of Venus, and the geocentric orbit of the moon.

58. Here the punctuation in Galileo (1632, 332) is more helpful than that in Favaro (7:368).

59. This paragraph is a good summary of the three main parts of this selection: the observational argument supporting the heliocentrism of planetary motions; the simplicity argument from the heliocentrism of planetary motions to the earth's annual motion; and the statement and criticism of three traditional observational astronomical objections to the earth's annual motion.

60. The passages omitted here correspond to Favaro 7:368.32–383.33 and Galileo 1967, 340–56. The first discusses the problem of explaining retrograde planetary motion: it formulates an argument in favor of Copernicanism, based on the greater simplicity and explanatory coherence of the Copernican as compared to the Ptolemaic explanation; and it advances some methodological reflections on the distinction between the mathematical instrumentalist and the philosophical realist approaches to astronomy. The second discusses the motion of sunspots across the solar disk: Galileo presents some of his empirical discoveries; he formulates a novel argument in favor of Copernicanism, based on its ability to explain this phenomenon in a simpler and less ad hoc manner than the Ptolemaic theory; and he reflects on the methodology of theoretical explanation. For more details, see Finocchiaro (1980, 40–41, 129–30, and 246–53).

# [12. The Role of the Bible]

[383] . . . SAGREDO. It is now time for us to listen to the objections from the booklet of conclusions or disquisitions, which Simplicio has brought back.

SIMPLICIO. Here is the book. And here is the place where the author first briefly describes the world system according to Copernicus's position [384] by saying: "so, Copernicus has the earth together with a moon and all this elemental world, etc."[61]

SALVIATI. Stop for a moment, Simplicio, for I think that in this initial statement this author shows to have very little understanding of the position he undertakes to refute, when he says that Copernicus has the earth together with the moon move along its orbit in a year from east to west. This statement is false and impossible, and it was never made by him; on the contrary, he has it move in the opposite direction, meaning from west to east, namely, in the order of the signs of the zodiac; it then follows that such appears to be the annual motion of the sun, positioned motionless at the center of the zodiac.[62] Look at the reckless arrogance of certain people! They undertake the confutation of someone else's doctrine but misunderstand its primary foundations, which support the greater and most important part of the whole construction. This is a bad beginning for gaining a reader's trust. But let us go on.

61. Cf. Locher (1614, 23–28). Strauss (1891, 557n.51) gives a fuller quotation in which Locher indeed states that according to Copernicus the earth revolves around the sun from east to west.

62. The apparent annual motion of the sun is in a direction that may be described as eastward, counterclockwise (as viewed from the north celestial pole), or in the order of the signs of the zodiac (Aquarius, Pisces, Aries, etc.). In the Ptolemaic system, which equates appearance and reality, the sun actually revolves along its annual orbit in the same direction. The Copernican system, which explains appearances in terms of some underlying reality, switches the annual revolution from the sun to the earth, but does *not* reverse the direction. The direction of the annual motion need not be reversed because the annual motion (whether by the sun or the earth) takes place inside, and is then projected onto, the (imaginary) zodiacal band of the celestial sphere; regardless of which body actually moves, their relative motion is counterclockwise and gets projected as counterclockwise onto the zodiac. By itself, this point may not be confusing; it becomes so when considered in the context of the diurnal motion, which is clockwise in appearance, and also in reality according to the Ptolemaic view, but which is really counterclockwise according to Copernicanism; the latter thus reverses the direction of diurnal motion, but not that of annual motion. The reversal of diurnal motion is needed because diurnal motion is an axial rotation and the two "bodies" in question are the earth and the rest of the universe. Cf. the introduction (3.1) and selection 6 ([143]).

SIMPLICIO. After explaining this world system, he begins to advance his objections to this annual motion; the first ones are the following, which he advances with sarcasm and scorn for Copernicus and his followers. This author writes that according to this fanciful arrangement of the universe, one must utter the most solemn nonsense; that is, that the sun, Venus, and Mercury are below the earth;[63] that heavy bodies go naturally upward and light ones downward; that Christ, our Lord and Redeemer, ascended into hell and descended into heaven when he approached the sun; that when Joshua ordered the sun to stand still,[64] it was the earth which stood still, or else the sun moved in a direction opposite to that of the earth; that when the sun is in the constellation of Cancer, the earth moves through Capricorn;[65] that the winter signs signal summer and the summer signs signal winter; that it is not the stars which rise and set on the earth, but the earth which rises and sets on the stars; that the east begins in the west, and the west in the east; and, in short, that almost everything in the universe is turned upside down.

SALVIATI. All this is proper, except for having mixed these childish and scurrilous inanities with passages of the Holy Scripture, which is to be treated always with reverence and awe; nor is it proper to try to wound with holy objects someone who, philosophizing for fun and in

63. This objection presupposes an absolute rather than relative understanding of the concepts of up and down; Aristotelian natural philosophy defines upward as away from the center of the universe and downward as toward it; since in the Copernican system the sun, Venus, and Mercury are closer to the center of the universe, Aristotle's definition implies that they are located "downward" from, or at a "lower" position than, or simply "below" the earth. For more details, see Galileo's "Reply to Ingoli" (Finocchiaro 1989, 175–80).

64. Cf. Joshua 10:12–13 (King James Version): "Then spake Joshua to the Lord in the day when the Lord delivered up the Amorites before the children of Israel, and he said in the sight of Israel, Sun, stand thou still upon Gibeon; and thou, Moon, in the valley of Ajalon. And the sun stood still, and the moon stayed, until the people had avenged themselves upon their enemies. Is not this written in the book of Jasher? So the sun stood still in the midst of heaven, and hasted not to go down about a whole day." This was one of the most commonly mentioned biblical passages alleged to contradict the earth's motion and used as the basis of the biblical argument against Copernicanism. This biblical objection was one of many religious arguments designed to undermine the new world view; the previous sentence embodies another one of these arguments. Galileo's criticism, including an examination of the Joshua passage, is found in his "Letter to the Grand Duchess Christina" (Finocchiaro 1989, 87–118); cf. Finocchiaro (1986) and Moss (1993).

65. Cancer is a northern constellation of the zodiac lying between Gemini and Leo, allegedly having the shape of a crab (which gives it its Latin name). Capricorn is a southern constellation of the zodiac lying between Sagittarius and Aquarius, allegedly having the shape of a goat (which gives it its Latin name). They are diametrically opposite to each other on the celestial sphere.

jest, does not affirm or deny anything, but rather makes certain assumptions or hypotheses and then reasons in a friendly manner.

SIMPLICIO. Frankly, I too was scandalized more than a little, [385] especially when he added that, although the Copernicans respond to these and other similar arguments (very tortuously, to be sure), they will not be able to satisfy and answer the things that follow.

SALVIATI. This is worst of all because it implies that there are things more effective and conclusive than the authority of the Holy Writ.[66] However, please let us revere this, and let us go on to natural and human reasoning. Indeed, if among natural arguments he does not produce more plausible things than those presented so far, we will be able to set aside this whole undertaking, for I certainly am not willing to waste words to respond to such foolish trifles. Finally, his saying that the Copernicans respond to these objections is most false; nor can one believe that anyone would spend his time so uselessly.

## [13. Stellar Dimensions and the Concept of Size]

SIMPLICIO. I also agree with the same judgment; so let us listen to the other objections, which he presents as being much more powerful.[67] Here, as you can see, is one conclusion he reaches by means

66. Here Galileo's attitude toward the Bible and its role in scientific inquiry is difficult to interpret; his language is evasive and ambiguous. On the one hand, he seems to suggest that it is not "proper . . . mixing . . . passages of the Holy Scripture" when one engages in "philosophizing . . . and . . . makes certain assumptions or hypotheses and then reasons," to quote selectively from Salviati's previous speech; this means that the Bible is not a scientific authority and its assertions about the physical universe carry no weight; and it corresponds to the methodological principle elaborated in Galileo's "Letter to the Grand Duchess" of 1615 (Finocchiaro 1986; 1989, 87–118). On the other hand, Salviati's present speech seems to assert that there is no source of scientific information which is "more effective and conclusive than the authority of the Holy Writ"; this statement is comprehensible as part of the book's religious rhetoric to give the impression that it was not defying religious authorities; for despite the ambiguities of the anti-Copernican decree of 1616 and the confusing status of his other restrictions, Galileo knew he was under prohibition to discuss the religious objections to Copernicanism. Besides this ambivalence, another puzzle is why he brings up the subject at all; although he does it indirectly by way of quoting from a book (Locher 1614) published before the anti-Copernican decree, and although he quickly drops the matter, the question remains why he does not avoid the topic altogether. In this context, it should be noted that the *Dialogue* was written and published while the anti-Copernican decree of 1616 was still in force. For more details, see the introduction (4) and the appendix (3).

67. The explicit source of these anti-Copernican arguments and the explicit target of Galileo's criticism is still Locher (1614).

of the most exact calculations: let us suppose, as Copernicus himself says we must, that the earth's orbit, along which Copernicus makes it revolve around the sun in one year, were imperceptible compared to the immensity of the stellar sphere; then one would have to say necessarily that the fixed stars are at an unimaginable distance away from us, that the smaller ones among them are larger than the whole annual orbit, and that some others are much larger than the whole orbit of Saturn; but such dimensions are too large, as well as incomprehensible and incredible.

SALVIATI. I have already seen a similar thing brought up by Tycho[68] against Copernicus, and it has been a while since I discovered the fallacy (or, to be more exact, the fallacies) of this argument; it is built upon extremely false hypotheses and upon an assertion by Copernicus himself which his critics take with the most narrow literalness, like those debaters who (being essentially wrong on the merits of the issue) focus on a single small word casually uttered by their opponents and clamor about it without end. For a clearer understanding, you should know that Copernicus first explained the wonderful consequences about the other planets which derive from the earth's annual motion; that is, in particular, the direct and the retrograde motions of the three superior ones, which are perceived to be greater for Mars than for Jupiter (due to [386] Jupiter being farther) and smallest for Saturn (due to its being farther than Jupiter); he then added that these apparent changes are imperceptible in the fixed stars, due to their immense distance from us compared to Jupiter's or Saturn's distance. Here the opponents of this opinion rise up: they first take the imperceptibility mentioned by Copernicus[69] as if he were assuming it to be really and absolutely nothing; they add that even one of the smaller fixed stars is perceptible since it is caught by the sense of vision; then, on the basis of other false assumptions, they make various calculations; and they conclude that according to the Copernican doctrine one must claim that a fixed star is much larger than the whole annual orbit.[70]

68. Cf. Brahe (1602, 481).

69. Copernicus, *On the Revolutions*, I, 10 (1992, 22).

70. This objection is the target of the critical analysis of this selection and is called the stellar dimensions argument; for it refers to the size of individual stars and their distance from the earth, and so there are *two* dimensions involved, stellar size and stellar distance; stellar distances are a measure of the size of the universe, and so we are dealing with the size of particular stars and of the whole universe. The objection is that the stellar dimensions implied by the earth's annual motion are impossible (or absurdly large) because in the course of a year apparent stellar sizes do not change (or change imperceptibly little). A complication of this discussion is that apparent stellar size is a function of both actual stellar size and actual stellar distance, increasing with the former and decreasing with the latter.

Now, to reveal the futility of this whole account, I will prove by means of a truthful demonstration the following conclusion: if a fixed star of the sixth magnitude is supposed to be no larger than the sun, then its distance from us becomes sufficient to ensure that it exhibits no noticeable effect due to the earth's annual motion, which causes such great and observable variations in the planets; at the same time I will show in detail the serious fallacies in the assumptions of Copernicus's opponents.

I first assume, with Copernicus and in agreement with his opponents, that the radius of the annual orbit (which is the distance from the earth to the sun) contains 1,208 radii of the earth;[71] second, I suppose, in agreement with the same and with the truth, that the apparent diameter of the sun at its average distance[72] is about half a degree, namely, 30 minutes or 1,800 seconds or 108,000 thirds.[73] Now, the apparent diameter of a fixed star of the first magnitude is no larger than 5 seconds (namely, 300 thirds), and the diameter of a fixed star of the sixth magnitude 50 thirds (and here lies the greatest error of Copernicus's opponents);[74] therefore, the diameter of the sun contains the diameter of a fixed star of the sixth magnitude 2,160 times. Consequently, if we assume that a fixed star of the sixth magnitude is in reality equal to the sun and no larger, which is the same as saying if the sun were moved so far away that its diameter would appear [387] 2,160 times smaller than it appears to us now, then its distance would have to be 2,160 times greater than is now the

71. Galileo takes this number from Locher (1614, 25), which corresponds to Ptolemy's estimate of 1210 earth radii (*Almagest*, V, 15) and Copernicus's of 1142 (*On the Revolutions*, IV, 21; 1992, 208). This is an underestimation by a factor of about 20, since the correct distance is about 23,439 earth radii.

72. This talk of average distance presupposes that Galileo would admit that the earth-sun distance varies, and so either the earth's orbit is not perfectly circular or the sun is not the exact geometrical center of that orbit.

73. A third (of a degree of arc) is 1/60 of a second of arc.

74. These estimates are literally correct in the sense that 5 seconds and 50 thirds are given here as the upper limits for the apparent diameters of fixed stars; but when used as approximate estimates, as Galileo goes on to do, they still represent great overestimations; the apparent diameters of stars are actually of the order of a fraction of a third. The point is that, despite his awareness of the irradiation effect and other disturbances in the observation of stars, his telescope did not enable him to perceive their apparent diameter due to its extremely small value and to the presence of other disturbances unknown to him (such as the motion of stars). Drake (1967, 487n) states that as late as the nineteenth century, astronomer Friedrich G. W. von Struve argued that it was still impossible to measure reliably the true angular diameter of a star; and Pagnini (1964, 124n.1) states that only in the last few years have astronomers been able to measure the apparent diameter of supergiant stars like Betelgeuse. Thus, here Galileo made a valuable beginning and was on the right track, despite the limitations of his instruments and his analysis.

case, which is to say that the distance of a fixed star of the sixth magnitude is 2,160 radii of the annual orbit. Moreover, because by common agreement the distance from the earth to the sun contains 1,208 earth radii, and because (as we said) the distance of a fixed star is 2,160 radii of the annual orbit, therefore the earth's radius in comparison to that of the annual orbit is much greater than (almost double) the radius of the annual orbit in relation to the distance of the stellar sphere; hence, the difference in the appearance[75] of a fixed star caused by the diameter of the annual orbit would be little more observable than what is observed for the sun as a result of the earth's radius.

SAGREDO. As a first step, this makes a big drop.

SALVIATI. Indeed it does. For, according to the computations of this author, in order to go along with Copernicus's statement, a fixed star of the sixth magnitude had to be as large as the whole annual orbit; but, by making it merely equal to the sun (which is much more than ten million times smaller than the annual orbit), we make the stellar sphere so large and distant that it suffices to remove the objection against Copernicus.

SAGREDO. Please do this computation for me.

SALVIATI. The computation is easy and very short. The solar diameter is eleven earth radii,[76] and the diameter of the annual orbit contains 2,416 of the same, as both sides agree; thus, the diameter of the annual orbit contains that of the sun approximately 220 times. Because spheres are to each other as the cube of their diameters, let us compute the cube of 220; this is 10,648,000, and so we will have the annual orbit ten million six hundred and forty-eight thousand times larger than the sun. This author was saying that a star of the sixth magnitude had to be equal to this annual orbit.

SAGREDO. Therefore, the error of these people consists in being very badly mistaken in estimating the apparent diameter of fixed stars.

SALVIATI. That is the error, but he is not alone.[77] Indeed I am greatly surprised that so many astronomers [388] have been so badly mistaken in

75. "Appearance" refers to apparent diameter or apparent position; the word is *aspetto* in the original.

76. This value is a mathematical consequence of the above mentioned estimates for the earth-sun distance and for the apparent solar diameter; it corresponds to Copernicus's value (*On the Revolutions*, IV, 20–21; 1992, 208). Since the distance had been grossly underestimated, so was this value for the solar diameter; its true value is about 20 times bigger, namely, 218 earth radii.

77. That is, Locher (1614) is not alone in committing the observational error of overestimating the apparent diameter of fixed stars. This is the issue here because the mathematical calculations just made are uncontroversial; they establish that *if* the apparent di-

determining the magnitudes of all stars, fixed as well as wandering, except for the two luminaries,[78] and that they have paid little attention to the adventitious irradiation which deceptively makes them appear more than one hundred times larger than they are seen without the hair; and I am referring to astronomers of great renown, such as Al-Farghani,[79] Al-Battani,[80] Thabit ibn Qurrah,[81] and more recently Tycho and Clavius,[82] in short to all the predecessors of our Academician. This inadvertence of theirs is inexcusable because it was within their power to observe them at will without the hair (it is enough to observe them at their first appearance in the evening or before their disappearance at dawn); at least Venus ought to have alerted them to their fallacy, for oftentimes during the middle of the day it is seen so small that we must strain our eyesight, and then the following night it seems to be a very large torch; and I cannot believe that they thought the true disk was the one which appeared in deep darkness rather than the one which is seen in a luminous environment because they could have been sufficiently alerted by our lamps, which at night from afar appear large but at close range show their real flame to be clear-cut and small. Indeed, if I must frankly state my opinion, I really believe that none of them ever made it a point to measure the apparent diameter of any star, except the sun and moon; I think this applies even to Tycho himself, who was so precise in handling astronomical instruments, and who built them so large and exact, without

---

ameter of a fixed star is 50 thirds, this could be explained by assuming its actual diameter to equal that of the sun and its actual distance to be 2,160 times greater; keeping the distance fixed, the actual diameter is directly proportional to the apparent diameter, and so overestimations of the latter lead to overestimations of the former; hence, difficulties about excessively large *actual* sizes may be answered by reducing *apparent* size. But the mathematical relationship among apparent diameter, actual diameter, and actual distance is such that if the apparent diameter is decreased we must either decrease the actual diameter or increase the actual distance; so other issues will arise in regard to the distance.

78. *Luminary* is a term sometimes used to refer to the sun or moon because they shine with much more light than other heavenly bodies.

79. Al-Farghani was a ninth century Arab astronomer, best known as the author of an elementary textbook on Ptolemaic astronomy that was popular in Europe until Galileo's time.

80. Al-Battani (c. 858–929) was an Arab astronomer and mathematician who made important contributions to the Ptolemaic system.

81. Thabit ibn Qurrah (c. 836–901) was an Arab mathematician, physician, and philosopher who was at one time the court astronomer in Baghdad and who translated and taught the works of Greek mathematicians and astronomers.

82. Christopher Clavius (1538–1612) was one of the leading mathematicians and astronomers of his time, a Jesuit professor at a Catholic university in Rome called Collegio Romano, and a friendly acquaintance of Galileo; Clavius never abandoned the geostatic theory.

worry for the very high expenses. On the contrary, I think one of the more ancient of them declared arbitrarily (at a glance, as it were) that this is the way it is, and then without any other checking those who came after accepted the earlier claim; for if any of them had devoted himself to making a further test, he would have undoubtedly discovered the error.

SAGREDO. But they lacked the telescope, and you have already stated that our friend has discovered the truth by means of this instrument; so, they must be excused and not accused of negligence.

SALVIATI. This would follow if the goal could not be accomplished without the telescope. It is true that this instrument renders the operation much easier by showing the disk of the star stripped and enlarged a hundred or a thousand times; but the same can be done without the instrument, even if not so precisely. I have done it many times, and the procedure I followed is this.[83] I let a string hang [389] in the direction of a particular star, using Vega,[84] which rises between north and northeast; then, by approaching and receding from the string hanging between me and the star, I found the place from which the thickness of the string exactly hides the star; after this I measured the distance from my eye to the string, which becomes one of the sides defining the angle formed at the eye and subtended[85] by the thickness of the string; this angle is similar, indeed identical, to the angle subtended by the diameter of the star; from the ratio of the string thickness to the distance from my eye to the string I immediately found the magnitude of this angle in a table of arcs and chords;[86] and I took the usual precaution in dealing with such acute angles, namely, to make the visual rays converge not at the center of the eye (where they would not go unless refracted) but beyond the eye (where the size of the pupil actually makes them converge).

83. This technique is practically not too useful, but the general idea and the motivation are methodologically important because they show an appreciation of the limitations of the human eye as a physical instrument; cf. selection 11, Galileo's *Operazioni astronomiche* (Favaro 8:449–56), and Brown (1985).

84. Vega is one of the brightest stars, located in the constellation Lyra.

85. An object under observation is said to *subtend* the angle formed by the lines drawn from its boundaries to the eye of the observer.

86. In a circle, a chord is a straight line connecting two points of the circumference; it subtends the angle formed at the center by the lines drawn from this center to the extremities of the chord, or the corresponding arc along the circumference; the chord, radius, and subtended angle or arc are related in such a way that from any two of them the third quantity can be calculated; appropriate tables listing such numbers can be compiled.

SAGREDO. I understand this precaution,[87] although I have something of a doubt about it. However, what troubles me most is that, if this operation is performed in the darkness of night, I think one would be measuring the diameter of the irradiated disk of the star, rather than its true and naked one.

SALVIATI. No, sir; for when the string covers the naked body of the star, it takes away its hair, which is not in it but in our eye and disappears from view as soon as the true disk is hidden; in fact, when you make the observation you will be surprised to see how a fine string covers a very large torch that seemed susceptible to being hidden only by a much larger obstacle. Then, to measure and determine very exactly the distance from the eye to the string in terms of its thickness, I do not measure a single diameter of the string; instead I take many pieces of the same, place them side by side on a table, and measure with a compass the width occupied by 15 or 20 of them; finally, I use this width to measure the distance from the string to the point of convergence of the visual rays, after this distance has been determined by means of some other finer string. Using this very exact technique, I find the apparent diameter of a fixed star of the first magnitude (commonly estimated at two minutes, and even three minutes by Tycho on page 167 of his *Astronomical Letters*)[88] to be no more than five seconds; this is 24 and 36 times smaller than they believed. Now, you see on what serious errors their doctrines are based.

[390] SAGREDO. I see and understand it very well; but, before going further, I should like to express the doubt I have in regard to the convergence of the visual rays beyond the eye when we observe objects that subtend very acute angles. My difficulty stems from my thinking that this convergence may take place farther or closer, not because of the greater or smaller size of the observed object, but because in observing objects of the same size I think that the convergence of the rays must take place at a greater or lesser distance from the eye due to some other reason.

87. The last few and next few speeches are a discussion of several precautions necessary in observing the apparent diameters of fixed stars, whose neglect has led astronomers to overestimate them and so to overestimate the calculated actual size of stars; the precautions involve not only the halo or irradiation effect (also discussed in selection 11), but also the measurement of very small thicknesses and the correction for the refraction of light rays by the eye. It is uncertain to what extent Galileo put into practice these precautions and to what extent they are practically useful, but the discussion conveys a good idea of the difficulties involved and of Galileo's observational ingenuity, sophistication, and skill. For more details, see Drake (1978; 1983; 1990).

88. Brahe (1596).

SALVIATI. I already see where Sagredo is going, with his perspicacity and great diligence in the observation of natural phenomena.[89] I would make any bet that, among a thousand persons who have noticed that cats contract and expand the pupils of their eyes a great deal, you cannot find two (and perhaps not even one) who have noticed a similar phenomenon in the human pupil, which depends on whether the medium through which men look is well or poorly lit; in clear daylight the circlet of the pupil contracts considerably, so that in looking at the solar disk it is reduced to a size smaller than a seed of panic grass;[90] but, in looking at objects that are not shining, in a less luminous environment, it expands to the size of a lentil or larger; in short, this expansion and contraction occur in a proportion greater than ten to one. From this it follows that when the pupil is greatly dilated the point of convergence of the rays is farther from the eye; this happens in looking at objects which are not very luminous. This was the account just suggested to me by Sagredo. It implies that, when we are making a very exact and very important observation, we should determine this convergence in the course of the same or of a very similar operation. However, here, to show the error of the astronomers, such accuracy is not necessary; for, even if in their favor we should suppose that this convergence is made at the pupil itself, their error is so great that this supposition would be of little consequence. I do not know, Sagredo, whether this is what you had in mind.

SAGREDO. It is exactly. I am pleased it was not unreasonable, as I am assured by my having met you. But this is also a good occasion to hear how one can determine the distance of the point of convergence of the visual rays.

SALVIATI. The procedure is very easy, and it is the following. I take two strips of [391] paper, one black and the other white, and I cut the black one to be one half the width of the white one; then I attach the white one on a wall and fix the other over a rod or some other support at a distance of 15 or 20 cubits. As I recede from the latter by an equal distance in the same direction, it is clear that this is the point of convergence of the straight lines which leave the boundaries of the width of the white strip and touch, in passing, the width of the other strip placed in

89. Drake (1967, 487n) states that in 1612 Sagredo wrote Galileo about refraction within the eye (Favaro 11:350). In this speech, Salviati attributes differences in where an image is focused to pupil size; this emphasis is misplaced since they are due chiefly to changes in the shape of the lens; but he is correct in attributing changes in pupil size to the intensity of light.

90. Panic grass is a grass in the millet family that produces various grains.

the middle; from this it follows that when the eye is placed at the point of convergence, the black strip in the middle should exactly hide the white on the opposite side if vision occurred at a single point; but if we find that the extremities of the white strip appear uncovered, this is a necessary argument that the visual rays do not converge at only one point. To have the white strip remain hidden by the black, the eye must get closer; when we get closer just enough to have the middle strip cover the remote one and we mark the amount we have had to approach, this amount will give us a precise measure of how far the true convergence of the visual rays is from the eye in such an operation. We will also have the diameter of the pupil, or rather of the hole through which the visual rays go; for the ratio of this diameter to the width of the black paper is equal to the ratio of that distance to the distance between the two papers, that (former) distance being the distance between the intersection of the lines produced along the boundaries of the papers and the place where the eye was when it first saw the distant paper hidden by the middle one. Hence, if we wanted to measure the apparent diameter of a star with some precision, after making the observation described above we would have to compare the diameter of the string with the diameter of the pupil; for example, having found the diameter of the string to be four times that of the pupil, and the distance from the eye to the string to be 30 cubits, we will say that the true convergence of the lines produced from the extremities of the stellar diameter along those of the string diameter takes place at a distance of 40 cubits from the string; for this is in accordance with the fact that the ratio of the distance between the string and the intersection of the said lines to the distance between this intersection and the position of the eye must equal the ratio of the diameter of the string to the diameter of the pupil.[91]

SAGREDO. I understand very well; but let us hear what Simplicio presents in defense of Copernicus's opponents.

[392] SIMPLICIO. Even if the very great and wholly incredible difficulty advanced by these opponents of Copernicus should be considerably modified in the light of Salviati's account, still I do not think it has

91. This passage is another good example of treating the human eye as a physical instrument, an important methodological development which Galileo helped to start. Strauss (1891, 560n.64) reconstructs this formula as follows; cf. Santillana (1953, 374n.55). Let $f$ be the diameter of the string, $p$ the diameter of the pupil, $d$ the distance between them, and $x$ the distance from the intersection of the rays to the eye; then $x/(x + d) = p/f$; and $x = d/[(f/p) - 1)]$. This formula is easily derived from the proportionality of the corresponding sides of similar triangles. In the example, $(f/p) = 4$, $d = 30$, $x = 10$ cubits.

been removed to such an extent as to lack the strength to destroy this opinion. For, if I understood properly the chief and latest conclusion, when a star of the sixth magnitude is assumed to be as large as the sun (which is, however, difficult to believe), it still remains true that the annual orbit should cause a change or variation in the stellar sphere like that which the radius of the earth produces in the sun (which is indeed observable); thus, since neither such a variation nor a smaller one is observed in the fixed stars, it seems to me to follow that the earth's annual motion remains devastated and destroyed.[92]

SALVIATI. Your conclusion would be right, Simplicio, if there were nothing else which Copernicus's side could produce; but many other things still remain. In regard to the reply you gave, nothing prevents us from supposing that the distance of the fixed stars is much greater than previously supposed; you yourself, and anyone else who may not want to deviate from the propositions accepted by Ptolemy's followers, will have to admit as a very appropriate thing that the stellar sphere may be taken to be very much larger than the estimate we just finished calculating.[93] For all astronomers agree that the greater slowness of a planet's revolution is caused by the larger size of its orbit,[94] so that Saturn is slower than Jupiter and Jupiter slower than the sun because the first describes a circle larger than the second and the second larger than the third, etc.; for example, Saturn's orbit is nine times bigger than the sun's, and because of this the period of Saturn's revolution is 30 times greater than that of the sun's revolution. Now, according to Ptolemy's doctrine a revolution of the stellar sphere is completed in 36,000 years,[95] whereas a revolution of Saturn takes 30 and that of the sun one. Then let us argue in accordance with a similar proportion and say: if Saturn revolves in a period 30 times greater because its orbit is 9 times greater, then by inverse reasoning[96]

92. This is a natural reformulation of the original argument at the beginning of this selection, and it uses Salviati's own estimate of the stellar dimensions; the variation in question is an annual change in the apparent diameter and in the apparent position of any one star. That is, the objection here is that if the earth had the annual motion, then fixed stars would appear now larger and now smaller as the earth approaches and recedes from them, and they would appear at different positions on the celestial sphere as we observe them from different places in the earth's orbit; such changes would be of the same order of magnitude as those seen in the sun by observing it from different places on the earth's surface; but no such stellar changes were observed.

93. That is, larger than a radius of 2,160 radii of the earth's orbit, for the distance from the earth to the fixed stars would define the size and radius of the stellar sphere; cf. [387].

94. Cf. selection 6.

95. This was Ptolemy's estimate for the period of the precession of the equinoxes.

96. In the original, *ragione eversa*. That is, the proportion would be $30:9 = 36{,}000:D$; then $D = 36{,}000 \times (9/30) = 10{,}800$.

how great is the orbit whose period is 36,000 times slower? We will find the distance to the stellar sphere to be equal to 10,800 radii of the annual orbit, which would be exactly 5 times [393] greater than what we calculated a little while ago from the assumption that a fixed star of the sixth magnitude is as large as the sun. Now you see how much smaller should be, in this regard, the variation caused in the stars by the annual motion of the earth.[97] Finally, if by an analogous proportion we wanted to calculate the distance of the stellar sphere from Jupiter or Mars, the former would yield a distance of 15,000 and the latter 27,000 radii of the annual orbit; that is, the former 7 and the latter 12 times greater than the distance yielded by the assumption that a fixed star is equal in size to the sun.

SIMPLICIO. To this I think one could reply that the motion of the stellar sphere has been observed after Ptolemy not to be as slow as he estimated it; indeed, I think I heard that Copernicus himself observed this.[98]

SALVIATI. What you say is very correct, but you are introducing something which does not favor at all the cause of the Ptolemaics. For they have never rejected the period of 36,000 years for the stellar sphere on the grounds that such slow motion would make it too vast and immense; and if such an immense size was not to be allowed in nature, they ought to have denied before now a revolution so slow that it proportionately corresponds to a sphere of intolerable size.

SAGREDO. Please, Salviati, let us waste no more time examining such proportions with people who are accustomed to allowing very disproportionate things, for with them in these matters it is absolutely impossible to gain anything. What more inappropriate proportion can one imagine than the following one, which they allow and pass over? That is, they write that there is no more convenient way of arranging the heavenly spheres than to be guided by the difference of their periods, gradually placing the slower ones around the faster ones; then, after placing the stellar sphere around all the others due to its being the slowest, they construct another farther and hence larger one, but they make it rotate

97. This variation is inversely proportional to the distance; for example, if the star's distance were 2,160 radii of the annual orbit, the change in distance and in apparent diameter would be at most 2 parts in 2,160 (about 0.1%); but if the star's distance were 10,800 radii of the annual orbit, the change would be at most 2 parts in 10,800 (namely, 0.000185, or 0.0185%, or about 0.02%).

98. Copernicus (*On the Revolutions*, III, 6; 1992, 129) calculates the period of the precession of the equinoxes to be 25,816 years.

in twenty-four hours[99] while the one just inside it turns in 36,000 years. However, these disproportions were sufficiently discussed yesterday.[100]

SALVIATI. Simplicio, I should like you to suspend for a moment the attachment you have for the followers of your opinion, and to tell me frankly whether you think that in their mind they comprehend the size which they then judge to be so immense that it cannot be [394] attributed to the universe. I, for one, think not; it seems to me that, just as with numbers, after one begins to exceed the thousands of millions, the imagination becomes confused and can no longer form a concept, so the same thing happens in conceiving immense sizes and distances. Thus, the effect that happens to reason is similar to what happens to the senses; that is, when on a clear night I look at the stars, my senses judge their distance to be a few miles, and the fixed stars appear to them to be no farther than Jupiter or Saturn, indeed no farther than the moon. We need go no further than to recall the past controversies between astronomers and Peripatetic philosophers in regard to the distance of the new stars in Cassiopeia and in Sagittarius;[101] the former placed them among the fixed stars, the latter regarded them as closer than the moon; such is the impotence of our senses to distinguish large from very large distances, even when the latter are many thousands of times greater than the former.

Finally, I ask you, O foolish man:[102] Does your imagination comprehend the size of the universe, which you then judge to be too vast? If you do comprehend it, do you want to say that your comprehension extends further than the divine power, and that you can imagine things greater than those which God can produce? And if you do not comprehend it, why do you want to pass judgment on things you do not understand?

SIMPLICIO. All these assertions are quite correct; no one denies that the heavens can exceed in size anything we can imagine, or that God could have created the universe a thousand times greater than it is. However, we must admit that nothing has been created in vain and is useless

99. This is the Prime Mobile, which accounts for the universal diurnal motion in some versions of the geostatic system.

100. See selection 6.

101. Cassiopeia is a constellation near the North Star with five bright stars forming the approximate shape of a *W* or *M*. Sagittarius is a constellation of the zodiac lying in the southern sky between Scorpio and Capricorn. A nova was observed in 1572 in Cassiopeia, and another in 1604 in Sagittarius.

102. As Drake (1967, 487n) suggests, this speech is addressed not to Simplicio but to the author of the book they have been examining (Locher 1614).

in the universe.[103] Now, when we see this beautiful order of the planets, arranged around the earth at appropriate distances to produce on it effects for our benefit, for what purpose should one then interpose between the farthest orbit of Saturn and the stellar sphere an extremely vast space that is without any star, superfluous, and in vain? For what purpose? For whose comfort and utility?[104]

SALVIATI. Simplicio, I think we arrogate too much to ourselves when we say that the only end of God's work ought to be our benefit, and that beyond this, nothing else ought to be done and brought about by the divine wisdom and power. However, I would not want to shackle His hands so much; instead, we can be certain that God and nature pay attention to the government of human affairs to such an extent that they could pay no more attention [395] to it even if they had no other concern but mankind. This can be explained, I think, by means of a very appropriate and noble example taken from the action of sunlight; that is, when it attracts those vapors or warms up this plant, it does so as if this were the only thing it is doing; indeed, in causing the ripening of a bunch of grapes, or even of a single grape, it does so in such a way that its dedication could be no more effective than if the purpose of all its actions were merely the ripening of the grape. Now, if this grape receives from the sun all it is capable of receiving, and if nothing in the least is taken away from it by the sun's producing simultaneously thousands of other effects, this grape would be guilty of envy and vanity if it should believe

103. This sentence formulates what I call the principle of teleology, which claims that everything that exists has a purpose. It is indirectly discussed in selection 6, but now the discussion gets more explicit. For example, it soon emerges that the Aristotelians are thinking of purposes definable in terms of human interests, which is also a form of what is called anthropocentrism.

104. This is a statement of a teleological argument against Copernicanism, which can be reconstructed as follows: if Copernicanism is correct, then the fixed stars are extremely far away (for example, at a distance of the order of 10,000 to 27,000 times farther than the sun) because this is the distance required to explain the lack of variation in apparent stellar sizes and positions; if the fixed stars are that far away, then there is a lot of empty space in the universe devoid of any bodies (namely, the space between Saturn's orbit and the nearest star); such empty space would be useless and superfluous; but nothing which is useless or superfluous can exist in nature; so Copernicanism is incorrect. The last premise is the teleological principle. The penultimate premise requires justification. One possible justification is to ground the superfluousness on lack of purpose for mankind, and lack of human purpose on human ignorance of it; doing so involves a form of anthropocentrism; and Galileo's criticism amounts to a criticism of such teleological anthropocentrism. But the superfluousness of the space between Saturn and the fixed stars could also be justified by means of metaphysical and theological considerations about cosmological order, ontological plenitude, and God's design; obviously Galileo does not consider such additional arguments.

or pretend that the action of the solar rays is being employed solely for its benefit. I am certain the Divine Providence neglects nothing that is required for the government of human affairs; but, as far as my reason can judge, I cannot bring myself to believe that no other things can exist in the universe which depend on its infinite wisdom; nevertheless, if the facts were otherwise, I would have no reluctance to accept reasons that might be adduced by a higher intelligence.[105] In the meantime, I repeat what I said to those who claim that it would be useless and vain to have an immense space interposed between the planetary orbits and the stellar sphere, without stars in it and superfluous, that is, to have the space belonging to the fixed stars be of such superfluous immensity as to surpass all our comprehension; I say it is rash to want our very weak reason to be the judge of God's works, and to call vain or superfluous anything in the universe which is of no use to us.

SAGREDO. I think it would be better to speak of *our not knowing that it is of any use to us*. I think one of the most arrogant, or rather insane, things which can be put forth is to say: "Because I do not know of what use Jupiter or Saturn is to me, therefore they are superfluous, or rather they do not exist in nature." O most foolish man![106] I do not even know of what use to me are the arteries, cartilages, spleen, or gall bladder; indeed, I would not know that I have a gall bladder, a spleen, or kidneys unless they had been shown to me in dissected corpses; only if my spleen were removed, could I understand what it does to me. To understand what this or that heavenly body does to us (in accordance with the idea that all their actions are aimed at us), it would be necessary [396] to remove that particular body for some time, and then we would be able to say that the effect on us which we would perceive to be absent depended on that body. Moreover, who will say that there are no other heavenly bodies in the space between Saturn and the fixed stars which those people regard as useless and too vast? Do those bodies not exist perhaps because we do not see them? So, did the four Medicean stars and the companions of Saturn[107] come into being when we began to see them, and not earlier? And did the other innumerable fixed stars not exist before men observed them? Were nebulas[108] at first merely patches of pale light,

105. This qualification is an expression of deference to religious authority.

106. Again, this is addressed to Locher, author of the book being examined (1614).

107. Galileo observed Saturn's rings but never developed an adequate explanation.

108. A nebula is a heavenly body seen by the naked eye as a faint, cloud-like patch of light in the night sky. With the telescope, Galileo discovered that many nebulas are collections of many stars, too many to be distinguished individually by the naked eye. These

and then with the telescope did we make them become collections of many luminous and very beautiful stars? How presumptuous, indeed reckless, is the ignorance of men!

SALVIATI. There is no need, Sagredo, to linger any more with these fruitless exaggerations. Let us, instead, continue with our task, which is to examine the weight of the reasons advanced by one side and by the other, without deciding anything; we will then defer to the judgment of those whose knowledge is greater than ours.[109] Going back to our natural and human arguments, I say that such terms as *large*, *small*, *immense*, *minute*, etc. are relative and not absolute;[110] thus, the same thing, compared to different things, can be called sometimes immense, and sometimes insensible, as well as small. Having said this, I ask in relation to what can Copernicus's stellar sphere be called too vast. In my opinion, it can be compared or called such only in relation to something else of the same kind. Now, let us take the smallest thing of the same kind, which is the lunar orbit; if the stellar sphere is to be declared to be too large relative to the orbit of the moon, then any other size that exceeds another one of the same kind by a similar or greater proportion must be called too large, and for this reason we must deny its existence in the world. Thus, elephants and whales will definitely be chimeras and poetical images; for the former are too large in relation to ants (which are land animals), while the latter are too large in relation to minnows (which are fish); and the reason for this is that undoubtedly an elephant and a whale exceed an ant and a minnow by a much greater proportion than that by which the stellar sphere exceeds the lunar orbit, assuming this sphere to be as large as required to accommodate the Copernican system; nevertheless, those large animals are certainly seen to exist in nature.

Moreover, how large is the orbit of Jupiter or of Saturn, which is designated as the receptacle of a [397] single body, and a rather small one in

---

nebulas correspond to what are now called galaxies, for example the Andromeda Nebula. Other nebulas are not galaxies but large clouds of gas and dust in interstellar space. Astronomers now tend to restrict the term nebula to such interstellar clouds and to use the term galaxy for nebulas of the type resolved by Galileo's telescope.

109. This is one of many disclaimers Galileo places in Salviati's mouth to ensure that his book would not be seen as a defense of Copernicanism or other act of defiance against the Church. For more details, see Finocchiaro (1980, 12–18).

110. Galileo is threading a fine line between absolutism and anthropocentrism: he wants to point out the many ways in which the concept of size is subjective and relative, and thus reject absolutism and dogmatism; but he also wants to steer clear of making the concept of size or actual sizes dependent on human interests or knowledge of purposes, as teleological anthropocentrism does.

comparison to a fixed star? It is certain that, if each fixed star were assigned as its receptacle such a part of the space in the universe, then the orb in which are placed the countless number of fixed stars would have to be many thousands of times greater than what suffices for Copernicus's needs.

Furthermore, do you not call a fixed star very small, whether it is one of the most visible ones, or one of those which are hardest to see? We call them this in comparison to the surrounding space. Now, if the whole stellar sphere were a single shining body, is there anyone who does not understand that in infinite space we can choose a distance so great that from there such a luminous sphere would appear as small as, and even smaller than, a fixed star now appears to us on the earth? Thus, from there we would judge small the same thing which from here now we call immeasurably large.

SAGREDO. I find extremely great the ineptitude of those who would want God to have made the universe more proportionate to the small capacity of their mind than to His immense, indeed infinite, power.

SIMPLICIO. What you are now saying is correct. However, what the opponents are objecting to is having to grant that a fixed star is not merely equal to the sun, but so much larger than it, even though they are both particular bodies placed within the stellar sphere. I think this author is posing very pertinent questions when he asks: "To what end and for whose benefit do such large bodies exist? Are they perhaps produced for the earth, that is, for an extremely small point? And why are they so distant as to appear so small and as to be able to effect absolutely nothing on the earth? For what purpose is there such a disproportionate and immense chasm between them and Saturn? Objectionable are all those things that are not supported by probable reasons."[111]

SALVIATI. From the questions this man asks, I think we can gather what he regards as the consequences of letting the heavens, stars, and distances remain at the dimensions and magnitudes he has so far believed them to be, although he has never determined with certainty any comprehensible magnitude; the consequences are that he grasps and understands very well the benefits coming to the earth from them; that the earth is not a minute little thing; that the stars are not so distant as to appear so very small, but are large enough as to be able to affect the earth; that the distance between them and Saturn has very good proportions; and [398] that he has very probable arguments for all these things. I

111. Cf. Locher (1614, 28).

would have been glad to listen to at least one of these arguments; but seeing that these few words of his contain confusions and contradictions, I am led to think that he is very lacking and wanting in these probable arguments, and that what he calls arguments are instead fallacies or, rather, shadows of empty images. For I now ask him whether these heavenly bodies really affect the earth and have been created of such and such a size and placed at such and such a distance in order to produce these effects, or whether they have nothing to do with terrestrial things; if they have nothing to do with the earth, it is very foolish for us terrestrials to want to be arbiters of their sizes and of their arrangement in space, when we are highly ignorant of all their affairs and interests; but if he says that they do affect the earth and exist for this purpose, then he is affirming what he himself had denied in another way, and he is praising what he had just now condemned when he said that heavenly bodies placed at such a distance from the earth as to appear so small cannot affect anything on it. But, my dear man, in the stellar sphere located at the distance at which it is and which you judge well proportioned for affecting these terrestrial things, there are very many stars that appear very small, and a hundred times more that are completely invisible to us (and this means that they appear even smaller than very small); therefore, contradicting yourself, you must now deny their effects on the earth; or you must admit that their appearing so small does not diminish their action (which again involves a self-contradiction); or you must admit and freely confess that our judgment on their sizes and distances is vanity, not to say arrogance or recklessness (which is a more sincere and modest admission).

SIMPLICIO. Indeed I, too, saw the clear contradiction immediately upon reading this passage; that is, it says that the stars of Copernicus (so to speak) could not affect the earth due to their appearing so small, but it does not realize having admitted the existence of some effect on the earth by Ptolemy's and his own stars, which not only appear very small but are for the most part invisible.

SALVIATI. However, I come to another point. On what grounds does he say that the stars appear so small? Perhaps because we see them like that? Does he not know that this derives from the instrument we [399] use in observing them, namely, our eye? The truth of this may be established by the fact that if we change instruments, we will see them larger and larger, as much as we please. Who knows? Perhaps to the earth, which looks at them without eyes, they appear very large and as they really are.

But it is time for us to set aside these trifles and come to more weighty matters. I have already proved two things:[112] first, how far one may place the firmament[113] so that the variation caused in it by the diameter of the annual orbit is no greater than that caused by the radius of the terrestrial globe in a body at a distance equal to that of the sun; second, that in order for a star in the firmament to appear to us to have the magnitude it does appear to have, it is not necessary to suppose it to be any larger than the sun. Having proved these things, I should like to know whether Tycho or any one of his followers has ever tried to determine in any manner whether in the stellar sphere we can perceive any phenomena on the basis of which we could more definitely confirm or deny the annual motion of the earth.

. . .[114]

## [14. Stellar Parallax and the Fruitfulness of a Research Program]

[404] . . . SALVIATI. I already said earlier that I do not think anyone has undertaken to observe whether during the various times of a year we can perceive any variation in the fixed stars that may depend on the earth's annual motion; moreover, I add that I doubt that anyone has properly understood what the variations should be and

112. Galileo has certainly proved these two things. It is also clear that this selection reflects on the methodology of the concept of size and raises the following issues, among others: the mathematical interdependence of apparent size, actual size, and actual distance; the precautions necessary for a reliable empirical observation of apparent size; the type of criteria for calculating estimates of the stellar distances and the size of the universe; and the relational but nonanthropomorphic nature of the concept of size. It is equally clear that this selection criticizes the stellar dimensions objection to Copernicanism. What is not so clear is exactly what the details and reasoning of this argument are, what Galileo's criticism is, and how this criticism affects the argument; the reconstruction of such details is a challenging problem.

113. *Firmament* is generally used to denote the sky or the heavens; here it refers to the stellar sphere.

114. The passage omitted here corresponds to Favaro 7:399.12–404.27 and Galileo 1967, 372–77. It discusses three peripheral issues: whether Tycho was clear about the changes in stellar appearances implied by the earth's annual motion, whether this motion would imply any change in the elevation of the celestial pole, and whether the resulting changes in stellar elevations would be comparable to the large ones resulting from moving around the spherical surface of the terrestrial globe. Galileo's answer to each of these questions is negative. For more details, see Finocchiaro (1980, 41–42 and 131–32).

which stars should be affected.[115] So we will have to examine this point diligently. The fact is that I have found written only the general argument that the earth's motion in the annual orbit should not be accepted because it is likely that we would see the apparent variations in the fixed stars it would cause, but then I have not heard in particular what [405] these apparent variations would have to be and in which stars; this makes me very reasonably judge that those who stop at this general pronouncement have not understood (and perhaps have not even tried to understand) how the business of these variations proceeds, or what are the things which they say should be seen. I am induced to make this judgment by the knowledge that if the annual motion attributed by Copernicus to the earth should have an observable effect on the stellar sphere, the apparent variation would not occur equally in all the stars; instead, this variation would appear greater in some, less in others, still less in others, and finally absolutely null in others, regardless of how large the circle of this annual motion is taken to be. Moreover, the variations which should be seen are of two kinds: one is a change in a star's apparent magnitude; the other is a change in its elevation at the meridian, which implies a change in the points where it rises and sets, in its distances from the zenith,[116] etc.

SAGREDO. It feels like I am being woven into such a knot of these variations that I pray to God to enable me to extricate myself from it. For, to confess my limitations to Salviati, I have occasionally thought about it but have never been able to find the loose thread; I am not referring so much to what regards the fixed stars, but to another more troublesome point which you reminded me of when speaking of these meridian elevations, rising latitudes, distances from the zenith, etc. The knot in my brain derives from what I am going to tell you now. Copernicus supposes the stellar sphere to be motionless and the sun at its center to be likewise motionless; therefore, any change that the sun or the fixed stars appear to undergo must belong to the earth, namely, to us; but the sun goes up and down on our meridian by a very large arc of

115. To support this claim, I would cite three things: (1) the misunderstandings Galileo exposed in the passage (here omitted) between selections 13 and 14; (2) the tortuousness of Aristotle's own version of this argument, first stated in selection 7, [150], and later criticized (Favaro 7:162.15–64.25; Galileo 1967, 136–38; cf. Finocchiaro 1980, 380–86); and (3) the relatively confusing character of the discussion of the stellar dimensions argument in selection 13.

116. The zenith is the point on the celestial sphere directly above the observer; it can be thought of as the point of intersection between the celestial sphere and the line drawn through the earth's center and the observer on its surface.

almost 47 degrees,[117] and the points of its rising and setting along oblique horizons change by even much greater arcs. Now, how can the earth tilt and rise so noticeably with respect to the sun but not at all with respect to the fixed stars (or rather so little that the phenomenon is imperceptible)? This is the knot I have never been able to untie; if you do that for me, I will esteem you more highly than Alexander.[118]

SALVIATI. These are difficulties worthy of Sagredo's intellect. The problem is such that Copernicus himself almost despaired of being able to explain it so as to render it intelligible; this is seen [406] both from his own admitting its obscurity and from his attempting to explain it twice in two different ways.[119] I myself frankly admit not having understood his explanation until after I rendered it intelligible in a different way, which is straightforward and clear but required a long and laborious mental concentration.

SIMPLICIO. Aristotle saw the same difficulty, and he used it to reproach some of the ancients who wanted the earth to be a planet. Against them he argued that, if this were so, it (just like the other planets) would have to have more than one motion; from the latter it would follow that there would be a variation in the points of rising and setting of the fixed stars, as well as in the meridian elevations. Since he advanced the difficulty without resolving it, it must be that it is hard (if not impossible) to resolve.[120]

117. This arc refers to the sun's apparent north-south motion in the course of a year; it is measurable in terms of the noon elevation of the sun, the eastern point of sunrise, or the western point of sunset; each of these three points moves north and south during the year, and the angular distance from one extremity to the other is 47 degrees. For more details, see the introduction (2.4).

118. Alexander the Great (356–323 B.C.) was one of the greatest military leaders of all time, who began as King of Macedon (north of Greece) and then conquered Greece, the Middle East, Persia, and parts of India; these conquests were important for the spread of Greek culture and the development of a cosmopolitan outlook. He became the subject of many stories, such as the one about the Gordian knot: a knot tied by King Gordius of Asia Minor which was so intricate that no one could untie it, and about which an oracle had declared that whoever untied it would become ruler of all Asia; when Alexander was confronted with the problem, rather than untie the knot, he cut it with a single stroke of his sword.

119. Cf. Copernicus, *On the Revolutions*, I, 12 (1992, 22–25).

120. Cf. Aristotle, *On the Heavens*, II, 14, 296a34-b7; see also selection 7, where it is presented as the second in Aristotle's list of four arguments against the earth's motion, and where I called it the two motions argument. Galileo criticizes it in a passage of the "Second Day" that is omitted here (Favaro 7:162.15–64.25; Galileo 1967, 136–38). Although there is no question that Aristotle's argument is reminiscent of stellar parallax, their relationship is unclear; for a critical analysis, see Finocchiaro (1980, 380–86).

SALVIATI. The seriousness and strength of the difficulty makes the resolution more beautiful and admirable; but I do not promise this for today and I beg you to excuse me until tomorrow. For now we will consider and explain the changes and variations that should be perceived in the fixed stars due to the annual motion, as we were just saying; the explanation of these will require the presentation of some points that will pave the way for the resolution of the greatest difficulty.[121]

Let us now recall the two motions attributed to the earth, namely, the annual and the diurnal (I say two because the third one is not a motion, as I will explain in due course).[122] By the annual we must understand the motion performed by the earth's center along the circumference of the annual orbit, namely, along a very large circle described in the plane of the fixed and unchangeable ecliptic. The other (namely, the diurnal) motion is performed by the terrestrial globe on itself and around its own center and its own axis; this axis is not perpendicular but inclined to the plane of the ecliptic with an inclination of about 23½ degrees. This inclination is constant during the whole year, and what is most remarkable is that it remains always oriented toward the same part of the heavens in such a way that the axis of the diurnal motion stays always parallel to itself. Thus, if we imagine this axis extended as far as the fixed stars, while the earth's center goes around the whole ecliptic in a year, the same axis describes the surface of an oblique cylinder having as one of its bases the said annual circle and as the other a similar imaginary circle described among the fixed stars by its extremity (or rather its pole); this cylinder is [407] tilted to the plane of the ecliptic with the inclination of the axis that generates it, which we have said is 23½ degrees; by remaining always the same, this inclination ensures that the terrestrial globe neither tilts down nor rises up but conserves its rotational orientation without change (except that in the course of many thousands of years there is a

121. This difficulty is called the stellar parallax argument against Copernicanism. The statement of this objection has just begun here; for unlike the other selections, this one mostly contains an elaboration of this argument. When Galileo later tries to answer the objection, his answer (unlike others) is not a criticism of it, but an outline of a research program to find the relevant evidence.

122. Copernicus (*On the Revolutions*, I, 11; 1992, 23) attributes to the earth a third motion, consisting of an annual rotation by the terrestrial axis around the axis of the earth's orbit in a direction opposite (westward) to that of its orbital revolution, in order to explain why the terrestrial axis always stays parallel to itself. As Galileo argues in a passage omitted here, this phenomenon is simply an instance of rest, and so no additional terrestrial motion needs to be postulated (Favaro 7:424–25; Galileo 1967, 398–99; cf. Galileo's "Reply to Ingoli" in Finocchiaro 1989, 191–92).

very small change, which is of no relevance to the present point).[123] From this it follows that in regard to the variation to be observed in the fixed stars as a consequence of the annual motion alone, the same thing will happen at any point whatever on the terrestrial surface which happens at the very center of the earth; hence, in the following explanations we will use the center, as representing any surface point whatever.

To facilitate the understanding of everything, we will draw a diagram [see fig. 10]. First, in the plane of the ecliptic let us draw the circle ANBO; let the points A and B be the extremities toward the North and toward the South, namely, the beginning points of Cancer and of Capricorn;[124] and let us extend the diameter AB through D and C indefinitely toward the stellar sphere. Now, I say, first, that any fixed star lying in the plane of the ecliptic will never change in elevation regard-

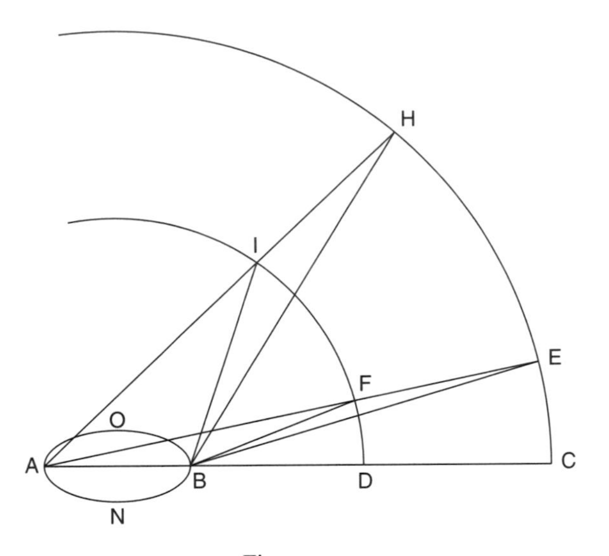


Figure 10.

123. This change refers to the precession of the equinoxes, which in a geokinetic system must be explained by making the axis of the earth's diurnal rotation rotate eastward around the axis of its orbital revolution; since the period is so slow, Galileo neglects this motion for the purpose of the present discussion.

124. To make sense of these directions, think of (1) the earth as revolving in a counterclockwise direction along ANBO, (2) the semicircle OAN as lying above or north of the celestial equator, and (3) the semicircle NBO as lying below or south of the celestial equator; the points O and N are then the points where the ecliptic intersects the celestial equator, namely, the two solstices.

less of any motion whatever performed by the earth in that plane, but rather it will be always perceived on that surface; on the other hand, the earth will approach and recede from it by a distance equal to the diameter of the annual orbit. This is sensibly seen in the diagram. For, whether the earth is at point A or at B, the star is always seen along the same line ABC; but the distance BC is less than CA by the amount of the whole diameter BA. Therefore, the most we can perceive about the star C, and about any other star whatever lying along the ecliptic, is an increase or a decrease in apparent magnitude due to the earth getting nearer or farther.

SAGREDO. Please stop for a moment because I am feeling a difficulty which bothers me; it is this. I understand very well how the star C is seen along the same line ABC when the earth is at A as well as when it is at B; moreover, I likewise understand that the same would happen from all points of the line AB if the earth should go from A to B by this line. However, as we assume, it follows the arc ANB, and so it is clear that when it is at [408] point N (and at any other except A and B) the star will no longer be seen along the line AB but along some other line; thus, if being seen along different lines causes an apparent change, then some difference would have to be perceived.[125] Indeed, I will say more, taking the philosophical liberty that ought to be allowed among philosophizing friends. I think you are contradicting yourself by denying now what earlier today you claimed, to our amazement, to be a remarkable and very true thing; I am referring to what happens to the planets, and in particular to the three superior ones; although they are always found in the plane of the ecliptic or very close to it, not only do they appear sometimes very close to us and sometimes very far, but their regular motions exhibit such variations that they sometimes appear motionless to us and sometimes appear to move backwards by many degrees; and the cause of all this is nothing but the annual motion of the earth.

SALVIATI. Although countless confirmations have made me certain of Sagredo's shrewdness, I wanted this other test to be more confident of how much we can expect from his intellect; the benefit to me is that, if my propositions can withstand the hammer and furnace of his judgment, I can be sure their consistency is good enough to withstand any

125. Sagredo is saying that star C is seen along line BC from position B but along line NC from N, and that there should be a way to distinguish these two lines or directions. The ensuing discussion reveals that these two lines cannot be distinguished if all stars are equidistant from the center, but that they can if different stars are located at different distances.

test. So I say that I had concealed this objection deliberately, but not with the thought of deceiving you or persuading you of any falsehood; this could have happened if the objection I concealed and you revealed were really how it appears to be, namely, truly powerful and conclusive; but it is not like this, and indeed I suspect that now you (in order to test me) are pretending not to know its worthlessness. However, on this particular I want to be more cunning than you by forcibly extracting from your mouth what you are artfully trying to conceal from us.[126] So tell me, How do you come to know about the phenomenon of the stopping and the retrograde motion of the planets which is due to the annual motion, and which is so significant that at least some trace of a similar effect should be seen in the stars lying on the ecliptic plane?

SAGREDO. Your question has two parts to which I must respond: the first regards your allegation that I am engaged in some kind of deception; the other involves what may appear in the stars, etc. As to the first, I will tell you, with your permission, that it is not true that I have lied about not knowing the worthlessness of that objection; and to prove this to you, I tell you that I now understand this worthlessness very well.

[409] SALVIATI. Now, I do not understand how it can happen that you were not speaking deceptively when you claimed not to understand this fallacy, which you now admit you understand very well.

SAGREDO. My very admission that I now understand it can assure you that I was not lying when I said that I did not understand it; for if I had wanted to lie or wanted to do it now, who could prevent me from continuing with the same deception by continuing to deny that I understood the fallacy? So I say that I did not understand it earlier, but that I understand it well at present; it was you who awakened my intellect, first by resolutely telling me that it is worthless, and then by starting to ask me in a general way how I came to know about the stopping and the retrograde motion of the planets. We come to know about this phenomenon by reference to the fixed stars, in relation to which the planets are seen to undergo variations (their motion being sometimes toward the west, sometimes toward the east, and sometimes coming to a standstill); but beyond the stellar sphere there is no other farther one

126. These remarks by Salviati along with Sagredo's ensuing comments primarily add rhetorical and dramatic interest to the discussion. They also suggest a methodological conclusion about the value of criticism which can be formulated as follows: it is advisable to test our arguments and theories against the criticism of persons of sound judgment and sharp intellect, for if they survive such criticism then our confidence in their correctness is enhanced. And they are also an indirect reference to the Socratic method.

visible to us by reference to which we can observe the fixed stars; therefore, we cannot perceive in the fixed stars any trace of those variations that we see in the planets. I believe this to be what you wanted to extract from my mouth.

SALVIATI. That is right, although you also gave us an example of your extremely subtle shrewdness. Moreover, if with one brief remark I fertilized your mind, you with another made me think it is not altogether impossible that sometimes something could be observed in the fixed stars, which would enable us to determine to which body the annual revolution belongs; thus they too, no less than the planets and the sun itself, would appear in court to testify in favor of such a motion for the earth. For I do not think the stars are lying on a spherical surface equidistant from the center, but rather I judge that their distances from us vary to such an extent that some may be two or three times farther than others; hence, if the telescope should reveal some very small and very distant star very close to one of the largest, it could happen that they would exhibit some perceivable variation corresponding to that of the superior planets.[127] Let this suffice for now in regard to the point about stars lying in the plane of the ecliptic.

Let us now consider fixed stars located outside this plane. Let us construct a great circle perpendicular to this plane [fig. 11], [410] for example, a circle on the stellar sphere corresponding to the solstitial colure,[128] which at the same time will be a meridian; let us label it CEH, where E is

127. This suggestion may be explained by reference to figure 10: assume that C and D are two stars lying in the plane of the ecliptic but at different distances (as indicated in the diagram); when the earth is at N, consider the angle formed at the earth by the two lines from the earth to the two stars, namely, the angle CND, which defines the angular separation between the two stars; as the earth moves from N to B, this angle decreases to nought, and as it moves from B to O, the angle increases to its maximum value; then as it moves from O to A the angle decreases again, and as it moves from A to N the angle again increases. Strauss (1891, 562n.76) states that this was the procedure followed in the discovery of annual stellar parallax. The German astronomer and mathematician Friedrich W. Bessel (1784–1846) first observed it for the star 61 Cygni and announced it in 1838. Aside from its scientific and historical significance, Bessel's discovery shows the fruitfulness of the research program Galileo outlines in this selection instead of directly refuting the parallax objection; this is methodologically important because fruitfulness is an important criterion for judging the merits of scientific ideas; for some general discussions, see Laudan (1977) and McMullin (1976).

128. The colures are two meridian circles at right angles to each other, which divide the celestial sphere into four equal parts in an east-west direction; the *equinoctial colure* goes through the poles and the equinoxes; the *solstitial colure* goes through the poles and the solstices.

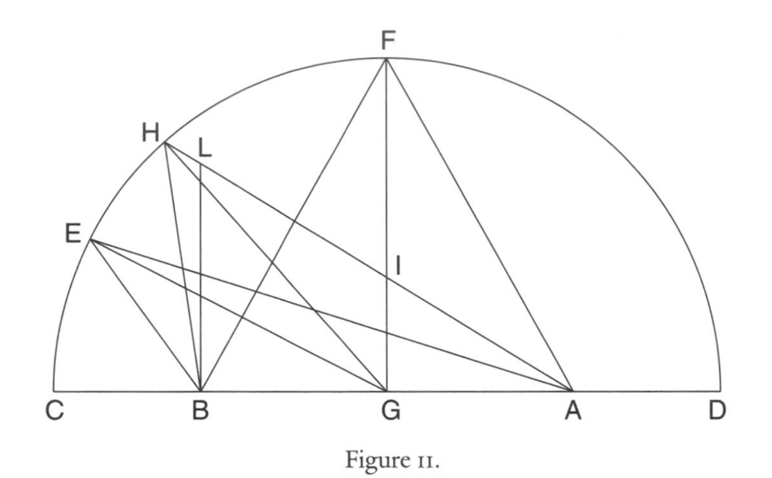


Figure 11.

a star outside the ecliptic plane.[129] Now, its elevation will clearly vary due to the earth's motion; for when the earth is at A, it will be seen along the line AE, at an elevation measured by the angle EAC; and when the earth is at B, it will be seen along the line BE, at an elevation measured by the angle EBC; but the angle EBC is greater than EAC since the former is an external angle[130] of the triangle EAB, and the latter is an internal and opposite angle; therefore, the distance of the star E from the ecliptic will be seen to change, and its elevation on the meridian will be greater when seen from B than from A. This change in elevation corresponds to the amount by which the angle EBC exceeds the angle EAC, which in turn is equivalent to the magnitude of the angle AEB; for, if in the triangle EAB the side AB is extended to C, the external angle EBC exceeds the angle A by the amount of the angle E since the angle EBC is equal to the sum of the two opposite internal angles E and A. If we take along the same meridian another star farther from the ecliptic (such as, for example, the star H), it will exhibit a greater variation when seen from the two places

129. This discussion is better followed by now disregarding figure 10 and looking only at figure 11, which corresponds to figure 10 but with some details added, with the left and right sides exchanged, and with minor irrelevant inconsistencies introduced. This becomes clear if we take points A, B, C, E, and H to represent the same things in both diagrams; the inconsistencies regard points D, F, and I; but D was not used earlier, and F and I were not even mentioned; so the substance of the present discussion is unaffected.

130. In a triangle, an external angle is the angle formed outside the figure by extending any side. This definition is useful because of the theorem that an external angle equals the sum of the two opposite internal angles; as a corollary, an external angle is greater than each of the opposite internal angles. These truths are used several times in this passage.

A and B, just as the angle AHB is larger than the other one at E; the new angle will keep on increasing as a result of the observed star being farther from the ecliptic until finally the maximum variation will occur for a star placed on the very pole of the ecliptic. To understand this fully, we can give the following demonstration.

Let AB be a diameter of the annual orbit and G its center, and let us extend it as far as the stellar sphere at points D and C; from the center G let us erect the axis of the ecliptic GF as far as the same stellar sphere; on this draw the meridian DFC, which will be perpendicular to the plane of the ecliptic; in the arc FC, take any two points, H and E, as the location of two fixed stars; finally, draw the lines FA, FB, AH, HG, HB, AE, GE, and BE. Thus, the angular variation, that is to say the parallax, of the star positioned at the pole F is AFB; that of the star positioned at H is the angle AHB; and that of the star at E is the angle AEB. Now, I say that the angular variation of the polar star F is the greatest; that, for the others, the closer a star is to this, the larger is its angular variation, [411] and hence that the angle F is larger than the angle H, and the latter is larger than the angle E. Imagine a circle circumscribed around the triangle FAB; the angle F is acute because its base AB is less than the diameter DC of the semicircle DFC; hence, the angle F will lie in the larger part of the circumscribed circle cut by the base AB; now, because AB is divided in half by FG and these two lines meet at right angles, the center of the circumscribed circle will be on the line FG; let it be point I. Moreover, of the lines drawn from point G (which is not the center) to the circumference of the circumscribed circle, the longest is the one which goes through the center, and hence GF will be longer than any other line drawn from point G to the circumference of the same circle; consequently, this circumference will intersect the line GH (which is equal to GF), and by intersecting GH it will also intersect AH; let the latter point of intersection be L and draw the line LB. Therefore, the two angles AFB and ALB are equal since they are both inscribed in the same portion of the circumscribed circle; but ALB, being an external angle, is larger than the internal angle H; therefore, the angle F is larger than the angle H. By the same procedure we can prove that the angle H is larger than the angle E; for the center of the circle circumscribed around the triangle AHB is on the perpendicular GF, and the line GH is closer to it than is the line GE, so that its circumference intersects GE as well as AE; from this the rest is clear. We conclude, therefore, that the change in apparent position (which with a proper technical term we could call the parallax of the fixed stars) is larger or smaller depending on whether the observed stars are more or less close to the pole of the ecliptic; thus, finally, for the stars which lie on the very plane of the ecliptic this change reduces to nought.

As regards the issue of the earth getting closer to and farther from the stars due to this motion, it approaches and recedes from those lying on the plane of the ecliptic by the amount of the whole diameter of the annual orbit, as we saw just now; but for the stars near the pole of the ecliptic this change of distance is almost null; and for the others this change is greater depending on their closeness to the plane of the ecliptic.

Third, we can understand how that change in apparent position is larger or smaller depending on whether the observed star is closer to or farther from us. To show this, let us draw another meridian closer to the earth, such as DFI in the diagram [fig. 12];[131] then, a star located at F would be seen along the ray AFE when the earth was at A, and would be perceived along the ray BF when observed from the earth at B; [412] the angular variation would be BFA, which is larger than the first mentioned AEB since it is an external angle of the triangle BFE.

SAGREDO. I have listened to your account with great pleasure and profit as well; to be sure that I have properly understood it, I will briefly state the gist of your conclusions. I think you have explained to us two kinds of different phenomena that could be observed in the fixed stars due to the earth's annual motion: the first is a variation in their apparent magnitude, depending on the extent to which we (carried by the earth) approach or recede from them; the other is their appearing to us sometimes at a higher and sometimes at a lower elevation along the same meridian (which also depends on the same approaching and receding). Moreover, you tell us (and I understand it very well) that these two kinds of variations do not take place equally in all the stars, but are larger in some, smaller in others, and nil in still others; the change in distance on account of which the same star should appear to us sometimes larger and sometimes smaller is imperceptible and almost null for the stars near the pole of the ecliptic, but is greatest for the stars lying in the plane of

131. Figure 12 is identical to figure 10, although in the earlier discussion relating to it Galileo did not mention points F and I and did not use point D. In relation to figure 11, there is some correspondence, insofar as points A, B, C, E, and H represent the same things but are reversed from left to right; there is also some inconsistency in regard to what the points D, F, and I represent. In the present discussion it is thus best to disregard figure 11 and focus just on figure 12. The point is that, though he could have used a single diagram for his three illustrations in this selection, his three diagrams add some confusion; the confusion is a nuisance but can be avoided.

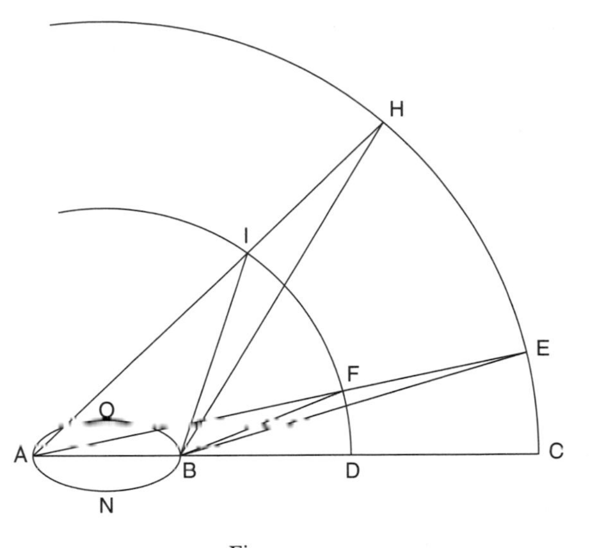


Figure 12.

the ecliptic, and intermediate for those in between; the opposite happens in regard to the other variation, that is, the change in elevation is null for the stars lying in the plane of the ecliptic, greatest for those near the pole of the ecliptic, and intermediate for those in between. Furthermore, both of these variations are more noticeable in the stars which are closer, less noticeable in those which are farther, and would vanish in the extremely remote ones. This is all I have to say. What is left now, as far as I can tell, is to see whether Simplicio is satisfied; I do not think he will easily agree to regard as imperceptible these variations deriving from a terrestrial motion so vast that it takes the earth to places which are distant from each other by twice the distance between us and the sun.

SIMPLICIO. Speaking frankly, I really feel great aversion to having to admit that the distance of the fixed stars should be so great as to make the variations just explained completely imperceptible.

SALVIATI. Do not despair so much, Simplicio, for perhaps your difficulties can be eased somewhat.[132] Consider, first, [413] the fact that the apparent magnitude of fixed stars is not seen to vary perceptibly; you should not find this at all surprising when you see that human judgment in regard to these matters is so seriously mistaken, especially in observing

132. So far in this selection the stellar parallax objection has been stated, elaborated, and clarified; now begin the answer and the criticism.

shining objects. For example, if you look at a flaming torch from a distance of 200 paces and then it is brought closer to you by 3 or 4 cubits, do you think you would notice that it appears larger? I for one would surely not notice even if it were brought closer by 20 or 30 cubits. Indeed, occasionally I have had the experience of observing a similar light at such a distance and being unable to decide whether it was moving toward or away from me, when in fact it was approaching. Finally, if the planet Saturn's approaching and receding (by an amount double the distance between us and the sun) is almost totally imperceptible, and that of Jupiter little noticeable, what should happen to the fixed stars which I believe you are inclined to place at a distance double that of Saturn?[133] As regards Mars, which approaches us . . . [134]

SIMPLICIO. Sir, you need not belabor this particular point, for I already comprehend how it may very well happen that, as you said, the apparent magnitude of the fixed stars does not change. However, what shall we say about the other difficulty, stemming from the fact that no change of apparent position is perceived?[135]

133. The following calculations and comparisons are approximate but instructive. Using the value given in selection 13 ([392]), the Ptolemaic estimate of Saturn's orbit is 9 times that of the sun; a fixed star is then 18 times farther from the earth than the sun; keeping the same estimates of distances but giving the annual motion to the earth, the distance between the earth and a fixed star changes by 2 parts in 18, or about 11%. Next, take a pace to be about 30 inches and a cubit about 20 inches; then 200 paces equal 6,000 inches, and 30 cubits equal 600 inches; thus, if a torch at 200 paces is brought closer by 30 cubits, the approach and so the change in apparent diameter would be 30 cubits:200 paces, or 600:6,000 inches, or 10%. Finally, let us use the most conservative Galilean estimate of stellar distance calculated in selection 13, namely, 2,160 times farther from the earth than the sun, or 2,160 radii of the annual orbit; the earth's annual motion changes this by 2 parts in 2,160, namely, by about 0.0009 or about 0.1%; this change is about 100 times smaller than either that of Saturn calculated from Ptolemaic estimates or that of the torch. Salviati is arguing that if we cannot notice an 11% change in Saturn or a 10% change in the torch, it is not surprising that we cannot notice a Copernican change of 0.1% or less in the apparent diameter of a fixed star.

134. This ellipsis is in the original to signal Simplicio's interruption of Salviati.

135. For all his alleged simplemindedness, here Simplicio shows some acumen; for other examples and a discussion of this issue, see Finocchiaro (1980, 46–47). That is, Simplicio raises a new issue which requires a different answer; he has understood that there is an important difference between changes in apparent magnitude and changes in apparent position. The difference is that changes in apparent magnitude are minute changes in minute quantities, whereas changes in apparent position are minute changes in much larger quantities. For example, let us use the same conservative estimate of stellar distance used above, namely, that a fixed star is 2,160 times farther than the sun; we have already seen that the earth's annual motion would cause a change of about 0.0009 (about 0.1%) in stellar distance and in apparent diameter; now, in selection 13, the Galilean estimate of apparent stellar diameter is 50 thirds of arc, namely, 5/6 of a second of arc; here we may ig-

SALVIATI. We shall say something that may perhaps satisfy you in this regard as well. To make a long story short, would you not be satisfied if we should really perceive in the stars those changes which you think must be perceived there as a consequence of the annual motion belonging to the earth?

SIMPLICIO. I would undoubtedly be, as far as this particular point is concerned.

SALVIATI. I would want you to say that if such a variation should be perceived, nothing would remain any longer to cast doubt on the earth's motion, given that no other explanation could be given for such a phenomenon. However, even if this were not perceived, its motion would not be thereby taken away, nor its immobility necessarily established.[136]

---

nore the fact that this is 24 to 100 times smaller than traditional estimates but still grossly overestimated (as we know today); the essential point is that we are dealing with a change of about 0.1% in 5/6 of a second of arc; the quantity to observe is then about 0.0008 of a second of arc. Let us now refer to figure 11 ([410]) to estimate the quantities involved in regard to position; suppose star F near the pole of the ecliptic is one of those stars that is 2,160 times further away than the sun; that means that GF = 2,160 × GA; by trigonometry, the tangent of the angle AFG is then 1/2,160; hence, AFG is the angle whose tangent equals 1/2,160; that is, angle AFG = arc tan 0.0004629, or about 1⅔ minutes; the full parallax (angle AFB) is then about 3⅓ minutes, or 200 seconds; this is a very large quantity compared to the change in apparent diameter, even though in relation to the quantity we are dealing with (about 90 degrees) the relative change is slightly smaller, namely, 200/(90 × 60 × 60) or about 0.0006 = 0.06%. As Galileo explained, the change in apparent position is greatest for stars near the pole of the ecliptic; so for a star like E in the same diagram the change would be less; but the quantities involved would still be much greater than for the case of apparent diameter. However, parallactic changes also decrease with distance; thus, if the stars are sufficiently far away, then even the change in apparent position may become too minute; this type of consideration corresponds precisely to one of the answers Salviati proceeds to give.

136. This latest exchange between Salviati and Simplicio raises some crucial issues in the methodology of theoretical explanation. First, Salviati seems to suggest that (1) we should accept a hypothesis (Copernicanism) if its consequences (stellar parallaxes) are observed to be true. Simplicio's response is basically "yes, other things being equal"; or more exactly, (2) a hypothesis becomes more acceptable insofar as its consequences are observed to be true. Simplicio's principle (2) is more plausible than Salviati's principle (1); as it stands, (1) commits the fallacy of affirming the consequent; that is, it corresponds to the following type of argument, which is formally invalid: "if *p* were true, then *q* would be true; *q* is true; therefore, *p* is true"; this point was famously made by Cardinal Bellarmine in his letter to Foscarini (Finocchiaro 1989, 67–69). But Salviati goes on to reformulate his principle to say that (3) we should accept a hypothesis if its consequences are observed to be true *and* these observations can be reasonably explained in no other way; this last clause is crucial; principle (3) is more restrictive than (1) but consistent with (2). However, principle (3) is inapplicable to the situation at hand since the stellar parallaxes were not observable in Galileo's time, although the point could be made "for future reference"; thus, Salviati goes on to state a fourth principle, which is more directly relevant: (4) we need not reject a hypothesis when its consequences fail to be observed *and* this failure can be

For, as Copernicus states, it may be that the immense distance of the stellar sphere renders such minute phenomena unobservable; it may also be, as we said, that they have not even been investigated, or (if investigated) not investigated [414] in an adequate manner, namely, with the exactness necessary for such minute details; but this exactness is difficult to achieve, both for lack of proper astronomical instruments (which are subject to many variations) and for the faults of those who handle them with less diligence than necessary. There is a necessarily conclusive argument for what little reliability such observations have: it is provided by the differences found among astronomers in assigning positions, I will not say to the new stars and the comets, but to the fixed stars themselves, and even to the polar elevations; in regard to these, most of the time they disagree among themselves by several minutes. Frankly, in using a quadrant or sextant[137] (whose sides are at most 3 or 4 cubits long), who can be so sure that the position of the perpendicular or of the line of sight is not off by two or three minutes, which along the instrument's circumference amount to no more than a grain of millet? Furthermore, it is almost impossible for the instrument to be built and maintained with ab-

---

explained to result from other causes; the relevant causes which he proceeds to enumerate in the case at hand are that stellar parallaxes have not been observed because they are extremely small, or because investigators did not know exactly what to look for, did not conduct a systematic search, lacked powerful or precise instruments, lacked the necessary skill, or failed to apply the proper care. This fourth principle should also be contrasted with another one not mentioned in this passage, namely, that (5) we should reject a hypothesis when its consequences fail to be observed; the difference between (5) and (4) is analogous to that between (3) and (1); but (5) has an advantage over (1), namely, that the following type of argument (called denying the consequent) *is* formally valid: "if *p* were true, then *q* would be true; *q* is not true; therefore, *p* is not true." Thus, from the viewpoint of strict logic, principle (1) is wrong and (5) is right; but methodologically speaking, in scientific practice principles (2), (3), and (4) are more useful. This is not the only place where Galileo discusses such issues (cf. selections 4, 6, and especially 15); he comes close to making all these distinctions and advancing this methodological doctrine about theoretical explanation in his "Considerations on the Copernican Opinion," answering Bellarmine's objection with these words: "It is true that it is not the same to show that one can save the appearances with the earth's motion and the sun's stability, and to demonstrate that these hypotheses are really true in nature. But it is equally true, or even more so, that one cannot account for such appearances with the other commonly accepted system. The latter is undoubtedly false, while it is clear that the former, which can account for them, may be true. Nor can or should one seek any greater truth in a position than that it corresponds with all particular appearances" (Finocchiaro 1989, 85).

137. A quadrant is an instrument used in astronomy, navigation, and surveying for measuring angles; it consists of a graduated arc of 90 degrees, to which is mounted a plumb line to determine the vertical direction or a spirit level to determine the horizontal direction, and a guiding sight to ascertain the direction of the observed object. A sextant is a similar instrument, except that it uses an arc of 60 degrees and so is easier to handle.

solute accuracy. Ptolemy does not trust an armillary instrument[138] built by Archimedes himself to observe the sun crossing the equinoctial circle.[139]

SIMPLICIO. If instruments are so suspect and observations so doubtful, how can we ever attain certainty and free ourselves from fallacies? I have heard great things attributed to Tycho's instruments (built at very great expense) and of his singular diligence in making observations.

SALVIATI. I grant you all this; but neither those instruments nor this diligence are sufficient to resolve an issue of such importance. I want to use instruments very much greater than Tycho's, extremely exact, and built at very little expense; their sides should have a length of 4, 6, 20, 30, and 50 miles, so that one degree would correspond to one mile, one minute to 50 cubits, and one second to little less than one cubit. In short, without spending anything, we can build them as large as we like.

While staying at a villa of mine near Florence, I have clearly observed the sun's arrival at and departure from the summer solstice. This happened one evening when, as it was setting, the sun went behind a cliff in the Pietrapana Mountains about 60 miles away, leaving uncovered toward the north a thin slice of itself whose width was less than one hundredth of its diameter; the following evening it likewise left uncovered a similar part, which however was noticeably thinner. This is a necessary argument [415] that it had begun to move away from the tropic.[140] It is certain that the sun's regression from the first to the second observation did not amount to even a second along the horizon. Then, if the observation is made with a good telescope, which magnifies the solar disk a thousand times, it becomes both easy and pleasing.

Now, I want us to make our observations of the fixed stars with similar instruments, selecting one of those stars in which the variation should be most noticeable. As we have already explained, these would be the ones which are farthest from the plane of the ecliptic; among these

138. An armillary instrument is an astronomical instrument consisting of an arrangement of interlocking rings and designed to represent the position and motion of the heavenly bodies.

139. The equinoctial circle is the celestial equator; the term refers to the fact that, when the sun is on it, night and day have the same duration (twelve hours) everywhere on earth.

140. That is, the Tropic of Cancer, which is a circle on the celestial sphere 23.5 degrees north of the equator marking the northern boundary of the sun's apparent annual motion. After the summer solstice the sun appears to move southward for six months until the winter solstice when it reaches the Tropic of Capricorn, which is the corresponding parallel at 23.5 degrees south.

Vega,[141] which is a very large star close to the pole of the ecliptic, would be a very appropriate one for countries far to the north. However, I selected another star and proceeded in the following manner, having found a very suitable place for such an observation. The place is an open plain to the south of which rises a very tall mountain; on its top stands a small church whose length runs from west to east, so that the ridge of its roof may intersect at right angles the meridian of some house located in the valley below. I want to install a beam parallel to the said ridge of the roof at a distance of about a cubit. After doing this, I will look for a place in the valley from which one of the stars in the Big Dipper[142] is hidden behind the already installed beam as it passes by the meridian; or else, if the beam is not thick enough to hide the star, I will find the spot from where the same beam is seen to cut the disk of that star in half; this effect can be discerned beautifully with a good telescope. If there is a dwelling in the place from which this phenomenon can be observed, that will be more convenient; if not, I will have a pole driven very firmly into the ground, with a permanent mark of where one should place the eye whenever one wants to repeat the observation. I will make the first of these observations around the summer solstice, continuing then from month to month (or whenever I should so wish) until the following solstice. By means of these observations we will be able to detect the star's change in elevation, however small it may be. If this procedure enables us to understand any variation, think of how much and how significant the gain will be for astronomy. For besides enabling us to ascertain the annual motion, this technique may enable us to determine the size and distance of the same star.[143]

141. Drake (1967, 488n) states that Vega was a good choice by Galileo; in fact, two centuries later German astronomer Friedrich G. W. von Struve (1793–1864) made it the subject of extensive observations for the purpose of detecting the parallax. Struve was a codiscoverer of the stellar parallax since he detected it in 1839 independently of Bessel's discovery a year earlier. Struve's work provides additional evidence that Galileo's research program outlined here was fruitful, and that his answer to the parallax objection is methodologically valuable despite the lack of effective refutation.

142. The Big Dipper is a group of seven stars in the constellation Ursa Major so arranged that four of them appear to form a bowl and the other three a handle.

143. Santillana (1953, 399n.72) states that Galileo's suggestions in this passage were closely followed a century later by the English astronomer James Bradley (1693–1762); but Bradley's observations led him to discover in 1729 a phenomenon called the aberration of starlight instead of the sought-after parallax. Like several earlier astronomers who had been searching for the parallax, he observed that the apparent position of fixed stars does vary annually; that is, when observed at six-month intervals, their apparent position shifts by about 40 seconds of arc. He concluded that this shift cannot be a true parallax partly because the change is too large, partly because it is the same for all stars, and partly for more

SAGREDO. I understand the whole procedure very well; I find it [416] so easy and appropriate for what is needed that one could very reasonably think that Copernicus himself or some other astronomer has already put it into practice.

SALVIATI. I think just the opposite. For, if anyone had done the experiment, it is likely that he would have mentioned whether the result was in favor of this or that opinion; and there is no record that anyone has undertaken this kind of observation, either for this or for some other purpose. At any rate, without an exact telescope it could not be easily brought about.

. . .[144]

---

technical reasons involving directions of the shift and times of the year, which are not the right ones for the phenomenon of parallax; later he understood that the shift results from the combination of the earth's annual motion and the finiteness of the speed of light. What happens is analogous to moving forward at high speed through vertically falling rain and seeing that the rain appears to slant obliquely backwards; the astronomical analogue of the rain is the starlight; as the earth moves in its orbit, the light coming from a given star appears to come from a direction shifted somewhat backwards; since at six-month intervals the earth's orbital revolution makes it move in opposite directions relative to the celestial sphere, at these times the light from the same star appears shifted in opposite directions. The aberration of starlight provided direct evidence for the earth's annual motion, but the detection of the parallax had to wait another century until Bessel in 1838. Bradley's discovery is yet another indication of the fruitfulness of Galileo's research program outlined in this selection and of the methodological value of his answer to the parallax objection.

144. The passages omitted here correspond to Favaro 7:416.10–441 and Galileo 1967, 389–415. There are two main topics. First, Galileo discusses the large changes in noontime solar elevation visible with the naked eye and related to the seasons of the year; why these changes are not like the minute stellar parallaxes, and so the contrast does not refute the earth's motion; and how they are explained by Copernicus. Then he discusses magnetism and the properties of lodestones in the context of two arguments: since lodestones have more than one natural motion, their properties undermine the objection that a simple body like the earth cannot have the several natural motions Copernicus attributes to it; and since there is evidence that the whole earth is a lodestone, this may explain why its axis of rotation stays always parallel to itself. For more details, see Finocchiaro (1980, 42 and 133–38).

# Fourth Day

. . .[1]

## [15. The Cause of the Tides and the Inescapability of Error]

[445] . . . SIMPLICIO. Salviati, these phenomena[2] did not just start to happen; they are very old and have been observed by infinitely many persons. Many have striven to explain them by means of some reason or other. Just a few miles from here, a great Peripatetic has advanced a new cause fished out of a certain text of Aristotle not duly noticed by his interpreters; from this text he gathers that the true cause of these motions derives from nothing but the different depths of the seas; for where the depth is greater, the water is greater in quantity and hence heavier, and so it displaces the more shallow water; once raised, this water wants to go down; the ebb and flow derives from this constant struggle. Then there are many who refer this to the moon, saying that it has special dominion over the water. Lately a certain clergy-

1. The passage omitted here corresponds to Favaro 7:442.1–445.21 and Galileo 1967, 416–19. It introduces the tides as a terrestrial phenomenon that can be shown to provide evidence of the earth's motion, and indicates that the aim of the Fourth Day is to present such an argument.

2. The tides have just been described briefly and presented as a fact to be explained; Salviati has proposed explaining them by the hypothesis that the earth moves.

man[3] has published a small treatise in which he says that, as the moon moves through the sky, it attracts and raises toward itself a bulge of water which constantly follows it, so that there is always a high tide in the part which lies under the [446] moon; but since the high tide returns when it is under the horizon,[4] he claims that to explain this effect one must say that the moon not only keeps this faculty naturally within itself, but also has the power of giving it to the opposite point of the zodiac. As I believe you know, others also say that the moon with its moderate heat has the power of rarefying the water, which rises as it expands.[5] We have also had someone who . . . [6]

SAGREDO. Please, Simplicio, do not tell us any more, for I do not think it is worthwhile to take the time to recount them or waste words to confute them; if you give your assent to these or similar trifles, you do an injustice to your judgment, which we know to be very seasoned.

SALVIATI. I am calmer than you, Sagredo, and so I will expend fifty words for the sake of Simplicio, should he perhaps think that there is any probability in the things he related. This is what I say. It is true, Simplicio, that waters whose exterior surface is higher displace those which are below them and lower; but this does not happen with those which are higher by reason of depth; and, once the higher ones have displaced the lower ones, they quickly calm down and level out. This Peripatetic of yours must be thinking that all the lakes in the world (which stay calm) and all the seas where the ebb and flow is unnoticeable have a bottom whose depth does not vary in the least; whereas I was so simpleminded as to believe that, even without any other sounding, the islands which emerge above the waters are a very clear indication of the variability of the bottom. To that clergyman you can say that the moon every day comes over the whole Mediterranean, but that the waters rise only at its

3. The clergyman was Marcantonio de Dominis (1566–1624), the archbishop of the Dalmatian city of Split, who had published a book on the subject (*Euripus, seu de fluxu et refluxu maris sententia*, Rome, 1624). Strauss (1891, 566–67n.3) states that a similar view can be found in Simon Stevin (*Les Oeuvres Matematiques de Simon Stevin de Bruges*, Leyden, 1634, 2:177–83).

4. It was then well known that normally there are two high and two low tides a day, the high and low alternating at approximately six-hour intervals.

5. This idea had been advanced by one of Galileo's teachers at the University of Pisa, Girolamo Borro (1512–1592), in *Dialogo del flusso e reflusso del mare* (Lucca, 1561).

6. Ellipsis in Galileo's text as given in the National Edition (Favaro 7:445), which corrected the 1632 edition in a few places such as here. Galileo wants to convey the impression that Sagredo interrupts Simplicio's speech.

eastern end and here for us in Venice.[7] To those who say that the moderate heat is capable of making the water swell, tell them to start a fire under a boiler full of water and keep their right hand in it until the water rises by a single inch due to the heat, and then to take it out and write about the swelling of the sea;[8] or at least ask them to teach you how the moon manages to rarefy a certain part of the water and not the rest, namely, the one here in Venice and not that at Ancona, Naples, or Genoa. One is forced to say that poetical minds are of two kinds: some adept and inclined to invent fables, others disposed and accustomed to believe them.

SIMPLICIO. I do not think anyone believes in fables while [447] knowing them to be such. In regard to the opinions about the causes of the ebb and flow (which are many), I know that the primary and true cause of an effect is only one, and so I understand very well and am sure that at most one can be true, and I know that all the rest are fictitious and false; and perhaps the true one is not even among those which have been produced so far. Indeed, I truly believe this is the way it is, for it would be strange if the truth produced so little light that nothing would appear

7. This is a relevant (but not decisive) objection to the lunar attraction explanation of the tides. Galileo also felt generally that attraction was an "occult" property involving a magical view of nature, and his attitude toward magic was as dismissive as his attitude toward astrology and alchemy (selection 5); for example, later in the "Fourth Day" (selection 16, [486]), even Kepler is chided for his "childish" belief in a lunar attraction theory. That is, Galileo thought it is methodologically inappropriate to explain tidal motion in terms of lunar attraction because he regarded such an explanation as an instance of the procedure of explaining by naming, and he found such a procedure uninformative and fruitless; for example, elsewhere in the *Dialogue* he objects on such grounds to explaining free fall by means of gravity (Favaro 7:260–61; Galileo 1967, 234–35) and magnetic phenomena by means of a magnetic force (Favaro 7:436; Galileo 1967, 410). He felt it was more appropriate to explain motion by means of an underlying mechanism, namely, by a type of explanation that soon thereafter was called mechanical and became the subject of a controversy still continuing today; his own geokinetic explanation of the tides is meant to be such an explanation. The irony (and the instructiveness) of this story is that the attraction explanation turned out to be physically correct, when Isaac Newton (1687) explained the tides in terms of the law of universal gravitation and the different gravitational forces exerted primarily by the moon (but also by the sun) on different parts of the oceans and of the land. Even in Galileo's time, the correlation between the moon and the tides was known to be very strong—too strong to be accidental: not only are high tides correlated with the moon crossing the local meridian, but they show a daily delay equivalent to the daily delay with which this happens; this point was raised by G. B. Baliani, a follower of Galileo who modified his theory accordingly; cf. Sosio (1970, lxxxi, n.2) and Favaro (14:342).

8. The wit of this remark is as aesthetically pleasing to a modern reader as it was psychologically offensive to Galileo's opponents; this is an example of rhetorical excess, in which he gets carried away by his immersion in the subject and his linguistic flourish and eloquence; cf. the appendix (3.4).

among the darkness of so many falsehoods. However, taking the liberty which we allow among ourselves, I will say that to introduce the earth's motion and make it the cause of the ebb and flow seems to me already to be an idea no less fictitious than the others I have heard; and if I were not offered reasons more in accordance with the nature of things, then without any reluctance I would go on to believe that this is a supernatural effect, and hence miraculous and inscrutable to the human intellect; this would be like the infinitely many others which are dependent directly on the omnipotent hand of God.[9]

SALVIATI. You speak very prudently and also in accordance with Aristotle's doctrine; as you know, at the beginning of his *Mechanical Questions*[10] he attributes to a miracle things whose causes are unknowable. However, as to whether the true cause of the ebb and flow is one of the impenetrable ones, I think the strongest indication you have for this is your seeing that, of all the causes which have so far been advanced as true ones, there is none from which we can reproduce a similar effect, regardless of whatever artifice we employ; for by means of the light of the moon or sun or temperate heat or differences of depth, we will never make the water contained in a motionless vessel artificially run back and forth and go up and down at one place but not another. On the other hand, if by moving the vessel very simply and without any artifice I can represent to you exactly all those changes that are observed in seawater, why do you want to reject this cause and resort to a miracle?

SIMPLICIO. I want to resort to a miracle if you do not convince me of natural causes other than the motion of the basins containing the waters

9. This appeal to divine omnipotence raises controversial issues and is historically and methodologically important. It is controversial because it amounts to an injection of theological considerations into the scientific discussion. From a historical viewpoint, this is related to (but not the same as) the favorite objection to Copernicanism of Pope Urban VIII, which Galileo was requested to include in a prominent place in the book and which he included on the last page (selection 16); but the manner of his inclusion was judged unsatisfactory, and this judgment was a factor leading to the Inquisition trial of 1633. Methodologically, Simplicio's suggestion of a divine miracle has a skeptical aspect insofar as it questions the possibility of any human knowledge, and it has a fallibilist component insofar as it suggests that human knowledge is uncertain and consists of contingent rather than necessary truths. For more details, see selections 1, 4, and 16, the introduction (4.2), Finocchiaro (1980, 6–22; 1985; 1986), and Wisan (1984b).

10. Pseudo-Aristotle, *Quaestiones mechanicae*, I, 847a11. This is a book wrongly attributed to Aristotle according to a tradition that lasted past Galileo's time. It was probably written a generation after Aristotle's death by one of his followers. Nowadays its unknown author is referred to as Pseudo-Aristotle.

of the sea, because I know that these basins do not move, the whole terrestrial globe being motionless by nature.

SALVIATI. But do you not believe the terrestrial globe could be made to move supernaturally, namely, by the absolute power of God?

SIMPLICIO. Who could doubt that?

[448] SALVIATI. Therefore, Simplicio, since we must introduce a miracle to produce the ebb and flow of the sea, let us make the earth move miraculously, and then this motion will naturally make the sea move. This operation will be all the simpler (and, I shall say, the more natural, among the miraculous ones), inasmuch as giving a turning motion to one globe (of the many we see moving) is less difficult than making an immense quantity of water go back and forth (in some places faster and in others slower) as well as rise and fall (more in some places, less in others, and not at all in still others), and having all these variations take place in the same containing vessel. Moreover, the latter involves many different miracles, the former only one. Finally, the miracle of making the water move implies another miracle as a consequence; that is, keeping the earth motionless against the impulses of the water, which are powerful enough to make it waver in this or in that direction unless it were miraculously restrained.[11]

SAGREDO. Please, Simplicio, let us suspend our judgment about declaring useless the new view which Salviati wants to explain to us, and let us not be too quick to place it in a pigeon-hole with the old ridiculous accounts. As regards miracles, let us resort to them after we have listened to discussions confined within the limits of natural reason, although I am inclined to find miraculous all works of nature and of God.

SALVIATI. My judgment is the same; to say that the natural cause of the tides is the earth's motion does not prevent this process from being miraculous.

Now, to resume our reasoning, I repeat and reaffirm that so far it is not known how it can happen that the waters contained in our Mediterranean basin undergo the motions they are seen to have as long as the containing basin or vessel remains itself motionless; what generates the difficulty and renders this subject inextricable are the things I will mention below which are observed every day. So, listen.

11. Here we have a simplicity argument, formulated and used in the context of answering the above mentioned theological objection. It would be instructive to reconstruct the details of this simplicity argument; that would also sharpen our understanding of the original objection. For an interesting reconstruction, see Wisan (1984b) and cf. Finocchiaro (1985).

We are here in Venice, where there is a low tide and the sea is tranquil and the air calm. The water begins to rise, and within five or six hours, it rises ten palms or more. Such a rise does not derive from the expansion of the water that was there before, but rather from new water that has come here, water of the same kind as the old, of the same salinity, of the same density, and of the same weight; [449] boats float on it, Simplicio, just as they did on the old water, without subsiding a hair lower; a barrel of this new water does not weigh a single grain more or less than an equal volume of the old water; it is as cold as the other, without any change; in short, it is new water that has visibly entered the bay through the narrows and mouth of the Lido.[12] Now, you tell me whence and how it has come here.

Are there perhaps around here some openings and caves at the bottom of the sea through which the earth inhales and regurgitates the water, breathing as if it were an immense and enormous whale?[13] If this is so, how is it that in a period of six hours the water does not rise likewise in Ancona, Dubrovnik, and Corfu, where the rise is very small and perhaps unobservable? Who will find a means of injecting new water into a motionless vessel and ensuring that it will rise only in a definite part of it and not elsewhere?

Will you say perhaps that the new water is supplied by the ocean, coming through the Strait of Gibraltar?[14] This does not remove the difficulties already mentioned and carries with it some more serious ones. First, tell me what must be the speed of the water which enters the strait and in six hours reaches the extreme shores of the Mediterranean (covering a distance of two or three thousand miles), and which then again covers the same distance in the same time when it returns? What will happen to the various ships at sea? What will happen to those which might be in the strait, where there would be such a constant and impetuous flow of an immense quantity of water that, by using a channel no more than eight miles wide, it would provide enough water to flood in six hours an area hundreds of miles wide and thousands of miles long?

12. The city of Venice, where this book's discussion occurs, is built on many small islands separated by canals and located in the middle of a shallow bay, the Lagoon of Venice; this bay is separated from the Adriatic Sea by a series of long and narrow islands, one of which is the Lido. The tides are especially noticeable there.

13. This explanation is distinct from the ones discussed above. According to Sosio (1970, lxxxiv), it is found in several authors of classical antiquity, the medieval Arab world, and the Renaissance (including Leonardo da Vinci).

14. The Strait of Gibraltar joins the Mediterranean Sea to the North Atlantic Ocean and separates southwestern Europe from northwestern Africa.

What tiger or falcon ever ran or flew at such a speed? I mean a speed of 400 and more miles per hour. There are indeed currents along the strait (I do not deny it), but they are so slow that rowboats outrun them, although not without a delay in their course. Furthermore, if this water comes through the strait, the other difficulty still remains; that is, how it manages to rise so much here in a region so remote, without first rising by a similar or greater height in the nearer regions.

In short, I do not think that either stubbornness or intellectual subtlety can ever find solutions to these difficulties or consequently uphold the earth's stability against them, as long as we confine ourselves within natural limits.

SAGREDO. I comprehend this very well already and am eagerly waiting to hear how these puzzling phenomena can without hindrance follow from the motions already attributed to the earth.

[450] SALVIATI. In regard to the manner in which these effects should follow as a consequence of the motions that naturally belong to the earth, not only must they find no repugnance or hindrance, but they must follow easily; indeed, not only must they follow with ease, but with necessity, so that it is impossible for them to happen otherwise; for such is the character or mark of true natural phenomena.[15] We have established the impossibility of explaining the motions we see in the water while simultaneously maintaining the immobility of the containing vessel;[16] so, let us go on to see whether the motion of the container can produce the effect and make it happen in the way it is observed to happen.

There are two kinds of motions which can be imparted to a vessel, and from which the water contained in it can acquire the power to flow alternately toward one of its extremities and toward the other, and alternately to rise and fall there. The first would occur when either one of the extremities is lowered, for then the water (flowing toward the inclined point) would be alternately raised and lowered, now at this extremity and now at that one. However, this rising and falling are nothing but a motion away from and toward the center of the earth, and hence this

15. Here Salviati expresses a controversial methodological principle, a version of what I call demonstrativism. One main interpretive question is whether he attributes necessity to natural phenomena by themselves or to their relationship to their causes. For more details, see the introduction (5.3) and the appendix (2.5 and 3.2). Demonstrativism was a key component of Aristotle's theory of scientific knowledge; this remark gives one of many indications of the Aristotelian background of Galileo's thought; see Wallace's works for the best elaboration of such an interpretation.

16. The previous criticism of alternative theories of tides, all of which are geostatic, may be regarded as the argument establishing this impossibility.

kind of motion cannot be attributed to the basins in the earth itself which contain the water; regardless of any motion attributed to the terrestrial globe, the parts of these containing vessels can neither approach nor recede from its center.

The other kind of motion occurs when, without tilting in any way, the vessel moves with forward motion at a speed that is not uniform but changing, sometimes accelerating and sometimes being retarded. The water contained in the vessel is not rigidly attached to it as its other solid parts are; instead, as a fluid, the water is almost separate, free, and not obliged to go along with all the changes of its container; it follows that, when the vessel is retarded, the water retains a part of the impetus already acquired and so flows toward the forward end, where it necessarily rises; on the contrary, if the vessel should acquire additional speed, the water would retain a part of its slowness and remain somewhat behind, and so (before getting used to the new impetus) it would flow toward the rear of the vessel, where it would rise by a certain amount. These effects can be more clearly explained and shown to the senses by means of the example of one of those boats that constantly come from Lizzafusina, full of the fresh water used by the city. [451] Let us then imagine such a boat moving at moderate speed across the lagoon and calmly carrying the water with which it is filled; suppose then that it is considerably retarded, either by running aground or due to some other obstacle in its way; the water contained in the boat will not thereby lose the already acquired impetus (as the boat itself will), but will conserve it and flow forward toward the bow, where it will noticeably rise while dropping astern; but if, on the contrary, while the same boat is on its quiet course it acquires additional speed by a noticeable amount, then before the contained water gets used to the new speed it will retain its slowness and remain behind, namely, toward the stern, where it will consequently rise while dropping at the bow.

This effect is indubitable and clear and can be experienced at any time. There are three particulars about it that I now want us to note. The first is that, in order to make the water rise at one end of the vessel, there is no need of any new water, nor need it flow there from the other end. The second is that the water in the middle does not noticeably rise or fall, unless the course of the boat is very fast and the collision or other restraining obstacle is sudden and very strong, in which case all the water could not only flow forward but even spill out of the boat for the most part; the same thing would also happen if, while going slowly, it should suddenly receive an extremely powerful impetus; but if its quiet motion undergoes a moderate retardation or acceleration, the water in the middle rises and falls imperceptibly (as I said), and for the rest, the closer it is to the middle the less it rises, and the farther it is the more it rises. The third is that, whereas the water near the middle undergoes little variation by rising and falling as compared with that at the end, on the contrary it flows a great deal forward and backward as compared with the same.

Now, gentlemen, what the boat does in relation to the water contained in it and what the contained water does in relation to the boat are exactly the same as what the Mediterranean basin does in relation to the water contained in it and what the contained water does in relation to the Mediterranean basin.[17] Next, we need to demonstrate how and in what manner it happens that the Mediterranean and all the other basins (in short, all parts of the earth) move with a significantly nonuniform motion, although only motions that are regular and uniform are assigned to the whole globe.

[452] SIMPLICIO. To me, who am neither a mathematician nor an astronomer, this seems at first sight like a great paradox; if it is true that while the motion of the whole is regular, the motion of the parts that remain always attached to the whole can be irregular, then the paradox will destroy the axiom affirming that the reasoning applying to the whole and to the parts is the same.[18]

SALVIATI. I will demonstrate my paradox and will leave to you, Simplicio, the task of defending the axiom from it or making them consistent;[19] my demonstration will be short and very easy and will depend on

17. Here the argument seems to use analogical reasoning, the analogy being between a moving boat and a moving earth; if this interpretation is correct, then this step in the argument does not necessitate the conclusion but at best makes it probable; and the reasoning can be questioned in the way arguments from analogy are usually questioned; for example, such an evaluation can use Galileo's own considerations on this analogy, advanced earlier (selection 8). But this argument step can also be interpreted as a new instantiation (to the case of the earth) of a generalization (that acceleration causes tidal-like motions); on this interpretation, the inference is necessary rather than merely probable; but then one can question the status of the generalization premise, namely, how and how firmly this proposition has been supported; the only supporting argument is the one based on the evidence of what happens when a boat carrying water is accelerated and retarded; and whatever its strength, this is an inductive argument whose conclusion is at best established with high probability rather than necessity.

18. As usual, Galileo writes this principle in Latin and in italics.

19. The parts-whole principle is also used in Aristotle's natural motion argument and is questioned by Galileo in his criticism (selection 2); that criticism shows that this principle is itself paradoxical. Whatever difficulties his discussion of tides may contain, it provides another counterexample to this principle: the earth's two Copernican motions (axial rotation and orbital revolution) are uniform for the whole earth but nonuniform for its parts.

the things discussed at length in our past arguments, when we did not introduce so much as a word about the tides.

We said that there are two motions attributed to the terrestrial globe: the first is the annual motion performed by its center along the circumference of the annual orbit in the plane of the ecliptic and in the order of the signs of the zodiac, namely, from west to east; the other is performed by the same globe rotating around its own center in twenty-four hours, likewise from west to east, but around an axis somewhat inclined and not parallel to that of the annual revolution. From the combination of these two motions, each of which is in itself uniform, there results, I say, a variable motion for the parts of the earth; I will explain this by drawing a diagram [fig. 13], so that it can be more easily understood.

First, around the center A, I describe the circumference of the annual orbit BC; on it let us take any point whatever B, and using B as a center let us describe this smaller circle DEFG, representing the terrestrial globe; then let us assume the center B to run along the whole circumference of the annual orbit from west to east, namely, from B toward C; and let us further assume the terrestrial globe to turn around its own center B in the period of twenty-four hours, also from west to east, namely, according to the order of the points D, E, F, and G. Here we must note carefully that as a circle turns around its own center, each part of it must move in opposite directions at different times; this is clear by considering that, while the parts of the circumference around the point

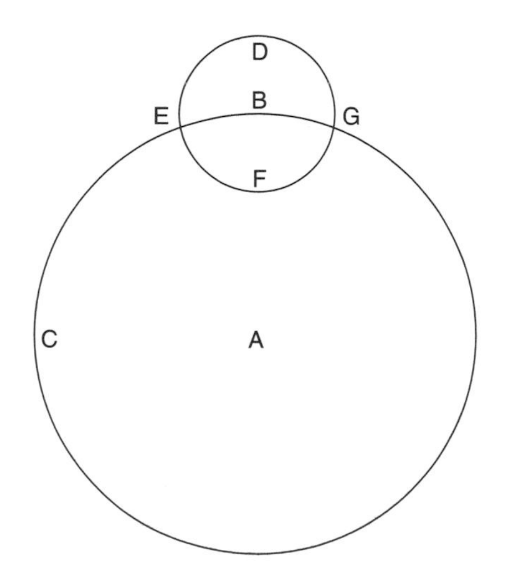


Figure 13.

D move toward the left (namely, toward E), those on the opposite side (which are around F) advance toward the [453] right (namely, toward G), so that when the parts D are at F their motion is contrary to what it was when they were at D; furthermore, at the same time that the parts E descend (so to speak) toward F, the parts G ascend toward D. Given such a contrariety in the motions of the parts of the terrestrial surface as it turns around its own center, it is necessary that in combining this diurnal motion with the other annual one there results an absolute motion[20] of the parts of the terrestrial surface that is sometimes highly accelerated and sometimes retarded by the same amount. This is clear from the following considerations: the absolute motion of the part around D is very fast since it originates from two motions in the same direction, namely, toward the left; the first of these is the annual motion common to all parts of the globe, the other is the motion of point D carried also toward the left by the diurnal rotation; hence, in this case the diurnal motion increases and accelerates the annual motion; the opposite of this happens at the opposite side F, which is carried toward the right by the diurnal rotation while together with the whole globe it is carried toward the left by the common annual motion; thus, the diurnal motion takes away from the annual, and so the absolute motion resulting from the combination of the two turns out to be greatly retarded; finally, around the points E and G the absolute motion remains equal to the annual alone, for the diurnal motion adds or subtracts little or nothing, its direction being neither left nor right but down and up. Therefore, we conclude that, just as it is true that the motion of the whole globe and of each of its parts would be invariable and uniform if it were moving with a single motion (be it the simple annual or the diurnal alone), so it is necessary that the mixture of these two motions together gives the parts of the globe variable motions (sometimes accelerated and sometimes retarded) by means of additions or subtractions of the diurnal rotation and the annual revolution. Thus, if it is true (and it is

20. The "absolute motion" mentioned here and in the rest of this discussion may be interpreted as motion in a reference frame with the sun as the origin, regardless of whether this solar frame is itself in motion; this solar frame should be contrasted with a terrestrial reference frame, namely, one attached to the earth; obviously in a terrestrial reference frame, there is neither axial rotation nor orbital revolution by the earth, nor are there accelerations and retardations deriving from their addition and subtraction. Some critics have sought to identify an internal logical error in Galileo's reasoning, lying in his confusion of solar and terrestrial reference frames; their point is that the tides take place in the terrestrial reference frame, but the acceleration which allegedly causes them takes place in the solar frame; for expositions of this criticism, see Aiton (1954, 46), Favaro (15:251–52), Koestler (1959, 464–66), and Shea (1972, 175); for a criticism of this criticism, see Finocchiaro (1980, 77–78).

most true, as experience shows) that the acceleration and retardation of a vessel's motion make the water contained in it run back and forth along its length and rise and fall at its ends, who will want to raise difficulties about granting that such an effect can (or rather, must necessarily) happen in seawater, which is contained in various basins subject to similar variations, especially in those whose length stretches out from west to east (which is the direction along which these basins move)?

[454] Now, let this be the primary and most important cause of the tides, without which this effect would not happen at all. However, there are many different particular phenomena that can be observed in different places and at different times, and that must depend on other different concomitant causes, although these must all be connected with the primary cause; hence, it is proper to present and examine the various factors that may be the causes of such various phenomena.[21]

The first of these is that whenever water is made to flow toward one or the other end of a containing vessel by a noticeable retardation or acceleration of that vessel, and it rises at one end and subsides at the other, it does not thereby remain in such a state even if the primary cause should cease; instead, in virtue of its own weight and natural inclination to level and balance itself out, it spontaneously and quickly goes back; and, being heavy and fluid, not only does it move toward equilibrium, but carried by its own impetus, it goes beyond and rises at the end where earlier it was lower; not resting here either, it again goes back, and with more repeated oscillations, it indicates that it does not want to change suddenly from the acquired speed to the absence of motion and state of rest, but that it wants to do it gradually and slowly. This is similar to the way in which a pendulum, after being displaced from its state of rest (namely, from the perpendicular), spontaneously returns to it and to rest, but not before having gone beyond it many times with a back-and-forth motion.

The second factor to notice is that the reciprocal motions just mentioned take place and are repeated with greater or lesser frequency, namely, in shorter or longer times, depending on the length of the vessels containing the water; thus, the oscillations are more frequent for the shorter distances and rarer for the longer. And this is exactly what happens in the same example of pendulums, where we see that the oscillations of those

21. This discussion of the "concomitant causes" introduces another crucial element in Galileo's theory of the tides. The earth's motion is for him "the primary and most important cause," but not the only one and not sufficient by itself to produce the two daily tides and other features. As the ensuing discussion reveals, the concomitant causes are mostly the fluid properties of water.

hanging from a longer string are less frequent than those of pendulums hanging from shorter strings.[22]

And here is a third important point to know: it is not only the greater or lesser length of the vessel that causes the water to make its oscillations in different times, but the greater or lesser depth brings about the same thing; what happens is that, for water contained [455] in vessels of equal length but of unequal depth, the one which is deeper makes its oscillations in shorter times, and the vibrations of less deep water are less frequent.[23]

Fourth, worthy of notice and of diligent observation are two effects produced by water in such vibrations. One is the alternating rising and falling at both ends; the other is the flowing back and forth, horizontally so to speak. These two different motions affect different parts of the water differently. For its ends are the parts that rise and fall the most; those at the middle do not move up or down at all; and as for the rest, those that are nearer the ends rise and fall proportionately more than the farther parts. On the contrary, in regard to the lateral motion back and forth, the middle parts go forth and come back a great deal; the water at the ends does not flow at all except insofar as by rising it goes over the embankment and overflows its original bed, but where the embankment stands in the way and can hold it, it only rises and falls; finally, the water in the middle is not the only part that flows back and forth, for this is also done proportionately by its other parts, as they flow more or less depending on how far or near they are relative to the middle.

The fifth particular factor must be considered much more carefully, insofar as it is impossible for us to reproduce it experimentally and practically.[24] The point is this. In artificial vessels which, like the boats men-

22. These references to the motion of a pendulum (here and in the previous paragraph) are not accidental but reflect Galileo's long-standing interest in the laws of falling bodies, to which he made important contributions. For example, he discovered that a pendulum's period of oscillation is independent of both the weight of the hanging bob and the amplitude of the swing (for small arcs), but increases with the length of the string.

23. The last two paragraphs contain an approximation to an important law in hydrodynamics. Combining the two points, Galileo says that the period of oscillation of water in a vessel varies directly with the length of the vessel and inversely with its depth. The relationship is actually more complicated insofar as the dependence on the length involves its *square root*. Cf. Strauss (1891, 568n.9) and Pagnini (1964, 3:243n.1).

24. This remark seems to contradict Salviati's claim in the penultimate sentence of this speech, namely, that he has under construction such a machine. Perhaps Galileo meant not that it is impossible but that it is very difficult, for in the "Discourse on the Tides" (of which this selection is a reworking) the corresponding sentence reads: "The fifth detail must be considered much more carefully, insofar as it is at least very difficult, if not impossible, to reproduce it experimentally and practically" (Finocchiaro 1989, 125). For an attempt to resolve the apparent contradiction, see Drake (1970, chapter 10).

tioned above, move now more and now less swiftly, the acceleration or retardation is shared to the same extent by the whole vessel and all its parts: thus, for example, as the boat slows down, the forward part is not retarded any more than the back, but they all share the same retardation equally; the same happens in acceleration; that is, as the boat acquires greater speed, both the bow and the stern are accelerated in the same way. However, in very large vessels like the very long basins of the seas, though they are nothing but certain hollows carved out of the solid terrestrial globe, nevertheless amazingly their extremities do not increase or diminish their motion together, equally, and simultaneously; [456] instead it happens that, when one extremity is greatly retarded in virtue of the combination of the diurnal and annual motions, the other extremity finds itself still experiencing very fast motion.

For easier comprehension, let us explain this by referring to the diagram drawn here [fig. 14]. In it, let us consider, for example, a portion of water spanning a quarter of the globe, such as the arc BC; here, as we explained above, the parts at B are in very fast motion due to the combination of the diurnal and annual motions in the same direction, whereas the parts at C are retarded insofar as they lack the forward motion deriving from the diurnal rotation. If, then, we take a sea basin whose length equals the arc BC, we see how its extremities move simultaneously with great inequality. The differences would be greatest for the speeds of an ocean a hemisphere long and situated in the position of the arc BCD, for

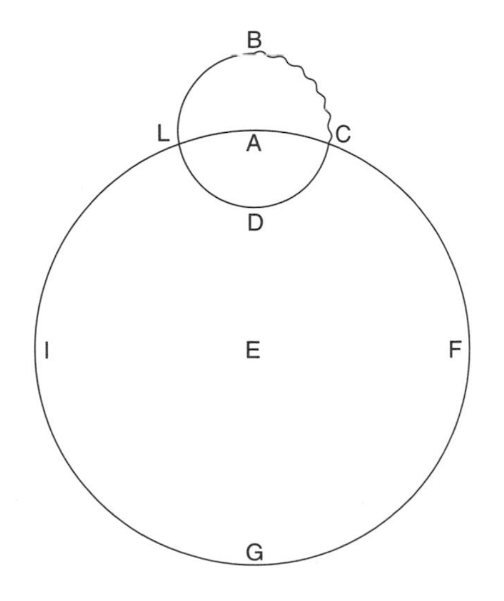


Figure 14.

the end B would be in very fast motion, the other D would be in very slow motion, and the middle parts at C would have an intermediate speed; further, the shorter a given sea is, the less will it experience this curious effect of having its parts moving at different speeds during certain hours of the day. Thus, if, as in the first case, we observe acceleration and retardation causing the contained water to flow back and forth despite the fact that they are shared equally by all parts of the vessel, what shall we think must happen in a vessel placed so curiously that its parts acquire retardation and acceleration very unequally? It seems certain we can say only that here we have a greater and more amazing cause of even stranger movements in the water. Though many will consider it impossible that we could experiment with the effects of such an arrangement by means of machines and artificial vessels, nevertheless it is not entirely impossible; I have under construction a machine in which one can observe in detail the effect of these amazing combinations of motions.[25] However, regarding the present subject, let us be satisfied with what you may have been able to understand with your imagination so far.

[457] SAGREDO. For my part, I understand very well how this marvelous phenomenon must necessarily take place in the sea basins, especially in those that extend for long distances from west to east, namely, along the course of the motions of the terrestrial globe; moreover, just as it is in a way inconceivable and unparalleled among the motions we can reproduce, so I have no difficulty believing that it may produce effects which cannot be duplicated with our artificial experiments.

SALVIATI. After these things have been clarified, it is time for us to go on and examine the variety of particular phenomena which experience enables us to observe in regard to the tides.[26] First, there will be no diffi-

25. I already noted that this sentence seems to contradict Salviati's claim at the beginning of the discussion of his "fifth particular factor" above ([455]), where he regards it as impossible to reproduce this effect experimentally. The corresponding sentence in the "Discourse on Tides" (1616) reads: "Though many will consider it impossible that we could experiment with the effects of such an arrangement by means of machines and artificial vessels, nevertheless it is not entirely impossible; I have under construction a machine, which I shall explain at the proper time, and in which one can observe in detail the effects of these amazing combinations of motions" (Finocchiaro 1989, 126–27). In the intervening period of sixteen years, Galileo apparently had not yet built the apparatus, and now he seems less optimistic than before. Again, for an attempt to explain the discrepancy, see Drake (1970, chapter 10).

26. Besides distinguishing the primary from the concomitant causes, Galileo distinguishes the main or primary effect from various secondary or particular effects; the primary effect is that every day seawater undergoes a number of up and down and back and forth motions in some but not all parts of a sea basin; the secondary effects are those which Salviati proceeds to list.

culty understanding why it happens that there are no noticeable tides in ponds, lakes, and even small seas; this has two very effective causes.

One is that, as the basin acquires different degrees of speed at different hours of the day, because of its smallness they are acquired with little difference by all its parts, and the forward as well as the backward parts (namely, the eastern and the western) are accelerated and retarded almost in the same way; moreover, since this change occurs gradually, and not by a sudden obstacle and retardation or an immediate and large acceleration in the motion of the containing basin, it as well as all its parts receive equally and slowly the same degrees of speed; from this uniformity it follows that the contained water too receives the same action with little resistance, and consequently it gives very little sign of rising and falling and of flowing toward this or the other end. This effect is also clearly seen in small artificial containers, in which the water acquires the same degrees of speed whenever the acceleration or the retardation takes place in a relatively slow and uniform manner. However, in sea basins that extend for a great distance from east to west, the acceleration or retardation is much more noticeable and unequal, for while one end is undergoing very retarded motion the other is still moving very rapidly.

The other cause is the reciprocal vibration of the water stemming from the impetus it also receives from the container, which vibration has very frequent oscillations in small vessels, as we have seen: for [458] the earth's motions can cause agitation in the waters only at twelve-hour intervals, since the motion of the containing basins is retarded and is accelerated the maximum amount only once a day, respectively; but the second cause depends on the weight of the water while in the process of reaching equilibrium, and it has its oscillations at intervals of one hour, or two, or three, etc., depending on the length of the basin; now, mixed with the former cause, which is very small in small vessels, the latter renders it completely imperceptible; for before the end of the operation of the primary cause with the twelve-hour period, the secondary one due to the weight of the water comes about, and with its period of one hour, two, three, or four, etc. (depending on the size and depth of the basin), it perturbs and removes the first, without allowing it to reach the maximum or the middle of its effect. From this contraposition, any sign of tides remains completely annihilated or much obscured.

I say nothing of the constant alterations due to air; disturbing the water, they would not allow us to ascertain a very small rise or fall of half an inch or less, which might actually be taking place in water basins that are no longer than a degree[27] or two.

Second, I come to resolving the difficulty of how tidal periods can commonly appear to be six hours, even though the primary cause embodies a principle for moving the water only at twelve-hour intervals, that is, once for the maximum speed of motion and once for maximum slowness.[28] To this I answer that such a determination cannot in any way result from the primary cause alone; instead we must add the secondary ones, namely, the greater or lesser length of the vessels and the greater or lesser depth of the water contained in them. Although these causes do not act to bring about the motions of the water (since this action originates only from the primary cause), nevertheless they have a key role in determining the periods of the oscillations, and this role is so powerful that the primary cause remains subject to them. Thus, the six-hour interval is no more proper or natural than other time intervals, although it is perhaps the one most commonly observed since it occurs in our Mediterranean, which for many centuries was the only accessible sea; however, such a period is not [459] observed in all its regions, for in some of the more narrow areas such as the Hellespont[29] and the Aegean,[30] the periods are much shorter and also much different from each other. Some say that Aristotle long observed these variations from some cliffs in Euboea[31] and found their causes incomprehensible, and

27. By *degree*, Galileo means 1 of 360 degrees in a circle; so, when the circle is the equator (which is about 24,000 miles), one degree is about 67 miles.

28. Note that Galileo mentions and tries to explain the fact that the interval between high and low tide is normally six hours, namely, that the period of one full tidal cycle is normally twelve hours, or again that normally there are two high and two low tides a day. His attempt is important because this fact constitutes prima facie evidence against his theory, as he admits in this paragraph and as his critics have always pointed out. The issue is whether the six-hour interval between high and low tide constitutes merely prima facie counterevidence or really refuting counterevidence. A related issue is whether the six-hour interval is a primary effect to be explained by the primary cause, or a secondary effect to be explained by the concomitant causes. He admits the prima facie refuting character of this fact; but he regards it as a secondary effect which his theory can explain by means of concomitant causes, and so as merely prima facie and not real counterevidence. For more details, see Shea (1972, 177–78) and cf. Finocchiaro (1980, 78).

29. The Hellespont, now called Dardanelles, is actually a strait separating southeastern Europe from Asia Minor and joining the Aegean Sea to the Sea of Marmara.

30. The Aegean Sea in the eastern Mediterranean separates Greece from Turkey and is joined in the northeast to the Sea of Marmara by way of a long and narrow strait called the Dardanelles.

31. In the original, *Negroponte*.

that because of this, (overcome with despair) he jumped into the sea and drowned himself.[32]

Third, we can quickly explain why it happens that although some seas are very long, for example the Red Sea, nevertheless they are almost entirely lacking in tides. This occurs because its length does not extend from east to west, but from southeast to northwest. For, the earth's motions being from west to east, the impulses received by the water always cross the meridians and do not move from one parallel to another; so, in seas that extend transversely in the direction of the poles and that are narrow in the other direction, no cause of tides remains but the contribution of some other sea, with which they are connected and which is subject to large motions.

Fourth, we can very easily understand the reason why, in regard to the rise and fall of the water, tides are greatest at the extremities of gulfs and smallest in the middle. This is shown by daily experience here in Venice, which is located at the end of the Adriatic[33] and where this variation amounts to five or six feet; but in areas of the Mediterranean far from the extremities, such a variation is very small, as is the case in the islands of Corsica and Sardinia and on the shores of Rome and Leghorn, where it does not exceed half a foot. We can also understand how, on the contrary, in places where the rise and the fall are very small, the flow back and forth is large. I say it is easy to understand the cause of these phenomena because we can make clear tests in all sorts of vessels we can artificially build; here the same effects are seen to follow naturally from our making them move with a motion which is nonuniform, namely, sometimes accelerated and sometimes retarded.

Fifth, considering how the same quantity of water that moves slowly through a wide area must flow with great impetus when passing through a narrow place, we shall have no difficulty in understanding the cause of the immense currents which flow in the narrow channel that separates Calabria from [460] Sicily;[34] for although all the water con-

32. This is not a true story; and later in the "Fourth Day," Galileo treats it as a mere legend (Favaro 7:472; Galileo 1967, 447). In fact, Aristotle discusses the tides only briefly in *Meteorology* (I, 1). The tides in the channel separating the island of Euboea from the Greek mainland are unusual; twice a month local conditions produce stationary waves that make the seawater ebb and flow every few hours (Strauss 1891, 568n.12; Santillana 1953, 440n.8). This is exactly the type of phenomenon to be explained by Galileo's concomitant causes.

33. The Adriatic Sea is a sea in the central Mediterranean on the northeastern side of the Italian peninsula; Venice is located at its northern end.

34. This is now called the Strait of Messina, named after a seaport city in northeastern Sicily.

tained in the eastern Mediterranean and bound by the width of the island and the Ionian Gulf may slowly flow into it toward the west, nevertheless, when constricted into this strait between Scylla and Charybdis,[35] it flows rapidly and undergoes very great agitation. Similar to this and much greater we understand are the currents between Africa and the large island of Madagascar,[36] as the waters of the North and South Indian Ocean,[37] which surround it, flow and become constricted in the smaller channel between it and the South African coast. Very great must be the currents in the Strait of Magellan, which connects the extremely vast South Atlantic and South Pacific Oceans.[38]

Sixth, to account for some more obscure and implausible phenomena observed in this subject, we now have to make another important consideration about the two principal causes of tides and then mix them together. The first and simpler of these is (as we have said several times) the definite acceleration and retardation of the earth's parts, from which the water would acquire a definite tendency to flow toward the east and to go back toward the west within a period of twenty-four hours. The other is the one that depends on the mere weight of water: once stirred by the primary cause, it then tries to reach equilibrium by repeated oscillations; these are not determined by a single period in advance, but they have as many temporal differences as the different lengths and depths of sea basins; and insofar as they depend on this second principle, some oscillations might flow back and forth in one hour, others in two, four, six, eight, ten, etc. Now, let us begin to join together the primary cause, whose fixed period is twelve hours, with one of the secondary causes whose period is, for example, five hours: sometimes it will happen that the primary and secondary causes agree by both producing impulses in the same direction, and with such a combination (a unanimous consent, so to speak) the tides are large; other times, the primary impulse being somehow opposite

35. This refers again to the Strait of Messina separating the island of Sicily from the Italian peninsula and joining the Ionian and the Tyrrhenian Seas. In classical Greek mythology, Scylla and Charybdis were two sea monsters who lived on either side of the Strait of Messina and who lured and devoured ships and sailors coming close to them while passing through the narrows; Scylla lived on the rocks on the Italian side, while Charybdis created a whirlpool on the Sicilian side; thus, the phrase *between Scylla and Charybdis* has come to mean between two dangers or evils that are difficult to avoid simultaneously because avoiding one brings us close to the other.

36. In the original, *S. Lorenzo*.

37. In the original, *Indico ed Etiopico*.

38. In the original, *Etiopico e del Sur*.

to that of the secondary cause, and thus one principle taking away what the other one gives, the watery motions will weaken and the sea will reduce to a very calm and almost motionless state; finally, [461] on still other occasions, when the same two causes neither oppose nor reinforce each other, there will be other variations in the increase or decrease of the tides. It may also happen that of two very large seas connected by a narrow channel, due to the mixture of the two causes of motion, one sea has tidal motions in one direction while the other has them in the opposite; in this case, in the channel where the two seas meet there are extraordinary agitations with contrary motions, vortices, and very dangerous boilings, as it is in fact constantly observed and reported. These conflicting motions, dependent on the different positions and lengths of interconnected seas and on their different depths, give rise sometimes to those irregular disturbances of the water whose causes have worried and continue to worry sailors, who experience them without seeing winds or any other serious atmospheric disturbance that might produce them.

These atmospheric disturbances must be significantly taken into account in other cases, and we must regard them as a tertiary accidental cause,[39] capable of significantly altering the occurrence of the effects produced by the primary and more important causes. For example, there is no doubt that very strong winds from the east can support the water and prevent it from ebbing; then, when at the appropriate time there is a second wave of flow (and a third), it will rise a great deal; thus, if sustained for a few days by the power of the wind, it will rise more than usual and produce extraordinary flooding.

Seventh, we must also note another cause of motion dependent on the great quantity of river water flowing into seas that are not very large. Here, in channels or straits connected with such seas, the water is seen flowing always in the same direction; for example, this happens at the Bosporus near Constantinople,[40] where the water always flows from the

39. Besides distinguishing between primary and concomitant causes of the tides, here Galileo is subdividing the latter into secondary and tertiary; both the atmospheric disturbances mentioned here and the water from large rivers mentioned in the next paragraph are examples of tertiary concomitant causes.

40. The Bosporus is a very narrow and very long strait separating southeastern Europe from Asia Minor and connecting the Black Sea and the Sea of Marmara. Constantinople was a city founded by and named after the Roman emperor Constantine in the fourth century A.D. to be the new capital of the Roman Empire; it later became successively the capital of the Eastern Roman Empire, the Byzantine Empire, and the Ottoman Empire; located on the European side of the Bosporus, it is now part of Turkey and is called Istanbul.

Black Sea to the Sea of Marmara.[41] For, in the Black Sea, because of its smallness, the principal causes of the tides are of little effect; on the contrary, very large rivers flow into it, and so with such a superabundance of water having to go through the strait, here the flow is very noticeable and always southward. Moreover, we must note that although this strait is very narrow, it is not [462] subject to perturbations like those in the strait of Scylla and Charybdis. For the former has the Black Sea on the north and the Sea of Marmara, the Aegean, and the Mediterranean on the south; and, as we have already noted, insofar as a sea extends in a north-south direction it is not subject to tides; on the contrary, because the strait of Sicily[42] is interposed between two parts of the Mediterranean that extend for long distances in an east-west direction, namely, in the direction of tidal currents, in the latter the disturbances are very large. Similarly, they would be larger between the Pillars of Hercules,[43] if the Strait of Gibraltar were less wide; and they are reported to be very large in the Strait of Magellan.

For now, this is all I can think of telling you about the causes of this first (diurnal) period of the tides and related phenomena;[44] if you want

41. The Sea of Marmara is a small sea in the eastern Mediterranean separating southeastern Europe from Asia Minor; it is joined by the Bosporus to the Black Sea in the northeast, and by the Dardanelles (or Hellespont) to the Aegean Sea in the southwest.

42. This is the same strait of Scylla and Charybdis, now called Strait of Messina.

43. This is the ancient mythological name for the Strait of Gibraltar, separating Europe from Africa, and connecting the Mediterranean Sea and the Atlantic Ocean; the "pillars" are the two mountains on each side of the strait, when approached from the Mediterranean.

44. The main interpretive issues raised by the tidal argument in this selection are (1) whether it is meant to be conclusive or merely probable; (2) what exactly is its structure, given that we take it to be a hypothetical, explanatory, and causal argument, namely, an argument whose essential premise is a statement of an effect to be explained and whose conclusion is a hypothesis that explains that effect by indicating its cause; and (3) whether there is an element of analogical reasoning. The main evaluative issues about the argument involve several aspects: the relevance of the correct scientific theory (known to us today) that the tides are caused by the gravitational attraction of the moon and sun; whether there is an illegitimate logical confusion of solar and terrestrial reference frames; whether the observation of the six-hour interval between high and low tide can be fairly taken to refute Galileo's theory, given that he tries to take it into account; whether he can be credited with groping toward the definition of a small tidal effect due to differences in rotational accelerations of different parts of the earth, which is a consequence of the laws of centrifugal force and is called the "tide of reaction"; and whether, despite the main scientific error, his theory has scientific value either because of the intuition of the tide of reaction just mentioned, or because of his method of seeking a mechanical explanation, or because of his treatment of the earth-moon system in his explanation of the monthly period (later in the "Fourth Day"). For more details, see Aiton (1954; 1963; 1965), Brown (1976), Burstyn (1962; 1963; 1965), Drake (1970, 200–213; 1979; 1983; 1986b), Finocchiaro (1980, 6–24 and 74–79; 1986; 1988b), Galileo's "Discourse on the Tides" (Finocchiaro 1989, 119–33), and McMullin (1967, 31–42).

to advance any comments, you can do it now, so that we can then go on to discuss the two other (monthly and annual) periods.[45]

. . .[46]

## [16. Ending]

[485] . . . SAGREDO. I think you have done a great deal in opening up for us the first door to such a lofty speculation. Even if you had given us [486] only the first basic proposition, in my opinion that alone so greatly surpasses the inanities introduced by so many others that merely thinking of them nauseates me; I mean the proposition (which seems unobjectionable to me) declaring very convincingly that if the vessels containing the seawater stood still, it would be impossible by the common course of nature for it to exhibit the motions we see, and on the contrary, given the motions attributed for other reasons by Copernicus to the terrestrial globe, such changes in the seas must necessarily follow. I am very surprised that among men of sublime intellect (of whom there have been many) no one has seen the incompatibility between the reciprocal motion of the contained water and the immobility of the containing vessel; this incompatibility seems very evident to me now.

SALVIATI. What is more surprising is that, while some have thought of finding the cause of the tides in the earth's motion (thus showing

45. The monthly period of the tides is the phenomenon that during certain times of the month the daily high tides are higher than usual and the daily low tides are lower than usual; similarly, the annual period is the phenomenon that during certain periods of the year the daily high tides are higher and the daily low tides lower than usual. This selection discusses only the diurnal period and is an adaptation of Galileo's "Discourse on the Tides" of 1616; later in the "Fourth Day," he elaborates his theory to include geokinetic explanations of the monthly and annual periods.

46. The passages omitted here correspond to Favaro 7:462.16–485.35 and Galileo 1967, 436–61 and contain three main parts. First, Galileo discusses whether on a moving earth its atmosphere should behave in a way similar to seawater, partly answering an objection to the tidal argument, and partly taking the opportunity to present a distinct argument in favor of Copernicanism based on the existence of the trade winds. Then he explains the monthly period of the tides by arguing that the earth's annual motion would undergo monthly changes in speed due to the fact that the earth-moon system would circle the sun in an orbit whose effective radius would increase when the moon is in opposition and decrease when it is in conjunction with the sun. Finally, he explains the annual tidal period by arguing that the inclination of the earth's axis of rotation would cause annual variations in the effective speed of diurnal motion of points on the earth's surface since this effective speed would be the diurnal speed projected onto the plane of the ecliptic. For more details, see Finocchiaro (1980, 43–44 and 139–40).

greater perspicacity than is common), when they then came to the point, they grasped nothing; they did not understand that it is not enough to have a single uniform motion (such as, for example, the mere diurnal motion of the terrestrial globe), but that we need an unequal motion, sometimes accelerated and sometimes retarded; for when the motion of a vessel is uniform, the water contained therein will get used to it and will never undergo any change. Moreover, it is totally useless to say (as an ancient mathematician[47] is reported to have said) that when the earth's motion encounters the motion of the lunar orb, such a contrast causes the tides; for it is neither explained nor self-evident how this is supposed to happen, but rather we can see its manifest falsity, given that the earth's rotation is not contrary to the moon's motion but in the same direction. Thus, what has been stated and thought so far by others is, in my opinion, completely invalid. However, of all great men who have philosophized on such a puzzling effect of nature, I am more surprised about Kepler[48] than about anyone else; although he had a free and penetrating intellect and grasped the motions attributed to the earth, he lent his ear and gave his assent to the dominion of the moon over the water, to occult properties, and to similar childish ideas.

SAGREDO. I am of the opinion that these better thinkers experienced what is now happening to me, too; that is, one cannot understand [487] how the three periods (annual, monthly, and daily) are entangled and how their causes appear to depend on the sun and moon without the sun and moon having anything to do with the water. For a full understanding of this business I need a longer and more focused application of my mind, which at the moment is very confused by its novelty and difficulty; but I do not despair of being able to grasp it if, in solitude and silence, I can return to chewing over what remains improperly digested in my mind.

Thus, the discussions of these four days provide strong indications in favor of the Copernican system. Among them, these three appear to be

47. In a marginal note Galileo identifies this author as Seleucus, a Babylonian who lived around 150 B.C. and was a follower of Aristarchus's geokinetic theory. Seleucus held that the interaction between the earth's daily rotation and the moon's monthly revolution produced disturbances in the atmosphere which then in turn agitated the oceans.

48. This is the only explicit mention of Kepler in the whole book. It seems uncharitable of Galileo to have mentioned him only in a negative context. But this fact is also revealing for it underscores that, while the two agreed on many specific conclusions, they also had deep philosophical disagreements. For example, regarding the explanation of tides, Galileo thought that a lunar attraction theory (such as Kepler's) was methodologically unsound and preferred a mechanical explanation. See also the relevant note to selection 15, [446], above.

very convincing:[49] first, the one taken from the stoppings and retrogressions of the planets and their approaching and receding from the earth; second, the one from the sun's rotation on itself and from what is observed about its spots; and third, the one from the ebb and flow of the sea.

SALVIATI. Soon we could perhaps add a fourth one and possibly even a fifth one. That is, the fourth one would be taken from the fixed stars, if the most exact observations were to reveal in them those minute changes which Copernicus assumes to be imperceptible.[50] There is now a fifth novelty from which one might be able to argue for the motion of the terrestrial globe. This refers to the extremely subtle things being discovered by the most illustrious Mr. Cesare Marsili,[51] member of a very noble family of Bologna, and also Lincean Academician; in a most learned essay he states that he has observed a constant though extremely slow motion of the meridian line. Having lately seen this essay with astonishment, I hope he sends copies of it to all students of the marvels of nature.

SAGREDO. This is not the first time I have heard of this gentleman's refined learning and of his great concern to be a patron of all scholars. If this or some other work of his comes out, we can be sure that it will be a thing of distinction.

SALVIATI. Now, since it is time to put an end to our discussions, it remains for me to ask you to please excuse my faults if, when more calmly going over the things I have put forth, you should encounter difficulties and doubts not adequately resolved. You should excuse me because these ideas are novel, my mind is imperfect, and the subject is a great one; finally, I do not ask and have not asked from others an assent which I myself do not give to this fancy, and I could very easily [488] regard it

49. This remark provides a very important indication of Galileo's judgment about the nature of his pro-Copernican case. For the first of the arguments mentioned, see the introduction (3) and selection 11; for the second argument, omitted here, see Favaro (7:372–83) and Galileo (1967, 345–56); for the third argument, see selection 15 and the rest of the "Fourth Day," which is omitted here.

50. This refers to the observation of the annual parallax of fixed stars, which was not detected until Friedrich W. Bessel in 1838; cf. selections 13 and 14, especially pp. [409–15].

51. Cesare Marsili (1592–1633) was a patron and friend of Galileo and an amateur scientist; he had been instrumental in helping one of Galileo's most distinguished students (Bonaventura Cavalieri) obtain the mathematics professorship at the University of Bologna. Marsili's essay has been lost and so it is impossible to evaluate his alleged discovery. But it is likely that the reported deviation of the meridian was due to observational error, and in any case it is unclear how the deviation would have supported the earth's motion.

as a most unreal chimera and a most solemn paradox.[52] As for you, Sagredo, although in the discussions we have had you have shown many times by means of strong endorsements that you were satisfied with some of my thoughts, I feel that in part this derived more from their novelty than from their certainty, and even much more from your courtesy; for by means of your assent you have wanted to give me the satisfaction which one naturally feels from the approval and praise of one's own creations.[53] Moreover, just as I am obliged to you for your politeness, so I appreciate the sincerity of Simplicio; indeed, I have become very fond of him for defending his master's doctrine so steadfastly, so forcefully, and so courageously. Finally, just as I express thanks to you, Sagredo, for your very courteous feelings, so I beg forgiveness of Simplicio if I have upset him sometimes with my excessively bold and resolute language; there should be no question that I have not done this out of any malicious motive, but only to give him a greater opportunity to advance better thoughts, so that I could learn more.

SIMPLICIO. There is no need for you to give these excuses, which are superfluous, especially to me who am used to being in social discussions and public disputes; indeed, innumerable times I have heard the opponents not only get upset and angry at each other, but also burst out into insulting words, and sometimes come very close to physical violence. As for the discussions we have had, especially the last one about the explanation of the tides, I really do not understand it completely. However, from the superficial conception I have been able to grasp, I confess that your idea seems to me much more ingenious than any others I have heard, but that I do not thereby regard it as truthful and convincing. Indeed, I always keep before my mind's eye a very firm doctrine, which I once learned from a man of great knowledge and eminence, and before which one must give pause.[54] From it I know what you would answer if both of you are asked

52. This is another instance of Galileo's disclaimers to ensure that his book would not be interpreted as an act of defiance against religious authorities, or as evidence that he accepted and was defending Copernicanism.

53. Note that Galileo's disclaimers are now applied to Sagredo's attitude, for his frequent siding with Salviati could convey a misleading impression.

54. This refers to Pope Urban VIII, who found the objection Simplicio proceeds to state a powerful and indeed unanswerable argument against Copernicanism; Galileo knew the pope's opinion from personal discussions during special audiences, first in 1624 after the pope's election, when Galileo went to Rome to pay him homage, and then again in 1630 after Galileo had finished his book and visited that city to get permission to publish it. The lengthy negotiations for this purpose included the stipulation that he would end the book with this argument; for example, in the letter dated 19 July 1631, which concluded these negotiations, the Vatican secretary wrote to the Florentine inquisitor that "at the end one must have a peroration of the work in accordance with this preface. Mr. Galileo must add the reasons pertaining to divine omnipotence which Our Master gave him; these must

whether God with His infinite power and wisdom could give to the element water the back and forth motion we see in it by some means other than by moving the containing basin; I say you will answer that He would have the power and the knowledge to do this in many ways, some of them even inconceivable by our intellect. Thus, I immediately conclude that in view of this it would be excessively bold if someone should want to limit and compel divine power and wisdom to a particular fancy of his.[55]

[489] SALVIATI. An admirable and truly angelic doctrine, to which there corresponds very harmoniously another one that is also divine.[56]

---

quiet the intellect, even if there is no way out of the Pythagorean arguments" (Finocchiaro 1989, 354n.57). See also Oregius's (1629, 194–95) report of one such discussion between Galileo and the pope, translated in Finocchiaro (1980, 10).

55. This is an important and controversial argument in the Copernican controversy; it may be descriptively called the divine omnipotence objection (besides being named after one its main proponents, Pope Urban VIII). As expressed here, it is a theological argument insofar as it uses as an unquestioned premise the idea that God is all powerful, which is part of the conception of God in Christianity and many other religions; a full evaluation of this argument must include a critical assessment of this idea; for some related points, see selection 4. This theological argument also has a nontheological analogue and its key point may be translated into purely methodological and logical terms; a key logical point is that reasoning of the form of affirming the consequent ("if P then Q; Q; therefore, P") is formally invalid; a key methodological point is the so-called problem of induction, a version of which would argue that, regardless of how much observational evidence there is in favor of a theory, we can always conceive a possible world in which the evidence is true and the theory is false; for such a methodological translation of Urban's theological argument, see Finocchiaro (1980, 9–11). But it is less clear what exactly the argument proves or tries to prove about Copernicanism, and how it does so; that is, it is unclear whether the conclusion is that Copernicanism *may be* false, or that it *is* false, or that it is *likely* to be false; or that Copernicanism is *not supported* by the evidence alleged in its favor (such as the tides), or that it is *contradicted* by this evidence; or that Copernicanism is less probable or less supported by evidence than the Ptolemaic view. Thus, another issue is whether the objection is any less effective against any other scientific theory, such as the Ptolemaic view. Urban may have had in mind the objection that God could have created a world in which the evidence suggested a moving earth despite its being motionless; this would lend support to Morpurgo-Tagliabue's thesis that Descartes' "*Discourse on Method* . . . is essentially an answer to the argument of Urban VIII" (1981, 104), which is a connection worthy of further investigation. Urban may also have thought that all pro-Copernican arguments are indirect, hypothetical, explanatory, and causal, like the tidal argument; thus, as long as that is the case, the earth's motion remains a hypothesis; this in turn would raise the issue of the meaning and status of hypotheses, and the problem of the ambiguity of the notion of hypothesis, namely, the equivocation between its instrumentalist and its fallibilist meaning; thus, Urban's favorite objection would connect with the issues raised in Galileo's preface, selection 1; for more details on hypotheses, see Clavelin (1964), Duhem (1969), Finocchiaro (1986; 1992b), Wallace (1981b; 1984), and Wisan (1984a).

56. This response to Pope Urban VIII's favorite objection against Copernicanism displeased him, as one can read in the report by the special commission appointed in the summer of 1632 to investigate the book's alleged transgressions; one of the many complaints was that Galileo had placed that objection "in the mouth of a fool and in a place where it can only be found with difficulty, and then he had it approved coldly by the other speaker, by merely mentioning but not elaborating the positive things he seems to utter against his

This is the doctrine which, while it allows us to argue about the constitution of the world, tells us that we are not about to discover how His hands built it (perhaps in order that the exercise of the human mind would not be stopped or destroyed).[57] Thus let this exercise, granted and commanded to us by God, suffice to acknowledge His greatness; the less we are able to fathom the profound depths of His infinite wisdom, the more we shall admire that greatness.

SAGREDO. This can very well be the final ending of our arguments over the last four days. Hereafter, if Salviati wants to take some rest, it is proper that our curiosity grant it to him, but on one condition; that is, when he finds it least inconvenient, he should comply with the wish, especially mine, to discuss the problems which we have set aside and which I have recorded, by having one or two other sessions, as we agreed. Above all I shall be looking forward with great eagerness to hear the elements of our Academician's new science of motion (natural and violent).[58] Finally, now we can, as usual, go for an hour to enjoy some fresh air in the gondola that is waiting for us.

---

will "(Finocchiaro 1989, 221). To get a glimpse of the variety of issues that make up the Galileo Affair, note that part of this complaint is that Urban's objection was placed in the mouth of Simplicio, the least intelligent of the three speakers, whose very name means "simpleton" and suggests simplemindedness; although the special commission report does not elaborate, Galileo's enemies had spread the rumor that Simplicio was a caricature of the pope himself. As for Salviati's allegedly cold reception of Urban's objection, it could be argued that Galileo had done something worse than that insofar as earlier in the Fourth Day (selection 15, [447–48]) he had criticized the objection by arguing that even if one invokes a divine miracle to explain the tides, the most likely miracle would be the simplest one, and the simplest miracle would be to move the earth (cf. Wisan 1984b and Finocchiaro 1985). But it could also be argued that Galileo had made an honest attempt to elaborate Urban's argument since the rest of Salviati's speech proceeds to connect it to the idea that God's creation can never be completely known or understood; that is, this selection may be interpreted as suggesting a Socratic lesson of epistemological modesty, which is a theme discussed elsewhere in the book (for example, selection 4) and corresponds to a genuine Galilean inclination (cf. Finocchiaro 1980, 141; Biagioli 1993, 301–11).

57. Ecclesiastes, 3:10–11 (King James Version): "I have seen the travail, which God hath given to the sons of men to be exercised in it. He hath made every *thing* beautiful in his time: also he hath set the world in their heart, so that no man can find out the work that God maketh from the beginning to the end."

58. In several places in the *Dialogue* Galileo has the speakers refer to him as the Academician, meaning member of the Lincean Academy. Here we also have a reference to a work on motion the research for which he had essentially completed at Padua but not published yet, and which eventually was published in Holland in 1638 as the *Two New Sciences* (Galileo 1974).

# Appendix

## 1. Critical Reasoning

1.1. *Reasoning* is the activity of the human mind that consists of giving reasons for conclusions, reaching conclusions on the basis of reasons, or drawing consequences from premises. More exactly, it is the interrelating of thoughts in such a way as to make some thoughts dependent on others, and this interdependence can take the form of some thoughts being based on others or some thoughts following from others. Reasoning is thus a special kind of thinking; all reasoning is thinking, but not all thinking is reasoning.

The occurrence of reasoning is normally indicated, and can always be explicitly indicated, by the use of *reasoning indicators*. These are words like the following (or phrases synonymous with such words): 'therefore,' 'thus,' 'so,' 'hence,' 'consequently,' 'because,' 'since,' 'for'. Reasoning indicators, however, are only hints, since it is possible to express simple reasoning without them and for them to have other meanings that do not indicate reasoning. Nevertheless, reasoning indicators enable us to formulate an operational definition: reasoning is the type of thinking that occurs whenever there is a high incidence of words such as 'therefore,' 'because,' and 'consequently'.

Reasoning is linguistically expressed in *arguments*. An argument is a basic unit of reasoning in the sense that it is a piece of reasoning sufficiently self-contained to constitute by itself a more or less autonomous instance of reasoning.

Reasoning indicators serve to interconnect the *propositions* of an argument. A proposition is any part of an argument that is capable of being accepted or rejected by itself. It is also capable of being stated as a complete sentence, so that it can stand by itself. Propositions may also be called assertions, statements, claims, or theses; for our purposes, we will not distinguish among these five terms. An argument may thus be conceived as a series of propositions some of which are based on others, where the interconnections are expressed by means of reasoning indicators.

The simplest possible argument contains two propositions and can always be expressed in either one of two standard forms which are logically equivalent: (1) A, therefore B; or (2) B because A. In both (1) and (2), B is the *conclusion* and A is the *reason* or *premise*. In other words, although both words 'therefore' and 'because' are reasoning indicators, they indicate different ways to express reasoning; the proposition preceding 'therefore' is the reason or premise, the one following it is the conclusion; whereas the proposition preceding 'because' is the conclusion, and the one following it is the reason. The conclusion of an argument is thus the proposition that is based on the others, whereas the reasons or premises are the propositions on which the conclusion is based. For our purposes, the words 'reason' and 'premise' will be used interchangeably.

Reasons and conclusions are mutually interdependent concepts: a proposition can be a conclusion only in a context where it is being based on some reason, and a proposition can be a reason only in a context where a conclusion is being based on it. The relationship between reasons and conclusion claimed in a given argument can be expressed by a number of terms that will be regarded as synonymous for our purposes. The conclusion may be said to be based or grounded on, to be justified or supported by, and to be inferred or derived from the reasons.

There are several standard ways of referring to arguments. Sometimes we speak of *the argument that S* or of the fact that someone *argued that S*, where S is a sentence; this refers to an argument whose conclusion is the proposition expressed by S, and whose premises are being left unspecified, perhaps because their identity is obvious in the context. Sometimes we speak of an *argument for N*, where N is a noun or noun phrase; this means an argument whose conclusion is some proposition obtained from N in a contextually obvious manner, and whose premises again are being left unspecified. Similarly, to speak of an *argument against N* is to speak of the argument whose conclusion is the denial or negation of such a proposition easily constructed from N. Sometimes it is contextually obvious what the conclusion of an argument is, and then one may want to identify a particular argument by a brief description of the most important premise; one would then speak of the *argument from N*, where N is again a noun phrase out of which one can easily form a proposition that serves as a premise from which that conclusion is drawn.

*Objections* are negative counterparts of arguments in the sense that they are arguments whose conclusions are the denial or negation of some controversial proposition. An *objection to or against N* is an argument whose conclusion is the negation or denial of some proposition constructed from N in a contextually obvious manner; that is, an objection to or against N is the same as an argument against N. A special and more complicated case arises when N refers to an argument rather than a proposition, for then the objection is an argument about another argument; that is, an *objection to an argument* is an argument whose conclusion is a proposition stating that the original argument has some flaw; this leads to the topic of the evaluation of arguments, which will be discussed presently. An *objection from N* means an argument whose main premise is a proposition easily formed from N and whose conclusion is both negative and easily identified; an objection from N is essentially the same as an argument from N, except that the 'objection' designation is more likely used by someone who rejects the conclusion of the argument, whereas the 'argument' designation is more likely used by someone who accepts it. The *objection that S*, where S is a sentence, means an argument whose main premise is a proposition expressed by S and whose conclusion is a proposition that is negative and easily identified in the context.

*Counterarguments* are special kinds of objections, namely, objections to the conclusions of the original arguments. That is, suppose we begin by considering an argument for N; then a counterargument to such an argument would be an objection to N or an argument against N. In other words, a counterargument to a given argument is an argument whose conclusion is the denial or negation of the conclusion of the given argument.

The differences among arguments, objections, and counterarguments are differences in perspective; that is, the differences relate to whether one is affirming or denying a given controversial proposition. However, normally there is nothing intrinsically positive or negative about a proposition; the same thought can usually be expressed either positively or negatively. The point is that in a normal controversial situation there are arguments for both sides of the dispute. Each side is affirmative from its own viewpoint and negative from the opposite viewpoint. Let us call two propositions, P and Q, contrary or inconsistent when they cannot both be true (though they could perhaps both be false); in such a case, the arguments for P are also arguments against Q, and the arguments for Q are arguments against P. In other words, the arguments for P are objections to Q, and the arguments for Q are objections to P. Or again, the arguments for P have counterarguments consisting of the arguments for Q, and the arguments for Q have counterarguments consisting of the arguments for P.

1.2. An *argument with serial structure* (for short, a *serial argument*) is made up of at least two subarguments combined so that the conclusion of one is simultaneously a reason of the other.[1] The simplest serial argument has the form: A because B, and B because C (or equivalently: C, therefore B; therefore A). Here B is the reason of the subargument "A because B" (or "B, therefore A") and also the conclusion of the subargument "B because C" (or, "C, therefore B").

Every proposition in a serial argument falls into one and only one of the following categories: intermediate proposition, final reason, final conclusion. An *intermediate proposition* in a given serial argument is a proposition that serves as the conclusion of one subargument and as a reason of another subargument. A *final reason* in a given serial argument is a proposition that is a reason of some subargument but not the conclusion of any subargument. The *final conclusion* in a given serial argument is a proposition that is the conclusion of some subargument but not the reason of any subargument. In the example here, A is the final conclusion, B is the one and only intermediate proposition, and C is the one and only final reason.

1. Here I adopt the terminology of Freeman (1991, 93–95) and Thomas (1986, 57–58); the concept, however, is the same as that elaborated in Finocchiaro (1980, 313–14) under the label (now dropped) of "complex argument."

The *propositional structure* of an argument or piece of reasoning refers to the interrelationships among its various elements, namely, among its subarguments and among its propositions. Such structure may be pictured in a *structure diagram* constructed in accordance with the following rules:[2]

1. Label each proposition with some number, letter, or symbol.
2. When one proposition is a reason supporting another, write the reason under the conclusion and indicate the fact by a solid line leading up from the first to the second.
3. Place at the top the proposition that is supported by one or more other propositions but that does not itself support any others; this proposition at the top is the final conclusion of the argument.
4. Place at the lowest level those propositions that support other propositions but are not themselves supported by anything else; such propositions are the final reasons of the argument.
5. Some structure diagrams may have propositions that both support and are supported by other propositions; such propositions are the intermediate propositions of the argument. That is, they are propositions which are logically placed between the final reasons and the final conclusion of the argument, and which are reasons from the viewpoint of what they immediately support and conclusions from the viewpoint of what they are immediately supported by. Intermediate propositions have some support lines leading up to them from below and some leading up from them to other propositions above them.

Sometimes it is useful to label propositions with a standard numbering system. This *standard labeling* is done in accordance with the following rules:

6. Label the final conclusion by some small number, for example 1.
7. Label reasons that directly support the final conclusion by adding decimals to the numbers denoting it. For example, if the final conclusion is denoted by 1, its directly supporting reasons are 1.1, 1.2, 1.3, and so on.
8. Reasons that directly support the same intermediate proposition are assigned the same digits, except that new decimals are added in accordance with the previous rule 7. For example, reasons supporting 1.2 are labeled 1.2.1, 1.2.2, 1.2.3, and so on.

Two reasons that immediately support the same proposition (and thus have the same number of digits in a diagram with standard labels) are *linked* when each depends on the other to support that proposition and each alone is insufficient or incomplete to provide that support. When reasons are linked, the rules for standard labeling apply without change. When two reasons are not linked,

2. Similar rules were adopted in Finocchiaro (1980) from Angell (1964); they are now common in the literature on informal logic and critical thinking.

they are *independent*. So two reasons are independent of each other when each does not depend on the other to support the conclusion based on them (whether or not each alone is sufficient to provide that support, that is, whether or not each is linked with *some other* reason). In other words, two reasons may be independent of each other and yet be linked with other reasons.

When an argument contains independent reasons, it may be useful to represent this fact in labeling the propositions and in drawing the structure diagram. The standard labeling of independent reasons is done as follows:

9. To distinguish one *set* of reasons from another independent set, the lowercase letters *a, b, c*, and so on, are placed after the label of the proposition they support; these letters are carried for all lower propositions supporting these independent reasons when these lower propositions are themselves labeled by the previous rules.

And the distinction between linked and independent reasons is represented in a diagram as follows:

10. The support lines from linked reasons always converge to some point below the conclusion they support; when there are no independent reasons, such lines converge directly to the symbol of this conclusion; when there are independent reasons, a horizontal bar is drawn between the supported conclusion and its immediately supporting reasons, and the various independent sets of linked reasons converge to distinct points on that horizontal line.

These rules will be illustrated and applied presently (see fig. 15 and fig. 16). One use of this idea of the independence of reasons is that it allows us to integrate separate arguments into a unified whole, if they have the same final conclusion.

The *latent* propositions of an argument are those propositions that are not explicitly stated in the argument but are implicitly assumed or taken for granted by the proponent of the argument. Latent propositions may also be called assumptions, presuppositions, missing premises, or implicit statements; for our purposes, such terms will be used interchangeably. The interrelationships among the latent propositions are called the latent propositional structure, or more simply latent structure; the latent structure may be contrasted with the explicit structure, namely, the interrelationships of the explicitly stated propositions. There are two main types of latent propositions. In one case, a proposition is latent when for a particular step of the argument it is needed, in addition to the other explicit propositions involved in that step, to fully justify or to better justify that step; in this case, the latent proposition is always linked with and never independent of the reason(s) explicitly present in that step of the argument. In another case, a proposition is latent when it is one of the reasons being implicitly used to justify one of the final reasons in the explicit argument. Because of their position in structure diagrams, in the first case the propositions are called *horizontally latent*, in the second case *vertically latent*. In a structure diagram all latent structure is drawn in dotted lines.

Finally, many arguments are such that what they need is not analysis but synthesis; that is, they need to be reconstructed. A *reconstruction* of an argument is a restatement of it such that no logically extraneous propositions are included and all logical interconnections among the stated propositions are explicitly and clearly indicated by means of reasoning indicators. A reconstructed argument normally makes explicit some of the propositions implicit or latent in the original formulation; however, the reconstruction cannot include all of them because their number is indefinitely large and the desirability of explicitness is subject to contextual constraints.

1.3. So far I have discussed the basics of how reasoning and arguments are expressed and stated, how they are described and referred to, how they are analyzed and what their structure is, and how they are reconstructed. Next, we must go on to the evaluation of reasoning and arguments. Evaluation is only one of a cluster of terms that are regarded as synonymous here; thus we may speak interchangeably of assessment, appraisal, judgment, and criticism.

Some caution is, however, needed in regard to the last term, for it can have a broad and a narrow meaning. Broadly construed, criticism is equivalent to evaluation; this is the relevant meaning when we speak, for example, of "critical reasoning." But criticism has also a negative connotation, for it can also mean negative or unfavorable evaluation; it provides, in fact, a handy term for the latter. Thus, I will often speak of criticism in its narrow meaning. However, I will use the term in both senses and, when needed, we will give appropriate clues to avoid misunderstanding.

This point leads immediately to a fundamental fact about the concept of evaluation, namely, that it can be of two opposite types: favorable, positive, approving, or appreciative; and unfavorable, negative, disapproving, or destructive. That is, the basic aim of evaluation is to determine whether an argument is good or bad, right or wrong, valid or invalid, correct or incorrect, sound or unsound, plausible or implausible, convincing or unconvincing, strong or weak, logical or illogical, and cogent or fallacious.

Another general point is that normally evaluation is a matter of degree, rather than an all-or-none affair. In a sense, evaluation is the judgment of value or worth, and value or worth is usually a nondiscrete or gradual notion, a spectrum embodying various shades of merit or demerit. One may, however, define particular evaluative categories referring to the extreme points along a particular dimension, and then one has a discrete evaluative concept.

Thirdly and perhaps most importantly, the evaluation of arguments is conceived here as involving two main things: an evaluative claim about some argument, and the articulation of the rationale for the evaluative claim. That is, evaluating an argument will be conceived as a special case of reasoning, namely, reasoning about the argument; in short, the evaluation of an argument is a higher-order or metalevel argument about the original argument.

Combining these three points, we may say that an evaluation of an argument consists of a new argument whose conclusion attributes some merit or flaw to the original argument, whose reasons are meant to justify such an attribution, and where it is understood that normally such a justification can only be more or less strong rather than completely right or completely wrong.

A fourth important point is that negative evaluation (criticism in the narrow sense) is much more common than favorable evaluation. The causes of this fact are unclear; I believe that part of the explanation is that criticism is instructive in a way in which positive evaluation is not.[3] At any rate, a fruitful way of proceeding is to look for flaws in an argument, which is then regarded as successful if and insofar as no flaws are found or demonstrated. Thus, it is useful to catalogue some basic types of criticism.

One type of criticism is to criticize the conclusion as an individual claim, namely, to try to refute it. This is usually done by means of an objection to the original conclusion, which, as explained above, is equivalent to advancing a counterargument. In short, suppose the original argument claims "C because R"; this type of criticism objects that "not-C because R′," where R′ is a proposition different from R. This is a common but not very perspicacious type of criticism because often the objecting reason has nothing to do with the original reason; so the criticism does not undermine the reason advanced in the original argument; and so it is of questionable relevance. As suggested earlier, the existence of an argument for a given proposition and of an objection against the same proposition provides the usual background for a controversy that generates argumentation; however, what is required for further progress in the discussion is to move to the level of evaluation by raising other questions, rather than by merely repeating the arguments that define the original controversy.

It is therefore not surprising that another common type of criticism of an argument is to criticize the reason adduced for the conclusion. One questions the individual truth of this reason by means of an objection against it, namely, an argument designed to show that the reason is not true. If successful, such an objection would undermine the conclusion of the original argument, for this conclusion was grounded on a reason, and this grounding disappears if this reason is not true. In short, to the original argument "C because R," this criticism responds by claiming "not-R because R′."

A third type of criticism tries to undermine an argument by undermining its reason, in the literal sense that this reason is alleged to be groundless or without foundation. The difference between this criticism and the previous one is the difference between claiming that a proposition is false and claiming that it is unsupported or not properly supported; this is the difference between saying that there is an argument concluding that the proposition is false, and saying that there is no argument concluding that the proposition is true. Thus, the proper answer to this type of criticism is to construct an argument supporting the truth of this proposition, or to strengthen or defend some existing argument supporting the same; whereas the proper answer to the former type of criticism is to point out what is wrong with the critical argument that tries to refute that proposition. In the simplest case, this type of criticism is a request that the argument "C because R" be strengthened by another argument to the effect that "R because R′." This type of criticism is related to the serial structure of an argument.

These three types of criticism share a common feature: none questions the relationship between the premise and the conclusion of the argument. In a sense,

3. For further details, see Finocchiaro (1980, 338–41; 1988c, 245–58; 1992a).

these criticisms involve the relationship between the conclusion or the premise and the world; or, if you will, the relationship between the conclusion or the premise and other propositions that were not part of the original argument. Other types of criticism focus on the relationship between the given premise and conclusion.

The most obvious way of criticizing the relationship between a reason and its conclusion is to question what might be called the relevance[4] of the reason. That is, this criticism questions how and why the conclusion can be inferred from the reason; it may take two different forms. In the weaker form, the criticism claims that it is unclear how the conclusion can be inferred from the reason; for example, given the original argument "C because R," the connection between the two propositions may not be obvious. In this case the critical claim is weak enough as to require little or no justification; the criticism is really a request for an elaboration of the original argument. In turn, the answer would be to provide other reasons which, together with the original one, make it clear, or clearer, how the conclusion would be inferred. In other words, the structure of the original argument needs to be complicated by adding other *linked* reasons; the new argument would read "C because R, R′, R″, and so on."

In the stronger form, the criticism claims that the conclusion cannot be inferred from the reason. In this case a supporting critical argument must be given. That is, this criticism takes the form that the conclusion does not follow from the premise(s) because of some specified reason. The type of such a supporting critical reason then generates various subtypes of this criticism. If we call *disconnection*[5] the general flaw in reasoning corresponding to this type of criticism, the various special cases may be regarded as types of disconnection.

Let us begin with the most familiar subtype. It may happen that the conclusion of an argument does not follow from the premises because it is possible for the premises to be true while the conclusion is false. This possibility is commonly shown by constructing an argument with the same form as the original one, but having obviously true premises and obviously false conclusion;[6] such a constructed argument is called a counterexample. So in this first type of disconnection, the conclusion does not follow because an appropriate counterexample exists. Other well-known ways of describing the flaw are to say that the conclusion does not follow for one of several reasons: because it does not follow *necessarily*; because it does not follow in virtue of the *form* of the argument; because it does not follow ("analytically") in virtue of the meaning of the terms involved; or because it does not follow in virtue of the rules of *deductive* inference. Correspondingly, this first type of disconnection could be labeled *formal, analytic*, or *deductive invalidity*.

A second reason why one may be entitled to say that the conclusion does not follow from the premises is that it does not follow with any greater probability than some other specifiable proposition. Here, the critic produces another argument which has the same premises as the original argument but a different con-

4. Cf. Freeman (1988; 1991), Govier (1985), and Johnson and Blair (1977).

5. I adapt *disconnection* from Perkins (1989) and Perkins et al. (1983).

6. It can also be shown by imagining a situation in which the premises of the original argument are clearly true and its conclusion is clearly false.

clusion, and which appears of equal strength as the original. This occurs primarily with explanatory arguments whose conclusion is an explanation of what is stated in the premises, and the criticism amounts to providing an alternative explanation. Occasionally it may happen that the explanandum occurs because of both factors mentioned in the two conclusions, but the point is that a given explanation has no force if there is no reason to prefer it to an alternative. This may be called *explanatory disconnection*.[7]

To appreciate the third type of disconnection we need to stress that we are dealing with natural language argumentation as it occurs in ordinary life, and that such arguments are always incompletely stated and have latent propositions or tacit presuppositions. Then it is easy to see that a reason why the conclusion does not follow from the premises may be that one of the presuppositions is false. What does this falsity mean in this context? It really amounts to the existence of some ("sound") argument constructible in the context, whose conclusion is the denial of the presupposition. Even the groundlessness of such a presupposition would create trouble for the original argument, at least as long as such groundlessness is not merely asserted but demonstrated, that is, as long as one gives contextually sound arguments to show there is no good reason to assert the presupposition. In short, to the original argument "C because R," this criticism objects that in the context this argument is equivalent to "C because R and R′," but it so happens that "not-R′ because R″." This third subtype may be said to involve a *presuppositional disconnection* and correspondingly *presuppositional criticism*.

Notice that a pattern is beginning to emerge. In the first type of disconnection the critical claim is grounded on the construction of an appropriate counterexample; in the second on the production of an alternative explanation; in the third on the construction of a presuppositional refutation; and these three entities are arguments different from but appropriately related to the original. This pattern allows us to define a fourth type of disconnection where the conclusion does not follow because what does follow from the premises is some specifiable proposition inconsistent with it. Such a proposition may be called a counterconclusion, and the new argument supporting it a counterargument; but this is a special type of counterargument whose premises contain some of the premises of the original argument and other propositions which are independently justifiable or contextually acceptable. In short, to the original argument "C because R," this criticism objects that "not-C because R, R′, R″, and so on." This type of flaw may be called *internal disconnection*, and the corresponding criticism *internal criticism*, because the crucial element of the criticism is an expansion of the original argument that accepts some of the original reasons.

The fifth type of disconnection may be called *semantical* and involves the problem of equivocation. It applies only to arguments having at least two linked reasons. Semantical disconnection occurs when the conclusion does not follow because the premises contain a term which has two meanings such that, if it is used in one sense, one of the premises is false (though they would imply the con-

7. This is reminiscent of inductive incorrectness, and so it may also be called inductive disconnection; still, since the connection between explanation and induction is problematic, it is advisable to avoid the latter label.

clusion), whereas if the term is used in the other sense, the premises clearly do not imply the conclusion (though admittedly the previously problematic premise becomes true); in short, in the context the conclusion cannot follow from true premises. This disconnection is intimately related to presuppositional disconnection since the semantical ambiguity in question is normally not a self-subsisting property of a term, but rather something that must be argued in the context on the basis of (often latent) inferential relationships affecting the term.

Finally, the sixth subtype also involves presuppositional disconnection but refers to the flaws of begging the question and circularity; it may be called *persuasive disconnection*. Here, the conclusion does not *follow* from the premises because it *is* one of the premises; this is not meant literally in the sense that the conclusion is identical to one of the explicit premises, but in the sense that it is identical to one of the latent propositions; moreover, this latency is frequently vertical and not always horizontal. That is, we seldom find arguments like "C because C" or "C because R, R′, C, and R″," where the circularity is too small and obvious; rather we find "C because R," with C distinct from R but included in the latent structure.

I have defined six types of disconnection: formal, explanatory, presuppositional, internal, semantical, and persuasive. All involve arguments whose conclusion may be said not to follow from the premises. Thus the corresponding evaluations involve criticism of the relationship between premises and conclusion. Each disconnection corresponds to a type of criticism where the critic argues that in the original argument the reasons given do not properly connect with the conclusion, and in each case the reason for the disconnection involves the construction on the part of the critic of some other argument, or some additional part of the original argument. These six types of criticism should be added to the first four, which were examined earlier and to which short labels may now be given: an argument against the conclusion of the original argument may be called *conclusion refuting criticism*; an argument against a premise of the original argument may be termed *premise refuting criticism*; a claim or argument that a reason of the original argument is groundless may be labeled *reason undermining criticism*; and a claim or argument that the relationship between the reason and the conclusion of the original argument is unclear may be named *reason relevance criticism*. The last one should be distinguished from the criticism that the reason-conclusion relationship is missing, namely, that there is no connection or that the reason is irrelevant; to deny the existence of the connection is a stronger criticism which in turn generates the six subtypes just defined.

1.4. It is now time to illustrate these general concepts by concrete examples. The observational argument for heavenly unchangeability is one of the most instructive because the original Aristotelian argument is very simple and the Galilean criticism very rich. This discussion may be found near the middle of the "First Day" of the *Dialogue* (selection 3).

The original argument reads quite simply: no heavenly changes have ever been observed; therefore, the heavenly region is unchangeable (selection 3, [71–72]). Its structure is so simple that its initial analysis is easily completed; we have just two propositions, the final conclusion (1) and the final reason (1.1). However, a few additional comments are in order.

First, there is the question of the latent structure, which in this case happens to be extremely interesting. This question may be motivated by asking how and why the reason is relevant to the conclusion, which is a way of proposing the fourth type of criticism mentioned above, the reason relevance criticism. That is, the original argument obviously presupposes that observation corresponds to reality, which is an extremely important epistemological principle. If such a missing premise is added, it constitutes a second reason linked to the first and should be assigned the label 1.2.

Note that this analysis implicitly rejects the interpretation that the original argument is an appeal to ignorance. That is, the argument might be taken to have the form "no A's have been observed; therefore, no A's exist"; and this might be regarded as tantamount to arguing that because we are ignorant of something, namely, because we do not know that P is true, therefore we can conclude that P is false. Now, this is certainly a fallacious manner of reasoning, which is easily recognized as such when made explicit, but which nevertheless is common. Thus, to interpret the original argument as an appeal to ignorance would immediately generate this criticism, depriving it of any worth. Such criticism in turn might be criticized as attacking a straw man, namely, as being based on an untenable interpretation (an interpretation made out of straw, so to speak). We might also add that such criticism would seem to violate the principle of charity, according to which, when evaluating an argument, one should interpret it in such a way that the argument avoids some of the most obvious errors. At any rate, the uncharitable interpretation would also involve taking the passage out of context, for the context makes it clear that the form of the argument is rather the following: "A's have been observed to have the property P; therefore, A's have property P"; that is, "the heavens have been observed to be devoid of (qualitative) change; therefore the heavens are devoid of (qualitative) change." Finally, it should be mentioned that Galileo does not take the argument as an appeal to ignorance; I mention the possibility merely for the sake of illustration.

Before proceeding to see how Galileo does criticize it, some more analysis will be instructive. As he points out, the argument for heavenly unchangeability is a step in a longer argument that seeks to support a further conclusion, namely, the earth-heaven dichotomy; that is, the proposition that the earthly and the heavenly regions of the universe are radically different. In other words, the proposition that the heavens are unchangeable would be linked to the proposition that the earth is constantly changing to support the dichotomy; the new argument (for the earth-heaven dichotomy) would then have two linked reasons supporting its final conclusion, and it would also have serial structure because heavenly unchangeability would no longer be the final conclusion (as it is in the original argument with which we started) but rather an intermediate proposition. The point is that the simple argument in the original statement can be grafted onto other arguments to generate bigger and more complex arguments, and in the process the original would become just a subargument of the new argument.

One other analytical point is worth making. The observational argument for heavenly unchangeability was only one of two independent arguments supporting the same conclusion; there was also an a priori or theoretical argument. That

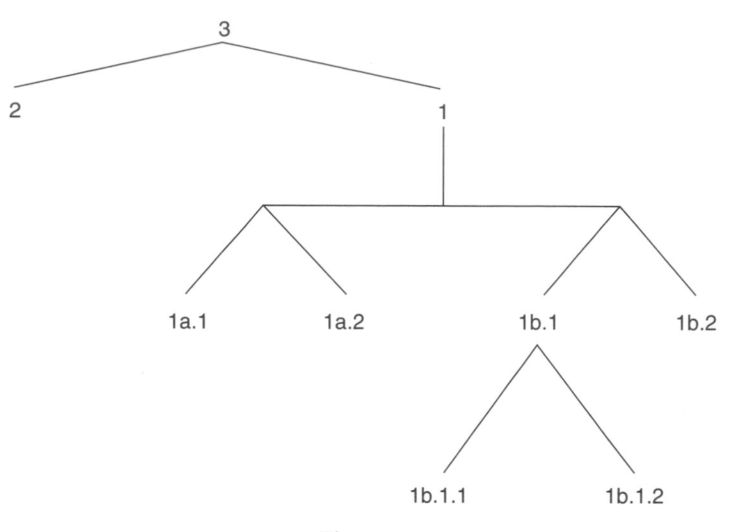


Figure 15.

is, to support heavenly unchangeability, the Aristotelians advanced another independent reason, namely, that there is no contrariety in the heavenly region.[8] Now, this reason immediately generates questions also illustrating many of the general concepts discussed earlier. Moreover, to keep proper track of the propositional structure, let us label this new reason (1b.1) and change the label of the previous reason from (1.1) to (1a.1), in accordance with the rules for the standard labeling of structure diagrams.

The first question about the alleged lack of heavenly contrariety (proposition 1b.1) is why should we accept it. The answer is because (1b.1.1) the heavenly region is the domain of natural circular motion, and (1b.1.2) natural circular motion has no contrary (the way, for example, that straight upward motion is contrary to straight downward). The other main question is why the lack of heavenly contrariety is relevant to heavenly unchangeability; the answer is in terms of the theory of change as deriving from contrariety, namely, because (1b.2) change can exist only if there is contrariety.

Finally, let us assign labels to the other propositions mentioned previously, namely, the numeral 2 to the proposition that the earth is constantly changing, and the numeral 3 to the earth-heaven dichotomy; and let us assign the standard label of (1a.2) to the latent proposition that observation corresponds to reality, in accordance with the rules that distinguish linked from independent reasons. Then we are in a position to picture all the arguments, subarguments, and propositions we have mentioned, and their interrelationships. The structure diagram would be as shown in figure 15. If proposition (3) were relabeled (1) and

8. Cf. Favaro (7:62–71); Galileo (1967, 38–47); and the earlier discussion in the introduction (2.2).

the rest relabeled accordingly, the structure would of course be unchanged, but the labels would be: 1 / 1.1, 1.2 / 1.2a.1, 1.2a2, 1.2b.1, 1.2b.2 / 1.2b.1.1, 1.2b.1.2. Thus, although the explicit structure of the observational argument for heavenly unchangeability is very simple, if we wish to consider its latent structure or its relationship to other ideas and other arguments, then it should come as no surprise that the structure becomes more complicated.

Galileo criticizes the observational argument for heavenly unchangeability in five ways. One is to argue that in his own time the premise is no longer true, in the light of, for example, the telescopic observations of sunspots and the naked-eye observation of novas. Although this criticism targets directly only the individual truth value of the premise, it is the one that takes him the longest to articulate, since he has to argue every inch of the way through all sorts of controversial issues in order to refute the premise (see selection 3, [75–83]).

It is also important to point out that besides refuting the premise, Galileo criticizes the argument by refuting its conclusion. For in this case, the refutation of the premise is not a purely negative affair; to deny that no changes have ever been observed in the heavens is to affirm that some changes have been observed. But the observation of heavenly changes can be combined with the same plausible version of the principle that observation corresponds to reality to yield the conclusion that the heavenly region is changeable.

Moreover, this principle is not only intrinsically plausible, but also it was accepted by the Aristotelians; indeed, we saw above that it is a missing premise of the argument being scrutinized. In short, here we have an example of what I called internal criticism: the original argument states "C because R"; to this we must add a latent and linked reason, say R′; so the argument is really "C because R and R′"; this may be reformulated as "C because R′ and other reasons"; Galileo's criticism amounts to the counterargument "not-C because R′ and R″," where R″ refers to observations of heavenly changes that had then become incontrovertible. What makes this criticism internal is that it is crucially based on a proposition which the proponents of the original argument accept, although this proposition is then combined with something else they do not accept or know anything about; however, since this additional critical premise is demonstrably true, the counterargument is successful.

Other criticisms are based on the contextual distinction of two meanings for the phrase "heavenly changes": a heavenly change can mean the generation or decay of a heavenly body as a whole, and it can mean a partial change within a heavenly body. Equivalently, the ambiguity might be taken to involve the term body, which could mean either a whole heavenly body or a part of it.

When interpreted holistically, the original argument amounts to the following: no one has ever observed any generation or decay of heavenly bodies in the heavenly region; therefore, the heavenly region is unchangeable. It is then subject to the criticism that this way of reasoning would lead one to the following absurd argument: no one has ever observed any generation or decay of terrestrial globes in the terrestrial region; therefore, the terrestrial region is unchangeable.

When the argument is interpreted the other way, Galileo objects that it is still wrong because no terrestrial changes would be noticeable to an observer on the moon before some particular very large terrestrial change had occurred, and yet terrestrial bodies are obviously changeable and would have been so even before that occurrence.

What are we to make of the last two criticisms? It seems they are trying to establish the invalidity or formal disconnection of these two versions of the observational argument. The technique used is that of the construction of counterexamples, namely, arguments of the same form but with true premises and false conclusion. Notice that, despite the ambiguity mentioned by Galileo, he is not charging equivocation. So what we have are two examples of criticism by counterexample, charging invalidity or formal disconnection, not semantical disconnection (see selection 3, [74–75]).

At this point someone might interject that the very fact that this particular argument was called an a posteriori argument shows that it was meant to be a probable or inductive rather than deductive argument, and hence the criticisms just made might be correct but of dubious force or relevance. Perhaps to anticipate this objection, Galileo has another criticism that can be taken to address precisely this issue (see selection 3, [72–74]).

His fifth criticism is directed to the more plausible particularistic (second) version involving parts of globes. It amounts to the following argument: if there were changes in the heavenly bodies, then most of them could not be observed from the earth, since the distances from the heavenly bodies to the earth are very great, and on earth changes can be observed only when they are relatively close to the observer; moreover, even if there were changes in the heavenly bodies large enough to observe from the earth, then they might not have been observed, since even large changes cannot be observed unless careful, systematic, exact, and continual observations are made, and no such observations have been made, at least not by the argument's proponents.

This criticism interprets the original argument as an explanatory argument; that is, it presents the conclusion about heavenly unchangeability as the explanation of the observational absence mentioned in the premise. Two other ways of explaining the fact are suggested: it may be due to the great distance between the earth and the heavenly bodies, and/or to the lack of sufficiently careful observations of the heavenly bodies. These alternative explanations do not *refute* the Aristotelian *explanation* but rather the Aristotelian *argument*; that is, this criticism does not prove the conclusion of the original argument false, but rather *weakens* the inferential link between premise and conclusion, for there is no reason to prefer the Aristotelian to the Galilean explanation. Thus, the point being made is a logical criticism, affecting primarily the premise-conclusion relationship in the original argument. The flaw being charged is explanatory disconnection. The type of criticism involved is criticism by construction of an alternative explanation.

To summarize, Galileo criticizes the observational argument for heavenly unchangeability by arguing in part that it has a false premise. He also shows that it has a false conclusion, which is to say he can demonstrate that the heavens are changeable. Moreover, he charges the argument with internal incoherence to some extent insofar as it presupposes that observation corresponds to reality, but in the light of new evidence this principle justifies a contrary conclusion. He argues further that the argument is formally invalid because the failure to observe heavenly changes shows merely that the heavenly bodies have undergone no changes so far, not that they are unchangeable by nature. Finally, he argues that even the probable version of the argument is problematic because it is questionable whether the failure to observe heavenly changes is due to their nonexistence, or instead to such factors as the great distance of the heavenly bodies, or the fact that past observations have been insufficiently precise and systematic. We may say that he criticizes the argument both factually and logically, as long as we understand that even factual criticism normally requires argument since part of what is in question is what the facts are; moreover, logical criticism should not be construed narrowly as referring merely to formal invalidity, but as referring to the evaluation of the relationship between premises and conclusion, and this evaluation again hinges crucially on argumentation.

1.5. Another instructive example of critical analysis is Galileo's discussion of the anti-Copernican argument from vertical fall. It is found near the beginning of the "Second Day" of the *Dialogue* (selection 8).

The original argument may be *reconstructed* as follows: (2) the earth does not rotate because (2.1) bodies fall vertically, and (2.2) this could not happen if the earth rotated; for (2.2.1) if the earth rotated and bodies fell vertically, then falling bodies would have a mixture of two natural motions, toward and around the center; but (2.2.2) this is impossible since (2.2.2a.1) every body can have only one natural motion, and since (2.2.2b.1) on a ship moving forward, rocks dropped from the top of the mast fall behind and land away from the foot of the mast (towards the stern). This is a reconstruction because it is meant to be not a mere synopsis of the relevant text but also an interpretation illustrating the concepts discussed earlier; for example, I used standard labels to reveal the propositional structure unambiguously. However, my interpretation is meant to be textually accurate.[9]

The first Galilean criticism (selection 8, [164–66]) is that if the phrase "vertical fall" is taken literally, then the argument begs the question because it presupposes that the earth is motionless, which is the conclusion it is trying to prove. The literal meaning of vertical fall is downward motion in a straight line along an extended terrestrial radius, namely, perpendicular to the earth's surface. This may be called *actual* vertical fall and should be distinguished from *apparent* vertical fall, which means fall that to an observer on the earth's surface appears to be vertical, as when a rock is dropped from the top of a tower and lands at its foot directly below with no visible deviation. The two would coincide on a motionless earth; but an important point easily agreed upon by both sides of the dispute is that *if* the earth were in axial rotation then the two would *not* coincide, since the appearance of vertical fall would imply an actually slanted path and thus an actual nonvertical fall.

This first criticism begins by noting that premise 2.1 of the original argument seems to refer to *actual* vertical fall; then there is no problem with the first inferential step from premises 2.1 and 2.2 to conclusion 2, since it is an instance of

9. For the documentation and the details, see Finocchiaro (1980, 115–17, 192–200, 208–13, 277–91, 329–30, and 387–91).

denying the consequent. However, the critic asks how the argument's proponents know that bodies do indeed fall vertically, namely, that premise 2.1 is true. The answer would be because (2.1.1) bodies are *seen* to fall vertically, namely, because of apparent vertical fall. Now, the truth of this observation is undeniable, but the critic next asks how (2.1.2) actual vertical fall follows from apparent vertical fall. This is a legitimate question since, as just explained, on a rotating earth apparent vertical fall would not imply actual vertical fall. It seems that the Aristotelians must assert this implication, and that the only way to justify it is to assume that (2.1.2.1) the earth is motionless, for indeed (2.1.2.2) if the earth is motionless, then apparent vertical fall does imply actual vertical fall. Unfortunately, this assumption (proposition 2.1.2.1) is precisely what the original argument is trying to prove. In other words, proposition 2.1.2.1 happens to be identical to proposition 2, and so the bigger argument (namely, the argument that includes the latent structure unearthed by the above analysis) is circular; thus the original argument, which overlaps with this bigger argument by sharing the subargument from 2.1 and 2.2 to 2 begs the question precisely at this point. This criticism is largely a presuppositional criticism, which exposes both presuppositional and persuasive disconnections.

However, perhaps the original argument was speaking of apparent rather than actual vertical fall. In that case the issue would reduce to the tenability of the impossibility of mixed motion, namely, proposition 2.2.2. This impossibility amounted to a denial of what later came to be known as the principle of the superposition of motions; this principle was then one crucial issue in the controversy. In the reconstructed argument the impossibility of mixed motion is explicitly supported with two independent reasons, and so its correctness largely depends on whether the corresponding subarguments involve any difficulty. Galileo criticizes both of these subarguments, thereby providing a good example of reason undermining criticism.

Consider first the subargument that (2.2.2) it is impossible for a falling body to have two natural motions, toward and around the center (i.e., vertical and horizontal) because (2.2.2a.1) every body can have only one natural motion. This may be called the theoretical argument for the impossibility of mixed motion since its premise was a basic principle of Aristotelian physics. As one might expect, Galileo explicitly questions the empirical correctness of this premise, but he does not do that in this passage, and so it need not concern us here. Let us note simply that, once again, the empirical issue requires argumentation, and that we would have an example of reason refuting criticism.[10]

Galileo also implicitly objects that the two "natural" motions of falling bodies on a rotating earth would be "natural" in different senses: the downward fall would be natural in the sense of spontaneous (not potentially everlasting), whereas the rotational motion would be natural in the sense of potentially everlasting (not spontaneous). Now, the premise here (proposition 2.2.2a.1) may be true of each of these two kinds of natural motion, but it would imply only that a body cannot have simultaneously two kinds of spontaneous motion or two

10. Cf. Favaro (7:281–89 and 423–42); Galileo (1967, 256–64 and 397–416).

kinds of everlasting motion; it clearly does not imply that it cannot have one of one kind, and one of the other kind.

This criticism involves a charge of equivocation and is an example of semantical criticism in the general classification elaborated earlier. But since it is, once again, not explicit in the passage being considered, the fuller demonstration of this point is left as an exercise for the reader.[11]

Galileo's explicit criticism is directed at the subargument that (2.2.2) a mixture of free fall and perpetual horizontal motion from west to east in accordance with diurnal rotation would be impossible because (2.2.2b.1) on a forward moving ship, rocks dropped from the top of the mast fall behind. This is clearly an observational argument insofar as the premise is an experimental report; it is also an explanatory argument insofar as the impossibility of mixed motion mentioned in the conclusion is advanced as the explanation of the experiment mentioned in the premise.

Galileo objects partly that the rock's failure to move simultaneously in two directions is not the only possible explanation of the alleged fact: it might happen because the horizontal motion imparted by the ship to the rock is violent motion, which would be dissipated after the rock is left to itself; or it might happen because of air resistance, which would oppose the horizontal motion acquired by the rock. This is a charge of explanatory disconnection (as defined earlier) and uses the critical technique of constructing an alternative explanation (selection 8, [166–69]).

He also criticizes this subargument (2.2.2b.1 to 2.2.2) by objecting to the claim about the results of the ship experiment (proposition 2.2.2b.1); he argues that this alleged experiment does not in fact happen this way. That is, he criticizes the truth of the premise. However, a simple appeal to the facts of observation was impossible in the context, which required instead an argument concluding with the denial of the proposition in question. This is a case of premise refuting criticism (selection 8, [169–74]).

In summary, Galileo criticizes the argument from vertical fall in several ways. He begins by distinguishing between actual and apparent vertical fall. Insofar as the argument refers to actual vertical fall, it is subjected to presuppositional criticism and shown to beg the question. Insofar as it refers to apparent vertical fall, it depends on a premise asserting the impossibility of mixed motion; he undermines this premise as groundless by criticizing a theoretical and an observational subargument supporting it. The criticism of the theoretical subargument involves premise refuting criticism and semantical criticism. The criticism of the observational subargument involves criticism by the construction of an alternative explanation and also premise refuting criticism. Again, we may say that he engages in a mixture of empirical and logical criticism, as long as we understand that the empirical is essentially dependent on argumentation and not just brute facts, and the logical involves not abstract irrelevancies but questions about the relationship between premises and conclusions.

11. This criticism is not explicit in the text of selection 8, but it can be derived from an earlier passage (Favaro 7:38–57; Galileo 1967, 14–32); cf. Finocchiaro (1980, 349–53 and 387–89).

1.6. The preceding discussion has provided a general elucidation of the relevant concepts of reasoning, analysis, and evaluation, followed by some concrete illustrations of reasoned analyses and evaluations of arguments. Now, reasoning aimed at the analysis and evaluation of reasoning may be considered to be the key element of critical reasoning. The discussion may then be seen as an elaboration of this definition of critical reasoning.

However, the two extended examples illustrate only part of this definition, for they are arguments that analyze and criticize *other* arguments and the definition of critical reasoning is not meant to have this limitation. That is, in referring to reasoning aimed at the analysis and evaluation of reasoning, the definition is meant to include both the case of reasoning aimed at the analysis and evaluation of *other* reasoning and the case of reasoning that is self-analytical and self-evaluative. If both of these notions are taken to be included in the concept of self-reflectiveness, then we can say that reasoning aimed at the self-reflective construction (or presentation, or formulation) of an argument is also a key element of critical reasoning.

It should be stressed that this definition does not equate critical reasoning with just reasoning, but rather with a special kind of reasoning; that is, according to this definition, not all arguments are instances of critical reasoning, only those arguments that analyze or evaluate other arguments or that show the proper degree of self-analysis and self-evaluation. But it should be equally stressed that the definition does include the self-reflective case; that is, it includes arguments that are not so explicitly directed outwardly, but rather are primarily directed inwardly, as long as they still contain the proper sensitivity to analysis and evaluation, so that they are indirectly aimed outwardly, so to speak.

What is required for reasoning to be self-reflective, and how do we distinguish reasoning that is self-reflective from reasoning that is not? We should not expect that, in the self-reflective construction of an argument, critical thinkers will analyze and evaluate it with the same degree of explicitness and formality as when they critically analyze the arguments of others. Here it is useful to contrast the analysis and evaluation of an argument such as a critical thinker might advance with the analysis and evaluation such as a logical theorist might produce. For example, although the illustrations given above are meant to be accurate representations and reconstructions of the relevant texts in the *Dialogue*, the actual texts are less explicit; they are also less systematic insofar as my illustrations are given in the context of a theoretical articulation of a conceptual framework which, while consistent with and inspired by the text, is certainly not contained there. Thus, two important distinguishing characteristics between a critical thinker and a logical theorist are conceptual explicitness and theoretical systematicity. Similarly, the self-reflective presentation of one's own argument will be less explicit and systematic in analysis and evaluation than the critical analysis of someone else's argument; but the degree of self-analysis or self-evaluation cannot be too low, otherwise we would have an instance of mere reasoning and not critical reasoning.

The sensitivity to self-analysis is normally shown by careful attention to such questions as what our conclusions are and what our reasons are, whether we are advancing just one reason or more than one, and how our reasons are meant to connect with our conclusions. These topics are discussed above under the headings of analysis and structure. The sensitivity to self-evaluation is normally shown by careful attention to such questions as whether we are advancing a conclusive or very strong or moderately strong or weak argument; this involves the degree of support the reasons lend to the conclusion. We should also pay attention to possible criticism, objections, and counterarguments against our own argument, and to ways of rebutting these.

An extended example will give more definite meaning to these imprecise stipulations and concrete content to the general concept of self-reflective reasoning. It consists of Galileo's argument for the earth's diurnal axial rotation at the beginning of the "Second Day" of the *Dialogue* (selection 6).

Galileo begins with a statement of the principle of the relativity of motion, namely, the idea that motion exists only in relation to things lacking it, whereas shared motion has no effect on the relationship of things sharing it; let us abbreviate this proposition by A (and then we will progressively label other key propositions in alphabetical order, unless explicitly labeled otherwise in the text under consideration). In the course of the discussion he gives a partial justification supporting this principle in terms of familiar examples of shared and unshared motion, although he does not stress this justification and regards the principle as relatively uncontroversial; let us label this support B. His primary interest is to apply this principle to the problem of explaining apparent diurnal motion; that is, the relativity of motion is taken to imply that the apparent westward diurnal motion can be explained by saying either that all heavenly bodies revolve westward daily around a motionless earth or that the earth alone rotates eastward on its own axis every day; let C refer to this consequence of the principle.

Given this premise (that diurnal motion can be explained either way), Galileo goes on to argue that therefore (D) terrestrial axial rotation is more likely than universal revolution around a motionless earth because (E) the geokinetic explanation of apparent diurnal motion is simpler than the geostatic explanation, and (F) nature usually operates by the simplest possible means. The argument so far is: A because B; because A, therefore C; and because C, E, and F, therefore D (see fig. 16).

Galileo makes it clear that he is asserting the final conclusion only with probability, and that his conclusion embodies an implicit comparison between the two contradictory views which generate the controversy. He also explicitly states that his conclusion needs to be qualified in another crucial way; that is, it depends on the assumption that all other relevant phenomena can also be explained either way, namely, the assumption that all the many anti-Copernican objections can be refuted and all the phenomena on which they are based could occur on a moving earth. These Galilean remarks are explicit indications of the self-evaluating element needed for critical reasoning.

Let us label the assumption just mentioned G. The best place for this assumption in the overall economy of this argument is as an additional premise directly supporting the final conclusion, alongside C, E, and F. We might ask at this point the more technical question whether this assumption is a reason linked with or independent of these other three. I would opt for a link on the grounds

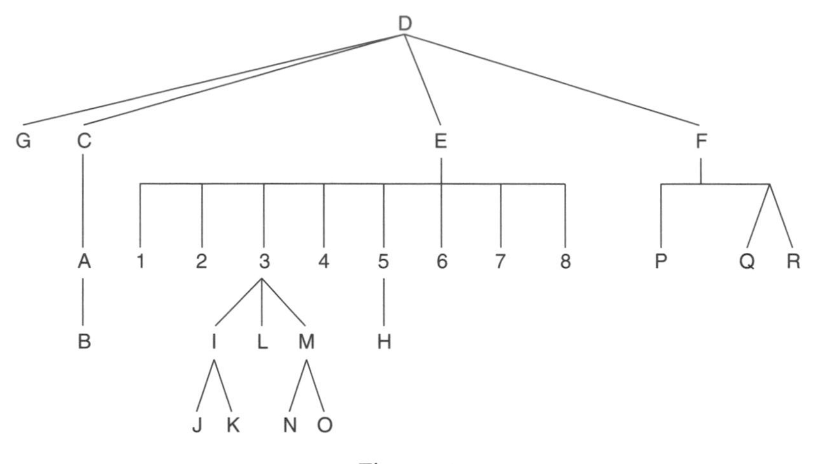


Figure 16.

that it makes a point which is similar to, though of course more general than, the specific claim that diurnal motion can be explained either way (C); moreover, the simplicity considerations made in the other two premises (E and F) apply as much to the general as to the specific point. Note also that this assumption is a generalization about anti-geokinetic arguments, and that the rest of the book supports it by critically examining each in turn and attempting to refute them all. However, here (selection 6) this assumption is unsupported and so constitutes a final reason.

The rest of this passage argues in support of the greater simplicity of terrestrial rotation (E) and in support of the principle of simplicity (F), which so far have merely been stated. In its central part, Galileo lists seven numbered reasons why terrestrial axial rotation is simpler than universal geocentric revolution; then at the end he adds an eighth reason, without labeling it as such. That is, the geokinetic system involves (1) fewer moving parts, smaller bodies moving, and lower speeds; (2) only one direction of motion (eastward) rather than two opposite ones; and (3) periods of revolution which follow a uniform pattern, namely, that of increasing with the size of the orbit. Moreover, the geostatic system involves (4) a complex pattern for the size and location of the orbits of the fixed stars; (5) complex changes in the individual orbits of fixed stars (due to [H] the precession of the equinoxes); (6) an incredible degree of solidity and strength in the substance of the stellar sphere which holds fixed stars in their fixed relative positions; (7) a mysterious failure for motion to be transmitted to the earth after it has been transmitted all the way down from the outer reaches of the universe to the moon; and (8) the postulation of an ad hoc Prime Mobile.

It is probably best if we leave these numerals alone and let them stand respectively for the propositions supporting the comparative simplicity claim (E). Although all of these eight premises are propositions which implicitly compare and contrast the two alternatives in regard to a particular property, it should be noted that only the first two seem to involve matters of degree, whereas the other six seem to involve discrete properties which one of the two world systems possesses but the other one lacks. An interesting question here is whether these eight reasons are linked or independent. I would say, first, that each strengthens the others, and so the total amount of support they give to their conclusion is a function of all of them taken collectively; this suggests that they may be viewed as linked. However, each reason provides some degree of support independently of the others; moreover, each proposition is a perfectly natural answer to the question whether there is another reason why one should accept the comparative simplicity claim; thus, I am inclined to regard them as basically independent. Note also that the first proposition has three parts, each of which would need separate support; that the fifth one is provided a brief justification; and that the third is especially important and is supported by a relatively lengthy, novel, and strong argument.

As phrased above, proposition 3 states that the periods of revolution have a uniformity in the Copernican system which they lack in the geostatic one. In the supporting subargument Galileo begins with a claim which I call the law of revolution: (I) it is probably a general law of nature that, whenever several bodies are revolving around a common center, the periods of revolution become longer as the orbits become larger. He then supports this by the well-known fact that (J) the planets revolve in accordance with this pattern, and by his own discovery that (K) Jupiter's satellites also follow the pattern. The important point is that, though this feature of planetary revolutions was known to the Ptolemaics and incorporated in their system, before the discovery of Jupiter's satellites it would have been rash to generalize a single case into a general law; however, the completely different and unexpected case of Jupiter's satellites suggested that this was not an accidental coincidence but had general systemic significance; thus, while this inferential step ("J, K, so I") is not conclusive, it cannot be dismissed.

Given the law of revolution (I), Galileo goes on to combine it with the claim that, whereas (L) the earth's diurnal motion in the Copernican system is consistent with the law of revolution, (M) the diurnal motion of the universe in the Ptolemaic system is not. The point would be that this difference gives the Copernican system a uniformity or regularity lacking in the Ptolemaic system, which is what is asserted by Galileo's third reason (proposition 3) why the former has greater simplicity than the latter (proposition E). He considers the consistency between Copernican diurnal motion and the law of revolution to be sufficiently obvious to need no support. The argument would be that in the Copernican system the diurnal motion is the axial rotation of the earth, and this axial rotation is not an orbital revolution and does not involve a member of a series of increasingly large orbits; thus, the law is not even meant to apply to terrestrial rotation. To justify the inconsistency between the law of revolution and Ptolemaic diurnal motion, he explains that (N) in the geostatic system the diurnal motion corresponds to the revolution of the outermost sphere (whether stellar sphere or Prime Mobile) around the central earth, but (O) this outermost sphere involves both the largest orbit and the shortest period. This completes the subargument supporting the greater Copernican uniformity in regard to periods and orbits of revolution (proposition 3), namely, the argument *from* the law of revolution, for short.

At the end of the passage, there is a discussion of the principle of simplicity (F), which was a crucial premise directly supporting the final conclusion about the greater likelihood of terrestrial rotation (D). Though there might be some question how this principle should be formulated, let us focus on the interpretation adopted above, namely, that nature usually operates by the simplest possible means. When so stated, Galileo in part suggests what might be called a teleological justification for the principle of simplicity, namely, the teleological principle that (P) it is useless to do with more means what can be done with fewer. However, he also recognizes that (Q) the principle of simplicity is subject to the following theological objection: given a God who is all-knowing and all-powerful, God can create and operate a more complicated system as easily as a simpler system, and so a more complex world system is as likely to exist as a simpler one; in other words, it is false that nature usually operates by the simplest possible means because nature was created by an infinitely powerful God and such a God would be as likely to use more power to operate a more complex universe as to use less power to operate a simpler world. Galileo tries to address this objection directly, although I do not find his answer too relevant or convincing; however, his presentation of this objection indicates the kind of self-evaluation that is part of critical reasoning. Finally, let us simply label his answer R, and let us note that we have labeled by the simple propositional label Q the whole admission of the existence of the theological objection, including a statement of its details as part of proposition Q; in terms of these labels, the last subargument has the structure "F because P, and because Q and R."

This leads to a final comment about this instance of critical reasoning. That is, although we have imposed some structure onto Galileo's argument for terrestrial axial rotation, his presentation of the argument is sufficiently self-reflective to make its overall structure relatively clear. The number of propositions is large, and the structure is relatively complex; but with the help of our technical apparatus and the labels we have progressively assigned, this structure may be represented in figure 16.

1.7. The three extended examples discussed above are especially incisive as illustrations of various aspects of critical reasoning. However, as I have mentioned several times, the *Dialogue* is full of critical reasoning of this sort; that is, the rest of Galileo's book has comparable significance. It is now time to substantiate this interpretive claim with a description from the viewpoint of critical reasoning of the various selections translated here.

The presence or absence of critical reasoning in an extended passage is a matter of degree, generalization, and overall synthesis; so I will talk about the primary content of each passage as a whole, and I do not exclude the existence of secondary elements. Moreover, the viewpoint of critical reasoning is not the only one we may adopt; as I said before, it is also important to appreciate the details of the methodological and the rhetorical content of the *Dialogue*, and I shall elaborate on this later.

I will say nothing further about selections 3, 6, and 8, which have already been discussed at length. Next, we should note that there are three selections which contain little critical reasoning: Galileo's preface (selection 1), the selection on the role of the Bible (selection 12), and the final selection on divine omnipotence and the limitations of human knowledge (selection 16). These have been included for other reasons, but even from the viewpoint of critical reasoning they have a function, namely, to provide instructive *contrasts* with the other selections containing critical reasoning. In other words, these three selections give examples of thinking *other than* critical reasoning.

Selection 2 is primarily a critical analysis of two interrelated arguments. The natural motion argument against Copernicanism states: the natural motion of the earth is straight toward the center of the universe because the natural motion of heavy bodies is straight toward the center of the universe, heavy bodies are parts of the earth, and whatever is true of the parts is true of the whole. This argument is criticized in several ways. The basic geocentric argument states: the natural motion of heavy bodies is straight toward the center of the universe because their natural motion is opposite to that of light bodies, and the natural motion of light bodies is toward the circumference of the universe; but the natural motion of heavy bodies is also straight toward the center of the earth; therefore, the center of the earth coincides with the center of the universe. This argument is criticized as begging the question. The connection is that these two arguments share the principle of natural motion for heavy bodies; this proposition is a final reason in the first argument and an intermediate proposition in the second. The passage also contains explicit reflections on the relative importance of observation and theory and the role of the logical authority of Aristotle; these reflections are important from the methodological viewpoint and will be examined later. From the viewpoint of critical reasoning, they are part of the critical analysis of the two arguments; however, in such reflections critical reasoning becomes, to some extent, the subject matter of the discussion, which thereby acquires a second relevant dimension.

Selection 4 may be interpreted primarily as the self-reflective formulation of an argument supporting epistemological modesty. This is a principle expressing a judicious, moderate, and balanced mean between the extremes of skepticism and dogmatism. This argument incorporates as subarguments the remarks on the cause of the moon spots visible with the naked eye, on the differences between the earth and the moon, on whether there is life on the moon, on the limitations and powers of human knowledge, and on the similarities and differences between human and divine understanding. Note that when the topics of discussion are epistemological questions like these, argumentation about them can hardly avoid being self-reflective, and so any serious thinking about them is almost inescapably critical reasoning.

Similar remarks apply to selection 5. The topics are methodological questions about the role of authority in general, the role of Aristotle's authority in particular, and the nature of independent-mindedness. Galileo self-reflectively argues for the importance of a judicious independence of mind, as contrasted with both uncritical subservience to and solipsistic disregard of what authorities say.

In selection 7, the primary argument is, again, methodological and thus self-reflective. It is especially relevant because the main conclusion is not only justified by self-reflective reasoning, but is also an explicit claim about the value of critical reasoning in the search for truth. Moreover, a key part of Galileo's argument is the statement of all objections against the earth's diurnal rotation; these arguments are not evaluated in this passage, but rather in later passages; however, their mere statement in this context involves a certain amount of analysis by way of classification and interpretation. The overall argument thus involves critical reasoning in three ways: the content of the conclusion, the preliminary analysis of the anti-Copernican arguments, and the self-reflectiveness.

Selection 9 is primarily a critical analysis of the anti-Copernican argument from the extruding power of whirling; this is subjected to several criticisms. Galileo begins with a constructive clarification or formal criticism: as usually formulated, the argument is improperly stated; its crucial step (subargument) should be stated to read that if the earth were rotating, then there would now be no loose bodies on its surface, since they would have all been extruded long ago. He then objects that the argument is quantitatively invalid for a number of reasons. First, a comparison of tangential extrusion with downward fall is needed because on a rotating earth bodies would have a tendency to be extruded along the tangent to the point of extrusion, but they would still have a tendency to move downward along the secant due to their weight; and the comparison shows that the downward tendency is greater than the extruding one. Second, the downward tendency not only happens to exceed the tangential one, but it necessarily does so for mathematical reasons; that is, he argues that extrusion would be mathematically impossible on a rotating earth. He tries to prove this mathematical impossibility on the basis of the geometry of the situation in the neighborhood of the point of contact between a circle and a tangent, and the behavior of the external segments (called exsecants) of the secants drawn from the center to the circle to the tangent (see diagrams on pp. [224] and [225]); he argues that as one approaches the point of tangency, (1) the ratio of an exsecant to the corresponding tangent segment tends toward zero, (2) the ratio of an exsecant to the corresponding speed of fall also tends toward zero, and (3) the ratio of an exsecant to the previous one (at twice its distance from the point of tangency) tends to zero as well; that is, the exsecants get smaller and smaller in relation to (1) the corresponding tangent segments, (2) the speeds of fall, and (3) each other; thus, the distances of fall required to prevent extrusion become infinitely smaller than the distances required to achieve extrusion, and they decrease more rapidly than the speeds of fall. He finally objects that the argument is quantitatively invalid because the cause of extrusion increases with the linear speed but decreases with the radius, and the linear speed at the equator is very small compared to the earth's radius. The passage is important for critical reasoning in another way, for there is little question that Galileo's criticism contains some errors; though it is hard to identify exactly where the difficulties lie, there seems to be at least one purely mathematical error and at least one physical misapplication of a mathematical truth. At any rate, the reader is given an opportunity to criticize the Galilean criticism, besides the usual opportunity to analyze and interpret his critical analysis. Finally, the passage contains a discussion of the methodological problem of the applicability of mathematical truths to physical reality, which suggests that his mathematical arguments have a certain amount of self-reflectiveness; this will be discussed in more detail later.

The main point of selection 10 is a critical analysis of the objection from the deception of the senses; as usual, several criticisms are advanced. Moreover, since this objection is a methodological argument, it should be no surprise that in the course of his critique, Galileo develops his own methodological position. The main issue is the proper relationship between observation or sense-experience and reason or reasoning. Thus, there is a second element of critical reasoning in the passage; the content as well as the form of the discussion involve critical reasoning.

In selection 11 is an example of self-reflective presentation of an argument, the pro-Copernican argument involving the heliocentrism of planetary motion. It has two main parts: a very strong argument showing that the planets (Mercury, Venus, Mars, Jupiter, and Saturn) perform revolutions whose center is the sun, not the earth; and a weaker but considerable argument, based on the heliocentrism of planetary revolutions but crucially involving simplicity considerations, and suggesting that the annual motion more likely belongs to the earth than to the sun. This passage also contains a critical analysis of three traditional objections to Copernicanism, involving the appearance of Mars, Venus, and the moon. Galileo's physical and critical arguments are both based on the new telescopic evidence, and so there is a discussion of the methodological issue of the role of the telescope and of artificial instruments; this in turn involves a discussion of the relationship between sensory experience, reason, and reasoning; thus, once again, critical reasoning is also the content of the discussion.

Selection 13 is primarily a critical analysis of the anti-Copernican argument from stellar dimensions. This argument is based on the premise that the sizes and distances of fixed stars implied by Copernicanism are absurdly great. Galileo undermines this premise by objecting that actual stellar sizes were traditionally overestimated due to overestimating *apparent* stellar diameters; that some Ptolemaic estimates of distances are of the same order of magnitude; and that the concepts of size and distance, while subjective and relative in several ways, are not anthropocentric or teleological. The third criticism involves a methodological claim, and its justification requires the usual sensitivity to self-reflection. There is also considerable analysis meant to clarify the original argument.

The next selection (14) deals with the following anti-Copernican argument suggested by the critical analysis in the previous selection: if the earth revolved yearly around the sun then the appearance of fixed stars would have to change in the course of a year; but there are no such changes; therefore, the earth does not revolve around the sun. The passage is primarily an analysis of this argument, together with an elaboration of a research program designed to evaluate the crucial empirical premise. Note that here, unlike in most other passages, there is no real criticism of the argument. Galileo's main point is twofold: (1) the first (conditional) premise needs further theoretical elaboration, designed to show exactly how the appearance of fixed stars would change, and he elaborates several conclusions about these changes; (2) because these changes would be extremely small, to test the second premise one needs considerable experimental sophistication, and he outlines one such experimental design. This type of evaluation may be labeled constructive analysis, to contrast it with critical analysis; the important point is that although this discussion has an evaluative and perhaps even a critical orientation, it consists primarily of the analytical elaboration which is a precondition for the later evaluation.

Finally, selection 15 is mostly a self-reflective presentation of an argument for the earth's motion based on the cause of the tides. Galileo regarded it as his strongest pro-Copernican argument. The main elements of self-reflection are an emphasis on the hypothetical and explanatory character of the argument; a brief discussion of alternative explanations; the explicit qualification that his primary cause is insufficient to account for all relevant details and must be combined with other concomitant causes; and a discussion of some possible counterevidence about whether the proper interval between high tides is or should be twelve or twenty-four hours. This argument contains difficulties that cannot be ignored, although it is hard to define the errors precisely; thus, it gives readers an opportunity to engage in that element of critical reasoning which is the evaluation of reasoning. That is, as in selection 9, from the viewpoint of critical reasoning, such errors are a blessing in disguise and give additional relevance to this passage.

1.8. We began with a general, abstract, and systematic elaboration of a number of concepts relating to critical reasoning. The large number of particular definitions can be grouped under some main headings. That is, we first explained what reasoning and argumentation are. Then we articulated what is meant by the structure of reasoning and argumentation, the description of such structure being the key task in the understanding, interpretation, and analysis of reasoning. This was followed by a clarification of the concepts of evaluation and criticism, and in the process we focused on their fundamental features and developed a taxonomy of ten common types of argument criticism. After this general discussion we illustrated most of these general ideas by two extended examples from the *Dialogue*, the critical analysis of the observational argument for heavenly unchangeability (selection 3), and the critical analysis of the anti-Copernican argument from vertical fall (selection 8). Then we gave a general elucidation of the notion of self-reflective reasoning by explaining how it relates to reasoned analysis and evaluation; and we reconstructed a concrete Galilean example consisting of his self-reflective presentation of the simplicity argument for terrestrial axial rotation. Lastly, all other selections translated here were described from the viewpoint of critical reasoning, thus substantiating our claim that the *Dialogue* is full of critical reasoning in both form and content.

All these general explanations and particular examples were meant to provide an elementary introduction to critical reasoning. The theme that unites them all is the definition that critical reasoning is reasoning aimed at the analysis and evaluation of reasoning, or equivalently, the reasoned analysis and evaluation of reasoning.

As suggested earlier, one immediate comment on this definition is that it includes two special cases. The first is reasoning aimed at the critical analysis of arguments formulated by others; this may be called critical analysis of reasoning, or argument analysis and evaluation, or just critical analysis. The second case is reasoning aimed at the self-reflective formulation of one's own arguments, which may be called simply self-reflective reasoning or argumentation; self-reflection refers to both self-analysis and self-evaluation, namely, the analysis and evaluation of one's own reasoning. In short, critical reasoning is reasoning aimed at the critical analysis of other reasoning or at the self-reflective presentation of one's own reasoning.

Another immediate comment is that critical reasoning so defined has three elements: reasoning, analysis, and evaluation. These are distinct but interrelated. Reasoning is the activity of basing one thought on another; analysis is the description of details to provide understanding; and evaluation is the assessment of worth. One relationship is that to be part of critical reasoning, analysis and evaluation have to be based on reasoning. Another is that normally evaluation has to be also based on analysis, or that normally analysis leads to evaluation. However, it is legitimate to focus on just one element of the analysis-evaluation pair; thus, if one is engaged in the reasoned analysis of an argument then to that extent one is doing critical reasoning; similarly, if one is focusing on the reasoned evaluation of an argument, one is thereby doing some critical reasoning. In short, critical reasoning is reasoning aimed at the analysis *and/or* evaluation of reasoning.

Another series of clarifications involves the interrelationships among thinking, reasoning, and critical reasoning. As here conceived, all critical reasoning is reasoning, but not all reasoning is critical reasoning; for example, reasoning which is relatively non-self-reflective is not critical reasoning. And all reasoning is thinking, but not all thinking is reasoning; for example, remembering and imagining are types of thinking but not of reasoning. That is, thinking is a broader notion than reasoning, and reasoning is broader than critical reasoning; conversely, critical reasoning is a special case of reasoning, and reasoning is a special case of thinking.

Finally, another relationship is worth discussing to point in the direction of further developments, namely, the relationship between critical reasoning and critical thinking. It is useful to introduce the notion of critical thinking as a type of thinking intermediate in generality between thinking and critical reasoning; it is useful to define critical thinking so that it includes critical reasoning as one important special case, but so as to leave open the possibility of other types of thinking to be subsumed under the same heading. In fact, there is an important type of thinking consisting of reflections on what one is doing when engaged in inquiry, the search for truth, or the quest for knowledge; that is, thinking aimed at understanding and evaluating the aims, presuppositions, and procedures of knowledge. Some call this metacognition;[12] I call it methodological reflection. Thus, methodological reflection is another special case of critical thinking. The point of grouping critical reasoning and methodological reflection together under the heading of critical thinking is that each corresponds to one of two important connotations of the notion of criticism, analysis and evaluation on the one hand and reflective awareness on the other.

## 2. Methodological Reflection

2.1. Any inquiry, search for truth, or quest for knowledge about nature or physical reality eventually leads to questions about the nature of

12. Cf. Battersby (1989), Kitchener (1983a; 1983b), Kitchener and Fisher (1990), McGuinness (1990), Meichenbaum (1986), and Paris and Winograd (1990).

inquiry, truth, and knowledge; the focus then temporarily shifts from natural science to epistemological or methodological reflection.[13] The same happens with any inquiry into any other topic; so these shifts occur in other fields, whether they study numbers, life, human nature, history, society, or the supernatural.

There are many causes for such methodological pauses. Sometimes investigators want to gain a deeper understanding of what they are doing, what their aims are, what procedures they follow, what rules they accept, and what their presuppositions are. Sometimes they are challenged by a critic about one of these things and want to determine whether the criticism is valid. Sometimes they want to resolve important differences that arise with fellow practitioners, and the resolution requires that such things be identified, analyzed, and evaluated.

Epistemological or methodological reflection is concerned with the identification, description, analysis, and evaluation of the aims, procedures, rules, and presuppositions of inquiry, truth-seeking, or knowledge-gathering. If we want to distinguish between the two, we could say that epistemological reflection focuses on the nature of inquiry, truth, and knowledge, and the meaning and relationships of these concepts; whereas methodological reflection focuses on the rules or procedures which investigators follow or should follow in order to be successful in inquiry, arrive at the truth, or acquire knowledge. So these two types of reflection are complementary aspects of the same business. Thus, here the two terms are usually used interchangeably; they are distinguished only when a special need arises.

As just defined, epistemological or methodological reflection is essentially context dependent and practically oriented; it arises in the course of inquiry about some other topic and ends when further reflection is no longer relevant. However, it is obvious that it can also be undertaken in a systematic, general, and theoretical manner, independent of the contextual and practical origin it had initially; indeed, it has become professionalized into a branch of technical philosophy. I refer to this branch of scholarship as systematic epistemology or systematic methodology. Normally, when I drop the adjective "systematic" and speak of epistemology, methodology, or epistemological or methodological reflection, I mean methodological reflection in the nonsystematic, context-dependent, and practically oriented version.

This relationship between methodological reflection and systematic methodology is analogous to that between critical reasoning and logical theory. The differences involve such characteristics as conceptual explicitness, theoretical systematicity, and generalizing drive. These are matters of degree, and sometimes it is difficult to determine whether we have an example of methodological reflection or systematic methodology; but usually this distinction will be immaterial. The point is that sometimes the differences of degree are great enough to generate a difference in kind. At any rate, methodological reflection is much more common than systematic methodology and can be fruitful without turning into the latter.

13. As previously noted, some scholars, especially those of critical thinking, call this metacognition, reflective thinking, or applied epistemology; cf. the references in the previous note.

Besides their contextual dependence and practical orientation, methodological reflection and critical reasoning share another formal characteristic. If we recall the definition of critical reasoning as the analysis, evaluation, and/or self-reflective formulation of arguments, then we can analogously define methodological reflection in terms of the formulation, analysis, and evaluation of methodological principles. However, this formal analogy must be seen in the light of the substantive difference between their respective subject matters (arguments versus methodological principles).

Before I discuss this substantive difference, note that the two formal similarities between methodological reflection and critical reasoning provide another reason for subsuming both under the notion of critical thinking. I have already given a semantic reason for this at the end of the previous section of this appendix when I said that critical reasoning and methodological reflection involve two connotations of the term "criticism," namely, criticism in the sense of analysis or evaluation and in the sense of self-reflective awareness. I can now add that they also share a contextual dependence or practical orientation and a focus on analysis or evaluation. These similarities should not lead to the confusion of critical reasoning and methodological reflection; they only mean that for certain purposes it is handy to have a single term (critical thinking) referring to either one of these two kinds of thinking.

A methodological principle is a general rule about the conduct of inquiry, search for truth, or quest for knowledge. Whereas an argument must have at least two propositions one of which is based on the other, a methodological rule is a single proposition. Thus, critical reasoning involves reasoning more directly than does methodological reflection.

Methodological principles are special kinds of propositions, defined partly by their content and partly by their form. I just described their content explicitly by reference to inquiry, truth, and knowledge; this content gives them special importance. I also described their form when I said that they are rules and general statements. To say that they are rules means that they are prescriptions about what we should do or what is desirable to do if we want to arrive at the truth, acquire knowledge, and conduct inquiry properly; this in turn means that they may be obeyed or disobeyed, followed or violated, acted upon or disregarded. Thus, another important point is the *application* of methodological principles, and this element should be added explicitly to the definition: methodological reflection aims at the formulation, analysis, evaluation, *and* application of rules for the conduct of inquiry. To say that methodological principles are general means that they convey information which can be applied to more than one particular case. To call a rule general does *not* mean that it is necessarily universal and categorical, or that it conveys information about each and every situation.

Besides contrasting methodological reflection with critical reasoning and with systematic methodology, it is useful to contrast it with method, or at least with one meaning of this word. Method is sometimes conceived as a set of rules that are exact, infallible, unchanging, and mechanically applicable, so that use of the method guarantees that we will arrive at the truth or solve the problem at hand. As conceived here, methodology is *not* meant to have this connotation. Instead a methodological principle is an inexact and fallible rule open to reformulation and requiring judgment for its application. Thus, when I speak of method I will usually construe it in such a way that it corresponds to a methodological principle so conceived, unless I indicate otherwise. In short, here a method is a procedure in accordance with some general but imprecise and fallible methodological principle.

This talk of procedure leads to the distinction between results and method or methodology. A method is a procedure used to arrive at certain results. The results are the conclusions, ideas, hypotheses, theories, or beliefs at which we arrive at the end of a particular investigation, or which we test by means of such an investigation. Thus, a methodological rule is a procedural rule—one about a procedure claimed to lead to the truth.

Another important distinction is that between *methodological reflection* and *methodological practice*. This is the distinction between what scientists say about the procedures they follow or should follow in their research and what they actually do when engaged in research. This is also called the distinction between theory and practice, between reflective pronouncements and practical involvement, and between words and deeds.

There is a widespread belief that the two do not correspond in science in general and for Galileo in particular.[14] This lack of correspondence is grounded on the indisputable fact that the skill required in the investigation of natural phenomena is not identical with that needed for reflection on one's own activities or the general study of the rules for the conduct of inquiry. However, I would argue that the correspondence or lack thereof depends largely on how we reconstruct scientists' methodological reflections and their methodological practices. Their methodological reflections are frequently taken out of context or injudiciously exaggerated and then compared with the actual procedure followed, not when those reflections were formulated but on some other occasion; when this happens the lack of correspondence is an inadequate interpretation rather than an accurate representation. The point I want to make is that words and deeds may or may not correspond. Moreover, even when they correspond, they should not be conflated; that is, they should be reconstructed independently of each other and then compared to see whether they correspond. It is as wrong to think that they never correspond as to think that they always do; and it is as wrong to conflate them as to dichotomize them; rather they should be distinguished *and* interrelated.

It is also important to distinguish between *methodology of discovery* and *methodology of justification*. The distinction stems from the difference between two contexts of inquiry. The context of discovery is the context in which scientists first make a discovery, conceive of a novel idea, or entertain a hypothesis; whereas the context of justification is the context in which scientists later explain the discovery to their peers, expound the novel idea in a professional article, or estab-

14. For example, there is Einstein's famous aphorism that "if you want to find out anything from the theoretical physicists about the methods they use, I advise you to stick closely to one principle: don't listen to their words, fix your attention on their deeds" (Einstein 1934, 30). In regard to Galileo, see Duhem (1969, 113–17) and cf. Finocchiaro (1980, 103–41; 1992b).

lish the hypothesis as a firm part of scientific knowledge. Here, it is indeed correct that the two do not normally coincide. But this is not to say that they never do; for example, in some cases a fact or idea is discovered in the process of proving its truth.

A more controversial point stems from the fact that some scholars disregard the context of discovery and deny the existence of a methodology of discovery; that is, they deny the existence of rules for generating novel ideas and making discoveries. When this is done, a method is conceived in the above mentioned special sense of an infallible mechanical rule; and we may agree that there is no methodology of discovery in this sense. However, in this case there would be no methodology of justification either, for outside certain special branches of mathematical logic there is no mechanical procedure for justifying propositions. By contrast, if we construe method in the looser sense of a methodological principle, then it is an open question whether certain principles are more conducive to generating fruitful ideas.

Moreover, these two contexts should not be regarded as exhaustive of the scientific enterprise. For example, if we construe the context of discovery as the context of first conceiving an idea, then what comes next is not the context of public justification, but rather the intermediate stage called the context of pursuit; here a scientist is not convinced that the idea is true but regards it as worthy of further investigation to determine whether it is true. In the same spirit, after the context of justification, one can define a context of clarification, in which scientists try to understand better what they already know to be the case. This would yield four rather than two importantly different contexts: discovery, pursuit, justification, and clarification, and four corresponding methodologies.[15]

2.2. In the *Dialogue* Galileo constantly engages in both critical reasoning and methodological reflection, as explained several times before. Thus, it is not surprising that the book contains a great deal of reasoning about methodological rules as well as many methodological rules about reasoning. Nor should it be surprising to find that one of his most overarching methodological principles is about the value of critical reasoning. These points are well illustrated in selection 7.

The passage occurs at the beginning of the "Second Day," before the detailed analysis and criticism of each argument against the earth's rotation. It may be subdivided into five parts: statements of Aristotle's five original arguments [7a]; statements of five more modern arguments, involving mostly gunshots [7b]; a contrast between Copernicans and Ptolemaics in regard to open-mindedness [7c]; a discussion of the importance of rational-mindedness [7d]; and statements of four other anti-Copernican arguments [7e].

In the third part of this selection [7c], Galileo mentions the fact that the Copernicans are acquainted with the arguments of the Ptolemaics but not vice versa. He regards this as an extremely important difference in their procedures, giving the Copernicans a methodological advantage over the Ptolemaics. Although he does

15. For more details on this sort of issue, see Finocchiaro (1973, 229–38), Nickles (1980), Popper (1959), and Wallace (1992b).

not label this particular procedure, we may speak of open-mindedness and define it as the willingness and ability to know, understand, and learn from the arguments against one's own view. Thus, part of what he is doing here is reflecting on the nature of open-mindedness.

The passage also advances the controversial argument that this fact suggests that the Copernican view is more likely to be true than the Ptolemaic one. In so arguing, Galileo presupposes the methodological principle that open-mindedness is an indication of the likelihood of an idea, at least in the sense that if one is open-minded one is more likely to arrive at the truth.

This argument was explicitly mentioned in a long list of points of censure compiled by the special commission appointed by Pope Urban VIII soon after the book's publication to investigate the many complaints against it.[16] One may oppose the Church authorities' censure, but one should not accept the argument uncritically. Still it deserves analysis and evaluation.

I suppose Galileo is thinking that open-mindedness is valuable partly because it may strengthen the view we hold, insofar as we can better defend it from critical objections; moreover, it facilitates the discovery of the truth, which may emerge more easily from the clash of opposing arguments. He may also be thinking that open-mindedness is useful if and insofar as we view scientific inquiry as an enterprise resulting in knowledge that is subject to revision, improvement, or falsification in the light of further evidence or reflection, and thus cannot be regarded as absolutely, categorically, or unconditionally correct; that is, open-mindedness may be justified as a consequence of the fallibilist view of knowledge. Finally, open-mindedness need not lead to mental confusion if we combine it with the art of analyzing and evaluating arguments (critical reasoning) and with the willingness and capacity to accept the views supported by the best arguments (which may be called rational-mindedness); this introduces an intrinsically important topic.

In fact, in the next part of this selection [7d], Galileo discusses a problem which I shall call the problem of *misology* (namely, hatred of logic) because this label is explicitly used in a similar passage in Plato's *Phaedo*; there Socrates dispels the confusions and discouragement felt by his interlocutors after hearing several arguments and counterarguments about the immortality of the soul. It is amazing that Galileo's solution is identical to the Socratic one; Socrates argues that the way out of the misologic attitude is to learn what one translator renders as "the art of logic" and explains as the skill of the "critical understanding" of arguments.[17]

In contrast to misologism, Galileo advances a version of what can be called rationalism; for such a label suggests a belief in the power of human reason to arrive at the truth, and Simplicio claims that there is a one-to-one correspondence between truth and good reasoning on the one hand and falsity and bad reasoning on the other. It should be noted, however, that the human reason in question is not apriorist speculation but critical reasoning; that is, Simplicio is not saying that one can discover the truth by pure thinking (without experience, observa-

16. Finocchiaro (1989, 222).
17. Plato, *Phaedo*, 88A-91D; Tredennick (1969, 144–45).

tion, and so on), but rather by collecting all the arguments, evidence, and reasons for and against a given view, and then critically analyzing them to determine which are good and which are bad.

On the other hand, Simplicio's statement of the principle seems oversimplified in two ways: (1) it gives the impression that distinguishing good from bad arguments is easier than it is, by saying nothing about the fact that arguments may appear to be good but in reality be bad, and vice versa; and (2) it gives the impression that arguments are simply good or bad, completely valid or fallacious, whereas they usually are better or worse, partly correct and partly incorrect, more or less strong. Thus, two qualifications are needed to make this principle viable because otherwise we could point to the Copernican controversy and the *Dialogue* as a counterexample to the principle, namely, a case where there are good arguments supporting two inconsistent propositions. I believe Galileo wants to qualify Simplicio's statement in precisely these ways; this is one reason why the oversimplification is put in Simplicio's mouth, and why at the end of this passage [7d] Sagredo makes precisely these qualifications.

Thus, instead of speaking of rationalism, I think it is preferable to speak of rational-mindedness (to avoid apriorist and intellectualist excesses) and of critical reasoning (to avoid dogmatist and demonstrativist excesses); by rational-mindedness I mean the ability and willingness to take the view supported by the best arguments to be true (or probably true, or provisionally true, or acceptable); and by critical reasoning I mean the skill of analyzing and evaluating arguments. Then Galileo may be interpreted as holding that rational-mindedness is difficult and problematic, but important in the quest for knowledge; and that it depends crucially on critical reasoning.

Thus, in selection 7 Galileo reflects on the nature and relationships among open-mindedness, rational-mindedness, and critical reasoning and their role in the search for truth. He tries to illustrate and clarify these concepts, and to formulate and justify several methodological principles corresponding to them. All these reflections arise naturally and directly in the context of the Copernican controversy.[18] We thus have a good illustration of that aspect of *Dialogue* I call methodological reflection.

2.3. The methodological reflections reconstructed above represent Galileo's response to the fact that for a long time during the Copernican Revolution reasonable people could take opposite sides on the issues by presenting arguments seeming to favor conflicting conclusions. His response is a plea to avoid both dogmatic one-sidedness and skeptical despair, both anti-intellectual misologism and rationalistic apriorism, and both simple-minded convergence and simple-minded divergence between physical truth and human reasoning. This Galilean attitude well illustrates a quality of mind I call judgment or judiciousness and define as the avoidance of one-sidedness and of extremes. As we shall see, judgment or judiciousness is the most fundamental trait of Galileo's methodological reflections.

18. Galileo began addressing these issues in 1615 and discussed them (among others) in his "Considerations on the Copernican Opinion" (Finocchiaro 1989, 71–73 and 85); these earlier writings contain other valuable points and may be usefully compared with selection 7.

Galileo also had to face the fact that the geokinetic worldview contradicted that of the most highly respected intellectual authority on the subject (Aristotle). This was especially problematic because Galileo did not want to reject authority in general or even all aspects of Aristotle's authority. Such total rejection would have been too extreme. Selection 5 and parts of selections 2, 3, and 11 represent his response to this situation.

Selection 5 is one of the longest and most sustained methodological discussions in the book. Negatively viewed, it is a critique of Aristotelian authority, namely, a criticism of the misuse and abuse of the authority of Aristotle; positively speaking, it is a plea for independent-mindedness.

It begins with the portrayal of an Aristotelian who witnesses the anatomical dissection of a human cadaver, who is shown that the nerves originate in the brain rather than in the heart, and who comments that he would have to believe what he saw if Aristotle had not said that the nerves originate in the heart. This memorable image is an example of an uncritical, undesirable, and excessive subservience to authority; here we have no argument but rather a vivid illustration of an undesirable methodological trait. The same is true for the equally memorable image of an Aristotelian's reaction to the first news of the invention of the telescope: he claimed that Aristotle anticipated the invention in the passage describing the phenomenon of the stars being visible during daylight if observed from the bottom of a well.

At the opposite extreme, there is the fictionalized story of a scholar who had written a book on the soul claiming, on the basis of certain passages, that Aristotle denied its immortality; when told that the book would not receive permission to be published (because of the theologically heterodox claim), he decided to elaborate the interpretation that Aristotle held the soul to be immortal, based on other passages. Such opportunism exemplifies an excessive and undesirable disregard for the words and texts of an authority.

Thus, for Galileo the proper use of Aristotle's authority involves a judicious independence of mind, as contrasted with both uncritical subservience and the arrogant self-centeredness of thinking we can make the authority stand for whatever we wish. So stated, this methodological rule is hard to deny; but, as with all judgment calls, the challenge is to apply it.

At any rate, selection 5 also contains a more specific suggestion: the proper use of authority involves studying the reasons underlying the authority's assertions, and not the assertions alone. That is, we need to understand, analyze, and evaluate the authority's arguments. In short, the proper use of authority involves critical reasoning in a crucial way.

In accordance with this principle, the selections in this book provide at least three examples of critical reasoning vis-à-vis authority. In the discussion of geocentrism in selection 2, instead of merely quoting Aristotle's conclusion, Simplicio does properly present his reasoning, namely, the basic geocentric argument ([59]). But the discussion then moves to the evaluation of this argument, and the issue becomes whether authority can play a role at that level. This issue arises in an especially natural way because Aristotle was also the founder of the science of logic; so the possibility arises that we may use his logical authority in the defense of the favorable evaluation of this argument. Thus, when Salviati faults it as question-begging, is it proper to object that this cannot be because the argument was advanced by the founder of logical science? Galileo's answer is to distinguish between logical theory and logical practice, to equate the science of logic with logical theory (the theory of reasoning), to equate argumentation with logical practice (the practice of reasoning), and to say that Aristotle's authority as a logical theorist cannot be used in the present context (which is one of concrete reasoning). The general methodological lesson is that when in an appeal to authority we use the authority's reasoning, we cannot also use its own evaluation, which must be always regarded as an open question; this is true even when the authority is also a genuine authority on the general theoretical evaluation of arguments.

In other cases, appeal to authority is subject to other principles. For example, in the discussion of the center of the universe at the beginning of selection 11, it is insufficient for an Aristotelian to say merely that the earth is at the center of the universe because Aristotle said so; one must also report his reasons. One of these involved the argument that the earth is at the center of the universe because it is at the center of the revolution of the planets. This argument presupposes that the center of the universe is the center of planetary revolutions; this may be regarded as the Aristotelian definition of the center of the universe. Now, Galileo thought that the new telescopic evidence enabled him to conclusively refute the minor premise of this argument and to prove instead that the sun is at the center of planetary revolutions. It followed that if one applied the Aristotelian definition to the new situation, one would have to conclude that the sun is at the center of the universe. The point is that if the reason for Aristotle's belief in universal geocentrism was his belief in the geocentrism of planetary revolutions, then insofar as one can show that planetary revolutions are heliocentric, this fact should become a reason for believing in universal heliocentrism instead. The general methodological lesson is that since argumentation involves basing relatively controversial conclusions on relatively better-known premises, appeal to an authority's reasoning opens up the possibility that when new premises displace old ones, the authority's own manner of reasoning will lead to new conclusions. That is, appeal to the argument of a past authority to help decide a current issue naturally leads to considering what conclusion that authority would reach in the current circumstances by reasoning in the same manner as before.

The same principle is illustrated (in selection 3) by Aristotle's observational argument for heavenly unchangeability, except that the situation there is complicated by the fact that the new different conclusion is also justified by other Aristotelian assertions. That is, we have the argument that the heavens are unchangeable because no heavenly changes have ever been observed; this argument presupposes that normal observation corresponds to reality; so in this case, the authority's manner of reasoning is to base the conclusion on observation. Now, when observation reveals instead that the heavens are changeable, as it did in Galileo's time, then the Aristotelian manner of reasoning leads to the un-Aristotelian conclusion that the heavens are changeable. Thus, following the authority in regard to reasoning implies disregarding it in regard to the particular physical conclusion in question. Obviously, the detection and resolution of such a tension requires both critical reasoning and the skill of discernment involved in judgment.

However, in this case the tension also emerges in another way. For Aristotle explicitly asserted that in the search for truth sensory observation should he given priority over intellectual theorizing—a methodological principle he asserted and used on many occasions. Thus, in examining whether the heavens are changeable, one might use Aristotle's authority by appealing to this principle. Now, an application of this principle in Galileo's time would yield that observation reveals that the heavens are changeable. By Aristotle's own principle, these observations should be given priority over his own theorizing to the contrary. Hence, in this case if we accept Aristotle's *methodological* authority, we are lead to reject his authority in regard to a particular physical theory. That is, in some cases, appeal to authority is rendered problematic by the fact that the authority's texts contain divergent elements, and so independence of mind (and the judgment and critical reasoning that go along with it) is required.

In summary, the problem of the role of authority in the search for truth is both generally important and constantly recurring in Galileo's own research. The *Dialogue* contains several methodological reflections on the problem. He is generally critical of appeals to authority but recognizes the legitimacy and importance of appealing to the authority's reasoning. However, there is always the question whether the authority appealed to is a genuine authority for the particular topic under investigation. Then, in using an authority's arguments, we cannot also use its own evaluation of these arguments but rather must arrive at our own evaluation. Moreover, in using its arguments, the implicit manner of reasoning may lead to a different conclusion when applied to the situation at hand. Finally, using its arguments may conflict with using other methodological principles which the authority itself may have advanced. Thus, appeal to authority is no substitute for doing one's own thinking, but rather it requires critical reasoning and independent judgment if it is to be properly used.

2.4. The problem of misology raises the specter of replacing critical reasoning with unreason in the quest for knowledge; and the problem of authority raises the possibility that the search for truth could be conducted without having to do our own independent reasoning, but instead use the reasoning of others and become "memory experts" (as Galileo cleverly puts it at the end of selection 5, [139]). There is, however, a third and more plausible alternative that must be considered, namely, observation.

In fact, one of the most fundamental problems in inquiry is that of the nature of observation, its relationship to thinking, and its role in the search for truth. Here observation is only one of a cluster of notions to which also belong the senses, experience, perception, and experiment; the label *empirical* may be used to refer neutrally to anything pertaining to them, while *empiricism* may be taken to mean an excessive emphasis on them. On the other hand, thinking belongs to a cluster to which also belong the intellect, intellectual activity, ideas, theorizing, speculation, conceptualization, reason, and reasoning; the label *conceptual* may be taken as the opposite of *empirical*, and *apriorism* or *intellectualism* as the opposite of *empiricism*.

In Galileo's work in general and the *Dialogue* in particular, there are indications of both an empirical approach and a conceptual approach. Thus, some scholars portray him as an empiricist, and some as an apriorist.[19] In my opinion, both interpretations injudiciously exaggerate what he said or did; and I deny that he was simply inconsistent in his methodological reflections and practices. He was not inconsistent vis-à-vis empiricism and apriorism because he did not universalize his reflections and procedures as required to generate an inconsistency. Nor is there a case of divergence between words and deeds because when his methodological remarks are examined in context, they usually correspond to his deeds. That is, he was what may be called a critical empiricist or a critical apriorist, which is to say neither a real empiricist nor a real apriorist; he was a judicious practitioner of both the empirical (not empiricist) approach and the conceptual (not apriorist) approach; this judiciousness may be detected in both his practice and his reflections.

One of the most instructive illustrations of these issues involves the fundamental principle of empiricism, which stipulates that observation has priority over theorizing. This principle is explicitly (and accurately) attributed to Aristotle in the discussion of natural motion (selection 2) and in the discussion of heavenly changes (selection 3). Moreover, the principle seems to be explicitly endorsed by Galileo when Salviati calls this stipulation "fitting" (selection 3, [75]), and he appears to use it in the same discussion when he argues that the heavens are changeable because heavenly changes can now be observed. Furthermore, Galileo's attitude toward Copernicanism before the telescopic observations may be construed as being in accordance with this principle; he admits this when, in the discussion of heliocentrism, he contrasts his own attitude with that of Aristarchus and Copernicus; he claims that they apparently gave priority to their geokinetic idea over the observational counterevidence, but that he would not have been able to join them without the telescope's help (selection 11, [355–56]).

On the other hand, there are indications of an opposite Galilean inclination, that is, toward apriorism. For example, in the discussion of vertical fall (selection 8), one of the issues is whether on a ship moving forward, rocks released from the top of the mast fall to the foot of the mast or some distance behind. It is striking that Galileo does not resolve this issue by actually doing the experiment or reporting the result of some actual experiment; instead he gives a plausible argument showing that the rock will fall to the foot of the mast. Or consider the discussion of stellar parallax (selection 14), where he admits both that the Copernican theory implies the existence of some annual changes in stellar appearances, and that observation (even telescopic observation) reveals none; but he does not thereby abandon the theory and instead chooses to give it priority by disregarding observations and explaining them away as due to various causes such as immense stellar distances, lack of systematic observation, and lack of sufficiently powerful instruments. Finally, in the discussion of the cause of tides (selection 15, [458–59]), after elaborating his geokinetic theory he admits that it may conflict with the observation of a six-hour interval between high and low tide; but he retains the theory and explains away the observation as due to other concomitant and secondary factors that are distinct from the primary cause; his explanation is

19. The classic example of an apriorist interpretation is Koyré (1966; 1978) and of an empiricist interpretation is Drake (1978; 1990).

plausible, but from the standpoint of the methodology of theory versus observation, the difficulty remains.

The last point deserves more elucidation. The difficulty remains because on this occasion Galileo gives priority to theory over observation, whereas in other cases he does the opposite; thus, in regard to the issue of priority of theory or observation, he does not seem to always follow the same principle. It should also be noted that in those cases where he disregards observations, he explains them away. But to use this locution (explaining away) is prejudicial and one-sided; such language presupposes that one is taking the side of observation, in the sense of simpleminded empiricism; so it is preferable to speak simply of explaining available observations, which is perfectly legitimate. To take such legitimate explanation into account complicates the situation, but also suggests that a more sophisticated principle is needed to make sense of such behavior, and that the fundamental principles of empiricism and apriorism are both oversimplifications. For example, one might hold that neither is observation prior to theory, nor is theory prior over observation, but rather sometimes the former holds and sometimes the latter, depending on certain conditions; that is, when there is a conflict between theory and observation, the theory should be abandoned if the conflict cannot be explained in any other way (for example, heavenly unchangeability and sunspots), but the theory need not be abandoned if the conflict can be explained as due to other causes (for example, earth's motion and stellar parallax). Galileo comes close to explicitly stating such a principle in the discussion of stellar parallax (selection 14, [413]).

Similarly Galileo's conceptual orientation in regard to the ship experiment can be taken, not as a sign of apriorism, but rather of the fact that sometimes one can give a justification of a truth based on other, more accessible premises, if one has already observed it empirically. This would give priority to observation chronologically speaking, or in the context of discovery; but theory would receive priority logically speaking, or in the context of justification. In selection 3 ([75–76]), he applies this distinction to Aristotle's thesis of heavenly unchangeability when he claims that probably Aristotle first observed a lack of heavenly changes and later formulated the theoretical justification based on contrariety. Therefore, to apply the distinction to Galileo's own conceptual approach to the ship experiment is in the spirit of his own methodological reflections.

More generally, refinements of the following sort would be necessary in pursuing a more systematic understanding of the methodology of theory and observation. One might begin with the fundamental principle that observation has priority over theory. This may be taken to mean that if there is a contradiction between a theory and an observation, the theory should be abandoned; this interpretation would make the priority a methodological one. It may also be taken to mean that in seeking knowledge, sensory experience should come first and theory later; this would give experience chronological priority over theory. The principle has great plausibility, but one can think of possible alternatives. For example, in regard to the chronological issue, sensory experience is potentially so rich and possible observations are so numerous that it is impossible or ineffective to begin with them unless one has done some prior thinking and is guided by some theory; but this means that the theory should have priority over experience. In regard to the methodological priority, if the theory is one which is well tested and has been able to explain many phenomena and solve many problems, and if the conflicting observation is one of questionable significance or reliability, then it would be more advisable to dismiss the observation rather than abandon the theory; even if the conflicting observation is robust and recurring, all that is required is that the theory be revised. Or we may want to weaken the requirement further and say that if there is a contradiction between theory and observation we should revise the theory, reinterpret the observation, or otherwise explain the conflict. The point is that a conflict between theory and observation merely creates a problem; but how serious the problem is, whether to give priority to either side, which side (theory or observation) should be given priority, and what the proper attitude should be toward the side with lower priority (abandonment, dismissal, revision, reinterpretation, and so on) are all issues that must be judged on a case-by-case basis.

Instead of pursuing such systematic methodology, let us look at another important illustration of the methodological problem of the role of observation which was crucial in the Copernican controversy. This is the principle that sensory experience is normally reliable, namely, that normally the human senses tell us the truth. This principle may be viewed as a species of empiricism insofar as it expresses an emphasis on observation and gives observation special importance in the search for truth. We have already seen that Copernicanism contradicted this principle insofar as the earth's motion cannot be directly seen, perceived, felt, or sensed. The so-called objection from the deception of the senses was a formalization of this difficulty.

Note that the principle of the reliability of the senses does not deny the existence of perceptual illusions. Everyone knew that a straight stick half immersed in water appears bent, and that the shore appears to move away from us when we look at it while standing on a boat which is moving away from the shore. But these deceptions were accidental, temporary, and easily corrected by further observations, whereas the sensation of the earth being at rest had none of these properties. Thus, the known perceptual illusions did not present a problem for the principle, whereas the earth's motion did.

Moreover, the principle did not deny the existence of exceptional cases when the senses were not reliable because they were diseased or disturbed by external factors such as drugs. But these were exceptions to the norm of healthy sense experience, whereas the earth's motion was not observable precisely under normal conditions.

Galileo confronts the problem head-on (selection 10) insofar as all his replies involve a partial abandonment of the original principle and a willingness to diverge from its naive empiricism. For example, he argues that experience with navigation shows that normally we can only perceive changes of motion and not mere motion; thus, the earth's motion is not something susceptible of being perceived by normal observation. While this example suggests that Copernicanism does not involve deception of the senses, it also suggests that the principle of the normal reliability of the senses is not true.

Galileo also argues that, insofar as there is a deception, it is more proper to speak of a deception of reason than of the senses. For the principle of the reliability of the senses is really a principle of reasoning that validates inferences from what is revealed to normal observation to what exists in physical reality. Now, if Copernicanism is right, then there are cases where something is revealed to normal observation (earth at rest) but does not correspond to reality. This means that such inferences are not always correct—that the principle of inference is invalid and so needs to be qualified. Still, the qualifications involve our reasoning, not our senses.

Finally, insofar as the senses are unreliable, that unreliability does not mean that they are totally unreliable and never tell us the truth. We must not go from one extreme (the senses are normally reliable) to the opposite extreme (the senses are normally unreliable). Normal observation is sometimes reliable and sometime unreliable. We must learn to live and to operate with this more qualified principle, which requires more of an exercise of our judgment.

Once again, what primarily emerges in Galileo's methodology is his interpretation of the situation in terms of critical reasoning and his judiciousness. The same is suggested by one last, important illustration of the methodology of theory and observation, namely, the telescope.

In a sense, Galileo's readiness to make scientific use of this artificial instrument indicated that he wanted to pursue the empirical method to a deeper level than his empiricist opponents; for telescopic observation is a special case of observation, and empowering the human senses with artificial instruments is still a case of using the senses. However, such an artificial instrument is essentially a product of human theorizing. He did not build his first telescope exclusively or primarily by a systematic application of optical theory, as he suggests in some of his writings; he largely followed a trial-and-error procedure; but even the latter procedure involved significant thinking that preceded the use of the eye in the new context. Thus, telescopic observation is observation guided by theory in an additional sense besides naked-eye observation. That is, observation with the help of artificial instruments is a telling example of the judicious combination of the senses and the intellect, of how the element of theory is present (presupposed) in an activity (observation) that at first seems very different from it.

To summarize, the methodological problem of the relationship between observation and theory arises on many occasions in the *Dialogue*. Galileo's various methodological reflections can and should be appreciated first of all on a piecemeal basis, independently of any systematic methodology. If we try to systematize them, they do not lend themselves to being interpreted in terms of empiricism or apriorism. This is due not to the incoherence of these reflections, but to the oversimplified and one-sided character of both empiricism and apriorism. A reconstruction is possible in terms of the idea that he avoided both of these extremes and achieved a judicious combination of the empirical and conceptual approaches. His scientific practice embodies this combined approach and thus reinforces this reconstruction.

2.5. Mathematical truths and techniques play such a central role in modern physical science that there can be no doubt that, as Galileo expressed it, "to want to treat physical questions without geometry is to attempt to do the impossible" (selection 9, [229]). Yet, as the term (geometry) used by him to refer to mathematics reminds us, mathematics is not a static discipline and, since Galileo, has come to subsume more and more novelties that were inconceivable to him and his contemporaries. Actually, even in his work, mathematics frequently involved quantitative, numerical, and arithmetical considerations which he himself would have easily distinguished from geometry narrowly conceived as the science of plane and solid figures such as triangles, circles, spheres, and cones; still, he harmlessly used the two terms interchangeably, and such is the case in the above quotation.

Not only is mathematics a dynamic discipline, but its development has been interwoven with that of physical science. Often, physical problems have forced scientists to invent new branches of mathematics, as happened soon after Galileo with the development of the calculus in order to come to terms with the phenomenon of motion, and with the development of probability theory to deal with questions of gambling and games of chance. Sometimes the interaction has gone in the other direction; that is, the availability of some independently developed branch of mathematics has led to the appreciation and understanding of novel phenomena and the creation of new branches of physical science.

It follows that the methodological problem of the role of mathematics must be taken to include the question, what is mathematics. Some commentators tend to equate mathematical thinking with apriorist conceptual thought since, after all, mathematics is an a priori science and observation has no role in mathematical inquiry. Given the nonempirical character of mathematics, this is a plausible conception. But if by mathematics we mean primarily its apriorist aspect, then the methodological problem of the role of mathematics becomes indistinguishable from the previous one of the role of observation, for both would be special cases of the problem of the relationship between empirical and conceptual elements in the search for truth; in that case, there would be little new to be added to the discussion in the preceding section.

However, by mathematics Galileo meant primarily deductive reasoning involving abstract entities such as numbers and geometrical figures. Here deductive reasoning means reasoning where the conclusion follows necessarily from the premises, so that if the premises are true then the conclusion must necessarily be true, and it is impossible for the premises to be true and the conclusion false.[20] At any rate, this is the relevant conception if we want to discuss the methodology of the role of mathematics as a distinct problem. However, although we can formulate this distinct problem in the methodology of physical inquiry, in regard to the proper reading of Galileo's text and interpretation of his methodology, we can draw a parallel with the situation discussed earlier regarding the problem of the role of observation.

To see this, let us go back to Galileo's claim that "to want to treat physical questions without geometry is to attempt to do the impossible" (selection 9, [229]). Although this claim is indisputable today (since we are inheritors of Galilean science), it was controversial at his time; this controversy, together with the inherent challenge of the mathematization of nature, was a key factor why the Scientific Revolution was so slow in coming. This Galilean methodological reflection is not an isolated example of his commitment to the mathematical

20. Cf. selection 4 [128–31], and selection 9 [229–33].

approach. One other formulation, which appears not in the *Dialogue* but in the *Assayer*, is worth mentioning. This is the eloquent image that the book of nature is written in mathematical language.[21]

On the basis of such remarks, some readers[22] attribute to Galileo a mathematicist methodology which is tantamount to an obsession with mathematics and the power of the mathematical approach in physical inquiry. I label such extremist methodology ***mathematicism*** and distinguish it from the mathematical approach as such, which may or may not involve such an excessive emphasis and universalization. This distinction is analogous to that between empiricism and the empirical approach and that between apriorism and the conceptual approach.

The distinction can be illustrated in terms of the Galilean reflection quoted above. It could be construed literally to mean that mathematical considerations are necessary in the treatment of many physical questions; then it would express a moderate, judicious commitment to the mathematical approach in natural science. But it could be taken to mean that mathematical considerations are necessary *and sufficient* in the treatment of *all* aspects of *all* physical questions; on this second interpretation, the claim is a good statement of mathematicism. It seems obvious to me that this mathematicist construal is an exaggeration and universalization on the part of the reader.

The mathematicist interpretation also involves taking the sentence out of context. There are several levels here. The most immediate one is that the sentence appears at the end of Galileo's attempt to prove the mathematical impossibility of extrusion on a rotating earth (selection 9, [223–29]). That is, he first explains that on a rotating earth bodies would have both a tendency toward extrusion along the tangent due to the rotation and a downward tendency along the secant due to gravity, and then he argues that the downward tendency (however small it might be) would always be able to overcome the extruding tendency (however large it might be). He tries to prove this claim on the basis of (1) the geometry of the situation in the neighborhood of the point of contact between a circle and a tangent, and (2) the behavior of the external segments of the secants drawn from the center of the circle to the tangent. The argument is interesting from several viewpoints,[23] but here the crucial point is that he justifies his claim by means of not just one, but *three* mathematical proofs. Now, one proof should have been sufficient since in valid mathematical reasoning the conclusion follows necessarily from the premises; so perhaps he is not sure about the validity and applicability of these proofs. This possible uncertainty suggests that the power of mathematics is not as great as claimed by the mathematicist interpretation of the Galilean sentence quoted above.

21. Drake and O'Malley (1960, 183–84). It can be argued that, like the claim from the *Dialogue* on which I am focusing, this statement in the *Assayer* also is not evidence of Galilean mathematicism; for when taken in context the statement is rather part of a plea for independent-mindedness; this is in accordance with the interpretation advanced in Biagioli (1993, 306–7).

22. Two such readers are Koyré (1966; 1978) and Shea (1972).

23. For more details, see Drake (1986a), Hill (1984), MacLachlan (1977), Wisan (1978), and the relevant notes in selection 9.

At the next level of context, it is important to stress that the sentence is uttered in the middle of the critique of the anti-Copernican objection from the extruding power of whirling. Galileo's substantial criticism of this objection is that it is quantitatively invalid due to its neglect of relevant quantitative considerations. One of these is the one just discussed, which involves the (problematic) mathematical impossibility of extrusion. Another consideration, indirectly suggested in the text just preceding it (p. [221]), is that the relevant physical parameters (terrestrial radius, rotating speed, acceleration in free fall) are such that the downward tendency happens to exceed the extruding tendency. A third consideration at the end of the passage (pp. [237–44]) is that the cause of extrusion increases with the linear speed but decreases with the radius, and so it would really be very small due to the fact that the linear speed at the equator is very small compared with the earth's radius. Thus, his main point is that without mathematical considerations of this sort it would be impossible to understand why the extrusion objection is ineffective.

However, this mathematical criticism does not apply to every aspect of the same problem. For example, the first thing he explains (pp. [214–16]) is that the extrusion objection as usually formulated is improperly stated and must be reformulated to see that there is a problem at all. This explanation involves logical criticism and constructive clarification, not mathematical or quantitative analysis.

Moreover, in the immediate context where the quoted remark is made, Galileo embarks on a long methodological discussion of the application of mathematical truths to physical reality. Unless this discussion is to be taken as a digression, its function must be to express some qualifications and precautions against a possible mathematicist misinterpretation of his use of the mathematical approach in his critique of the extrusion objection. The complex and sophisticated nature of the content of these methodological reflections about the relationship of mathematics and physics constitute the best evidence of his mathematical (but nonmathematicist) methodology.

These reflections may be reconstructed as follows. First, mathematical truths are *about abstract entities* in the sense that they are statements about the necessary consequences of certain definitions and axioms; for example, the proposition that a sphere touches a plane at a single point is about abstract spheres and planes insofar as it is a necessary consequence of the definition of a sphere and the axiom that a straight line is the shortest distance between two points. Second, mathematical truths are also about physical reality, although only *conditionally*; that is, a mathematical proposition is *physically* true if and only if the abstract entities to which it refers happen to exist as material entities in physical reality; for example, the proposition about spheres and planes is physically true in the sense that *if* there happen to be material spheres and planes *then* they touch in only one point. Third, mathematical truths are *applicable* to physical reality because and insofar as material entities instantiate or approximate abstract ones, for when material entities do not approximate one type of abstract entity they are bound to approximate another type; for example, if and to the extent that material spheres touch in more than one point, they instantiate abstract spheres and planes that are imperfect, and for these it is equally true in mathematics that they touch in more than one point. Finally, the real challenge is to find the proper type of abstract entity in terms of which to interpret physical entities; although one can be sure that the latter must correspond to *some* type of abstract entity that can be treated mathematically, one cannot be sure of which one; for example, this may be the difficulty with the relevance of the earlier mathematical analysis of extrusion.

At a more general level of context, we can point out that the *Dialogue* contains many other critiques of anti-Copernican objections, and most of these do not involve mathematics. Indeed, the book is largely qualitative and thus belies the mathematicist interpretation of Galileo's commitment to mathematics. Thus, to attribute mathematicism to him, we must also attribute to him the proverbial inconsistency between theory and practice: his words express mathematicist pronouncements, but his deeds do not live up to them.

However, it is more proper to construe the remark literally rather than to exaggerate it out of proportion; to consider it in context; and to make it correspond to what he did. When all this is done, the remark becomes evidence of a moderate, judicious commitment to the mathematical approach.

There are other important indications of Galileo's judiciousness about mathematics. An especially important one involves the discussion with which the *Dialogue* begins,[24] which is not included in our selections. The first thing he does in the book is to dissociate himself from various abuses and misuses of the mathematical approach that was popular among the followers of Pythagoras; the key practice he wants to reject is the confusion and conflation of the world of numbers with the world of nature. He not only wants to express his commitment to the mathematical approach but also to ensure that this is not misunderstood.

In summary, Galileo was indeed committed to the mathematical approach to the study of physical phenomena, and his commitment was shown not just in words but also in deeds. Accordingly, one of his key goals was the mathematization of motion, whose accomplishment is found in his *Two New Sciences*. However, his mathematical methodology was judicious, moderate, and balanced; it did not go to the mathematicist extremes of wanting to use nothing but quantitative considerations, or using them when qualitative considerations were more appropriate, or confusing the world of numbers with the world of nature. For example, his contributions to astronomy were largely qualitative and nonmathematical. His methodological reflections on the role of mathematics show the same avoidance of one-sidedness and extremes found in his actual use of the mathematical approach. By contrast, the mathematicist interpretation tends to conflate the mathematical with the apriorist approach, to exaggerate and universalize his assertions beyond the literal meaning of the text, to take his remarks out of context, and to focus one-sidedly on his undisputed mathematical analyses to the exclusion of other ways of proceeding.

2.6. Other methodological problems and principles are discussed in the *Dialogue* besides those reconstructed so far, namely, besides the problem of misology and value of critical reasoning, the problem of authority and importance of independent-mindedness, the problem of the relationship of observation to the-

24. The discussion can be found in Favaro (7:33–38) or Galileo (1967, 9–14).

ory and need of judiciously combining them, and the problem of the role of mathematics and avoidance of mathematicism. Those other topics are intrinsically less important, less extensively discussed in the book, or less well represented in our selections. Thus, we need not examine them in equal depth or detail. But it is useful to give a brief overview of them.

Although Galileo's preface (selection 1) cannot be taken at face value, it does raise the issue of the nature and role of hypotheses. This methodological problem overlaps with that of the theory-observation relationship, insofar as a hypothesis may be construed as a theoretical principle. It also overlaps with the problem of the role of mathematics, insofar as a hypothesis may be conceived as an instrument of mathematical calculation. But a large part of the problem is what is or should be meant by *hypothesis*. Is a hypothesis a description of physical reality, capable of being true or false, although we may not yet know which it is? This is what fallibilist realism holds. Or is it merely an instrument of calculation, capable of being more or less useful but neither true nor false, and capable of yielding testable descriptions of physical reality without itself being a description? This is what instrumentalism claims. Or is a hypothesis a description of physical reality that refers to processes that cannot be directly observed and functions primarily as an instrument of prediction of phenomena that can be directly observed? On this last conception, the issue becomes whether the impossibility of observation is temporary and accidental or permanent and essential. Obviously the question of what role hypotheses have in the quest for knowledge depends on what we mean by a hypothesis. As mentioned earlier, this problem arose partly because the earth's motion could not be directly observed (and indeed seemed to be physically impossible), and yet by assuming it one could explain many observable phenomena and calculate predictions that could be verified. The problem was also part of the methodological background to the *Dialogue* because to regard the earth's motion as a hypothesis, instrumentalistically understood, was one compromise position in the theological controversy in the Galileo Affair. The problem also gets explicitly discussed in at least one other passage in his book,[25] not included here. In general, it is clear that he rejects instrumentalism and accepts some version of epistemological realism; but it is more difficult to determine exactly what kind of realist he was. I think he was a realist in the fallibilist sense defined here rather than in a demonstrativist sense; that is, his realism holds not that scientific hypotheses must always be demonstrably true, but rather that they need be only probably true.

Selection 4 discusses various topics that should be appreciated, reconstructed, analyzed, and assessed on their own individual merits. But there are two main clusters, centering on the comparisons between the earth and the moon and between the human and the divine mind. Both discussions are elaborations of the principle of epistemological modesty; by epistemological modesty I mean an attitude which recognizes the value of the human mind and senses in providing true and useful knowledge about the world, but which also recognizes that such knowledge as we may acquire is at best limited, incomplete, revisable, and infinitesimally small compared to the knowledge we lack. Epistemological

25. Favaro (7:368–72); Galileo (1967, 340–45).

modesty represents a balanced mean between skepticism and dogmatism, and is thus an instance of judiciousness. To substantiate this interpretation, we would have to see that in this passage Galileo is claiming the following: that the moon spots visible with the naked eye may be explained by a number of causes besides the existence of lunar oceans, and so we are in no position to say that they are definitely caused by oceans; that differences between the moon and the earth involve the length of the night-and-day cycle, the existence of significant seasons, the range of temperature, and the existence of clouds and rain, and hence the analogy between the earth and the moon (elaborated in the preceding passage, omitted here) should not be carried too far; that the powers of human understanding are real, important, and undeniable, and they may be illustrated by the fact that in mathematical reasoning we can achieve an objective certainty equal to that of God; but that the limitations of human knowledge are equally real, important, and undeniable, and they may be illustrated by the fact that even in mathematics we have to follow a slow step-by-step reasoning process and we can discover only a few of the infinitely many truths possible.

The methodological principle of simplicity is discussed in selection 6, which I reconstructed earlier (appendix 1.6) as an example of critical reasoning in the sense of self-reflective argumentation. We saw that in the simplicity argument for terrestrial rotation, the principle appears as the major premise in the subargument that the geokinetic explanation of apparent diurnal motion is better than the geostatic explanation because the former is simpler than the latter; we also saw that there were questions about exactly how the principle should be formulated and justified. Other passages in the *Dialogue* (omitted from these selections)[26] also indicate that the principle of simplicity is far from simple; that is, they give further evidence for the complexity of the concept of simplicity, so to speak. I shall add only the following point about Galileo's notion of simplicity: on the one hand, when two theories are otherwise equivalent, if one is simpler than the other, then that is a methodological advantage in favor of the first; on the other hand, the greater simplicity of one theory than another is no guarantee that the former is true. That is, we must avoid saying both that simplicity is epistemologically irrelevant and that it is decisive.

At the end of selection 11 ([365]), Galileo begins a discussion of some details of the methodology of observation which he continues intermittently in selections 13 ([387–91]) and 14 ([414–16]). The issues pertain to such questions as the nature of the process of vision, the interpretation of the human eye as a physical instrument, the interaction between the eye and the telescope, and the difficulties faced and precautions necessary when the eye and telescope are used to make very small observational discriminations. These reflections provide clear evidence of his powers for keen observation, skillful building and use of instruments, and self-reflective awareness. They repay careful rereading and deserve further analysis and evaluation. But that is beyond the scope of the present work.

Selection 13 ([393–99]) also contains a methodological discussion of teleology and anthropocentrism. The problem of teleology is the question of the role,

26. See, for example, Favaro (7:416–25) and Galileo (1967, 389–99); and cf. Finocchiaro (1980), 133–34.

if any, of considerations about the purpose of natural phenomena; whether, for example, something can be ruled out of existence if no purpose can be found for it. By anthropocentrism is meant the view that mankind is the center of everything that exists, where the notion of center may have a physical, biological, epistemological, spiritual, or other type of meaning. Combining teleology and anthropocentrism yields the view that we can discover the truth about natural phenomena by speculating about their purpose defined in terms of human interests. This was a common way of thinking before the Scientific Revolution. An example would be to argue that the universe cannot be as large as required by Copernicanism because then the space between the outermost planet and the fixed stars would be superfluous and useless. Galileo objected to this manner of thinking, but he was careful not to go to the opposite extreme of completely denying all elements of subjectivity and relativity. He held that the concept of size is indeed subjective and relative in several ways, but not anthropocentric or teleological. For, on the one hand, normally the observation of the apparent size of a fixed star is disturbed by the way the human eye works; further, when we imagine very large sizes, we cannot clearly distinguish them from one another; and in general the criteria for "large" and "small" sizes are relational. On the other hand, it is questionable to think that the size of the universe or the distance of the fixed stars must have some purpose, or that their purpose is to serve some human interest, or that such a human purpose must be knowable to us.

Selection 12 refers to the crucial problem of the relationship between biblical interpretation and physical investigation. Because of the restriction under which Galileo was writing, it cannot be taken at face value. His real views on the subject are rather found in the "Letter to Castelli" and the "Letter to the Grand Duchess Christina."[27] The only thing I want to stress here is that there was a genuine methodological problem. Today the problem may have been solved and the controversy may have ended (although this may not be entirely true if one pays attention to the recent controversy in the United States over the status of evolutionary theory and the viability of scientific creationism). However, for Galileo the problem of the role of the Bible could be regarded as parallel to the methodological problems of the role of critical reasoning, appeal to authority, observation, and mathematics.

Analogous remarks apply to the book's ending (selection 16). Its content cannot be taken at face value because it presents without criticism the favorite objection of Pope Urban VIII, and Galileo had been ordered to include it in a prominent place in the book. But recall that this objection begins with the premise of divine omnipotence and reaches a conclusion about the epistemological status of Copernicanism. Thus, its analysis and evaluation would raise important methodological issues and yield useful lessons relating to fallibilism, realism, hypotheses, the relation between the human and the divine mind, and the certainly and applicability of mathematics; and Galileo explicitly discusses these in other passages.[28]

2.7. To recapitulate, aside from other readings, the *Dialogue* may be read from the point of view of methodological or epistemological reflection; by method-

27. See Finocchiaro (1989, 49–54 and 87–118) and cf. Finocchiaro (1986).

28. For more details, see Finocchiaro (1980, 8–12; 1985), Margolis (1991), Wisan (1984b), the introduction above, and the relevant notes in selection 16.

ological or epistemological reflection is meant the identification, description, analysis, evaluation, and application of the aims, procedures, principles, and presuppositions of inquiry, the search for truth, or the quest for knowledge. The book is in fact full of such methodological reflections about issues which were important in the Copernican Revolution and have some relevance even today. It is best to read such reflections in a contextual manner, without any commitment to any particular systematic methodology. If one wants to attempt a systematic reconstruction of sorts, two interrelated possibilities appear promising. One is to interpret his various reflections in terms of critical reasoning: principles about its nature and value; about its relationship to such things as unreason, authority, sense experience, and mathematics; and about inferring conclusions from premises in accordance with considerations of simplicity, teleology, and anthropocentrism. The other is to do it in terms of the notion of judgment or judiciousness, by which I mean the avoidance of one-sidedness and of extremes; for this is a quality of mind which constantly recurs in the book, whether the issue is the problem of misology, authority, observation versus theory, the application of mathematics, the extent of human knowledge, the import of simplicity, or the relativity and subjectivity of the concept of size.

Finally, it should now be clearer how methodological reflection is distinct from, but related to, critical reasoning. Methodological reflection focuses on the formulation, analysis, evaluation, and application of truth-seeking or knowledge-gathering procedures and principles; whereas critical reasoning focuses on the formulation, analysis, and evaluation of arguments. Methodological reflection can take the form of reasoning about methodological principles, in which case it becomes critical reasoning pure and simple. And critical reasoning can sometimes be about methodological principles, in which case it is a form of methodological reflection. I have already suggested that both be subsumed under critical thinking; as long as such association does not lead to confusion, this semantical move is fruitful.

## 3. Varieties of Rhetoric

3.1. Rhetoric is a term that is more ambiguous than most, and so discussions of the role of rhetoric in scientific inquiry tend to focus on many disparate things.[29] Few would deny that rhetoric encompasses such human activities

29. This may be seen from the variety of approaches and concerns in recent rhetorical studies. For example, Feyerabend (1975) and Hill (1984) seem to focus on what they perceive as the art of deception and of making the weaker argument appear stronger. A historian of rhetoric extracts from the *Dialogue* what he calls an "epideictic" rhetoric (Vickers 1983); this seems to be primarily a verbal redescription of elements which I describe differently; but to say this is to say that this author engages in a rhetoric of epideictic rhetoric, whereas I tend to engage in a rhetoric of "critical reasoning" and "scientific methodology." Other dimensions of Galileo's rhetoric are examined by Moss (1993). Aspects of Kepler's rhetoric in the sense of literary structure are examined by Jardine (1984, 72–79), who shows that Kepler's "Defense of Tycho" has the structure of a judicial oration in the style of Cicero. Other valuable general works are Gross (1990), Pera (1994), and Prelli (1989).

as speech communication, persuasive argumentation, written composition, verbal eloquence, emotive language, beautiful language, imaginative description, bare assertion, nuanced assertion, repetition, wit, satire, humor, and ridicule. Here the focus will be on three of these: persuasive argumentation, eloquent expression, and the projection or communication of images or impressions. I plan to use various passages from the *Dialogue* to help us understand what these rhetorical concepts mean; and I plan to use these concepts to convey a sense of the possibility of appreciating the book from a viewpoint which is different from that of its historical connections to the Copernican Revolution, the Scientific Revolution, and the Galileo Affair, and is also different from the philosophical viewpoints of critical reasoning and methodological reflection. The rhetorical standpoint is of course *related* to all these others, just as they are related to one another; but such relationships do not make them the same and cannot be defined without presupposing their distinction.

Scientists tend to have an aversion to rhetoric, for there is a tradition of anti-rhetoric in science. On this point, as for so many other issues about the nature of scientific inquiry, the *Dialogue* provides classic illustrations. For example, in the middle of the passage on heavenly changes (selection 3), there is a discussion of the nature of sunspots where the issue is whether they are part of the body of the sun like clouds on the earth, or are swarms of planets revolving around the sun and partially obscuring it when a number of them get in our line of sight. The latter was the Aristotelian interpretation, which retained the doctrine of the unchangeability of the heavenly bodies, whereas the former was Galileo's (correct) view. Toward the end of this discussion, Simplicio expresses his confidence that even if the planetary interpretation should turn out to be incorrect, others will be able to come up with a theory compatible with heavenly unchangeability. To this Salviati replies by making a contrast between science and rhetoric (see first paragraph of his speech in selection 3, [78]). And in the margin of the page Galileo writes the postil that "in the natural sciences the art of oratory is ineffective."[30] Although he does not use the word *rhetoric* here, the connection with rhetoric is unmistakable because of his reference to Demosthenes, his talk of "the art of oratory," and the fact that a contemporary critic (Antonio Rocco, about whom more below) explicitly used the term "rhetoric" in his comments.[31]

Such a methodological pronouncement cannot be taken at face value. As discussed earlier in this appendix (part 2.1), practicing scientists are not always the best interpreters of their own activities, and their words and deeds in regard to scientific method do not always correspond. This does not mean that their words and deed *never* correspond; indeed, Galileo's *usually* do, as I have argued. But in this particular case they do not, and we have an exception to the usual Galilean correspondence. However, these are not the points I wish to stress here.

30. Favaro (7:78); cf. Galileo (1967, 54). I have omitted such postils from my selections because most of the time they would just distract the reader.

31. Rocco wrote a book against the *Dialogue* entitled *Esercitazioni filosofiche* (meaning *Philosophical Exercises*), now reprinted in Favaro (7:569–750); this particular criticism occurs on pp. 628–29.

Rather, the main purpose of referring to this passage is to introduce a number of relevant meanings of rhetoric.

One of these is a meaning which Galileo does *not* explicitly have in mind but implicitly uses, and which corresponds to a common present-day conception. That is, in making the anti-rhetorical pronouncement just mentioned and in other passages where he expresses a similar sentiment, he himself uses a kind of rhetoric: he attempts to convey or communicate a particular message or impression. That the *content* of this message is "anti-rhetorical" is not relevant at the moment. The point is that there are many occasions when scientists want to project a certain image; and it is important to recognize that they are engaged in such image projection, and that such an image does not necessarily correspond to what they are actually doing, although the fact that they cherish such an image may itself be important.

This is the sense of rhetoric in which one speaks of "campaign rhetoric" in politics, when candidates for office make statements meant to create certain impressions in the minds of voters. This sense of rhetoric also corresponds to the domain of public relations, in which an organization tries to create a favorable impression with the public. Be that as it may, my main point is that in the passage mentioned above Galileo expresses a rhetoric of anti-rhetoric, so to speak, and that this is a typical attitude in science.

However, if the rhetoric of anti-rhetoric is to avoid being a contradiction in terms, the senses of the two occurrences of "rhetoric" must be different. The rhetoric that is explicitly criticized by this type of rhetoric must be different from the rhetoric that expresses the criticism. In fact, part of what Galileo is against in this passage is oratory such as that in which Demosthenes excelled, namely, the art of eloquent expression. This is a second common meaning of the term, and one problem is whether the rhetoric of verbal eloquence has no function in scientific inquiry.

This passage also suggests a third sense of rhetoric. For besides expressing misgivings about the scientific effectiveness of "verbal fluency" and Demosthenes' talent, it expresses misgivings about intellectual subtlety of the type in which Aristotle excelled. It is not completely clear what Galileo refers to, but let us take this to refer to persuasive argumentation.

Thus, there are at least three senses of rhetoric relevant to the question of the role of rhetoric in science. One is only implicitly used but not explicitly mentioned in this passage, while the other two are explicitly mentioned. They pertain, respectively, to the art of effective communication, the art of eloquent expression, and the art of persuasive argumentation.[32]

3.2. To elaborate, let us begin with the first sense of rhetoric: the projection of images, production of impressions, or communication of messages. Far from

32. These seem to correspond to the three present-day disciplines of communication studies, English composition, and philosophical rhetoric in the sense of the "new rhetoric" (Perelman and Olbrechts-Tyteca 1969). They also correspond to the three senses articulated, in a different order, in the first three chapters of Finocchiaro (1980), the rhetoric of effective communication corresponding to chapter 1, the rhetoric of persuasive argumentation to chapter 2, and the rhetoric of eloquent expression to chapter 3.

having no role in scientific inquiry, this rhetoric is unavoidable because science is a human and social enterprise done by human beings and addressed to other human beings. Even if science were completely independent of the knowing subject, there would be a need for a corresponding rhetoric, namely, for projecting the image that images are irrelevant, so to speak.

However, the important point regarding this type of rhetoric is not its pervasiveness, but that we must learn to see beneath this rhetoric to appreciate the more important things that go on. The reason stems from the very definition of this rhetoric: it involves appearances, impressions, images, and messages sent (in short, public relations) in contrast to reality, substance, actuality, and messages received; indeed, calling things rhetoric in this sense is a way of diminishing their reality or importance.

On the other hand, to say that such rhetoric is *relatively* unimportant is not to say that it is *completely* unimportant. It has a purpose, for even appearances have their own reality. The error is to take it at face value or to mistake appearance for reality. Let us look at more examples.

The effective content of the *Dialogue* is, as we have seen, a justification of the geokinetic hypothesis (advanced by Copernicus a century earlier). By justification I mean an argument showing that the geokinetic theory is more likely to be true than the geostatic view, and that the former should be accepted in preference to the latter. The justification proceeds by criticizing all pro-geostatic and anti-Copernican arguments as incorrect and flawed for one reason or another, and by favorably presenting several pro-Copernican and anti-geostatic arguments of various degrees of strength.

Despite this content, the book also contains a great deal of rhetoric trying to convey different impressions. To understand this rhetoric, it is important to recall the book's historical background. It was written at a time when, as a Catholic, Galileo was bound by a decree issued by the Church in 1616 to the effect that one could neither hold nor support nor defend the physical truth of Copernicanism; and perhaps he was also bound by more stringent restrictions about the kind of discussion he could engage in.

One of these rhetorical impressions is conveyed by the long title, which translates as follows: *Dialogue by Galileo Galilei, Lincean Academician, Extraordinary Mathematician at the University of Pisa, and Philosopher and Chief Mathematician to the Most Serene Grand Duke of Tuscany; where in meetings over the course of four days one discusses the two Chief World Systems, Ptolemaic and Copernican, Proposing indeterminately the Philosophical and Natural reasons for the one as well as for the other side.*[33] The message here is that the book merely *discusses* all the arguments for and against the two views, without defending either one. This is reinforced many times by appropriate reminders interspersed in the body of the work.

The point of this rhetoric is clear—Galileo wanted to make sure he was not seen as violating the anti-Copernican decree of 1616. This rhetoric did not work; it was seen as mere rhetoric because the book actually criticized geostatic arguments and favorably portrayed geokinetic ones. That is, appearances to the contrary, the book

33. Galileo (1632); cf. Salusbury (1661) and Santillana (1953, liii, n.1).

amounted to a defense of Copernicanism. The ineffectiveness of this rhetoric was a factor leading to the trial of 1633.

Another rhetorical impression is conveyed by the preface (selection 1) together with a number of related passages, such as those on the role of the Bible (selection 12) and on divine omnipotence at the end of the book (selection 16). The preface tries to give the appearance that the book is a work of religious apologetics. It makes it seem as if the book's aim is to show the world that Catholics know all relevant scientific arguments; that they are aware that the Copernican arguments are better; that they hold that these arguments prove only that Copernicanism is more useful mathematically (not more likely to be true) than the geostatic view; and therefore that the decree of 1616 was the result not of scientific ignorance but of religious motivations. One of these was awareness of divine omnipotence, and so the book's ending states the favorite objection of Pope Urban VIII.

However, such rhetoric was no more effective than the rhetoric of the title. The difficulty was that although the scientific arguments admittedly did not establish the physical truth of Copernicanism with certainty, they did show not only that it was more mathematically convenient, but also that it was more likely to be physically true than the geostatic view; that is, Galileo defended not merely the mathematical usefulness but also the physical truth of Copernicanism. Hence, though perhaps the decree of 1616 did not result from ignorance, he violated it and implicitly criticized it as erroneous.

A third rhetorical impression derives from a few scattered remarks Galileo makes to the effect that what he is doing in regard to the earth's motion is giving a strict demonstration or rigorous proof or conclusive argument, and to the effect that this is what is generally required in scientific inquiry. An example of this is Salviati's opening paragraph in his presentation of Galileo's explanation of the tides (selection 15, [450]). Analogous remarks are expressed at the beginning and at end of the discussion of the argument for the earth's annual motion based on the explanation of the apparent motion of sunspots,[34] although this is not included in our selections.

This rhetoric derives partly from the Aristotelian ideal of science as demonstration, which Galileo inherited from his cultural milieu[35] but was in the process of revising. The rhetoric of demonstration may be partly an attempt to evade the anti-Copernican decree of 1616, insofar as a rigorous proof of a conclusion (if it is really rigorous and really a proof) is not really a defense of the conclusion, but only an exhibition of mathematical and logical relations. Moreover, such rhetoric is not always expressed by Galileo's main spokesman (Salviati); for example, it is Sagredo who at the end of the argument from the explanation of sunspot motion evaluates it as demonstratively necessary; and in the passage explicitly discussing critical reasoning (selection 7), it is Simplicio who expresses the demonstrativist ideal; thus, if the dialogical aspects of the discussion are taken into account, the rhetorical appearances may not correspond to

34. Favaro (7:372 and 383); Galileo (1967, 345 and 356).

35. The depth of this Aristotelian legacy has been demonstrated anew by Wallace (1977; 1981a; 1984; 1992a; 1992b).

Galileo's real intentions. Even when the speaker is Salviati, the rhetorical appearance may not survive critical analysis; for example, at the beginning of the exposition of the tidal theory (selection 15, [450]), the alleged necessity refers more likely to the *relationship* between the geokinetic hypothesis and the tidal effects than to this hypothesis per se. Finally, the rhetoric of strict demonstration is in part rhetorical exaggeration by Galileo, when he was advancing an argument which he felt to be especially strong.

At any rate, such rhetoric failed because Galileo's arguments for the earth's motion, however ingenious and strong, do not in fact amount to a strict demonstration, and this was readily perceivable. They are not as weak or fallacious as his contemporary enemies and modern critics claim,[36] but they are not demonstrative proofs. One important rhetorical incoherence is that, as suggested in the introduction (5.3), if any of his arguments were really conclusive (even to him), he would not have given so many of them.

3.3. Another especially relevant example of this type of rhetoric, which I call image projection and which involves public relations, will pave the way for the examination of the other two types of rhetoric. The example is the rhetoric of anti-rhetoric mentioned above for another purpose, namely, to introduce the distinction among these three types. We saw that that initial passage (selection 3, [78]) contains rhetoric critical of the arts of eloquent expression and persuasive argumentation. The purpose now is to determine the effectiveness of this rhetoric of anti-rhetoric.

We can first analyze more deeply the internal coherence of the impression projected by that passage, to determine whether any message gets sent after all. One question we could raise concerns what is being contrasted to Demosthenes' eloquent expression and Aristotle's persuasive argumentation. Presumably it is necessary demonstration. But if so, then (using Galileo's own words) "intellectual subtlety" is often required to discover or grasp it, and so the "average intellect" is unlikely to be "fortunate" enough to do it.

It is also useful to examine how other readers of the book perceived such rhetoric, even though their perceptions need not be accepted uncritically and may have their own difficulties. They were not impressed. The response of one reader is especially relevant because it contains a good statement of the essential argument for the need of some rhetoric in scientific inquiry (though also some exaggeration). The respondent was Antonio Rocco, who wrote a whole book against Galileo's *Dialogue*. Addressing Galileo, Rocco says:

> To support your views, you say that, since this dispute deals not with some point of law or other human studies, but with conclusions which are about nature and are necessary, there is no place for the human will, for intellectual subtlety, and so on. And I say that in every controversy there is only one truth, and that because the present one deals with natural phenomena which are very remote in a thousand ways from us and our knowledge, its resolution is more uncertain and complex than is the case for the enigmas of the Theban sphinx; therefore, to assert something about it is more akin to guessing than to philosophizing,

36. The nonconclusiveness and nondemonstrative character of Galileo's arguments has encouraged some scholars to advance the claim that they were mostly unsound and sophistical. For a discussion of this issue, and a criticism of this claim, see Finocchiaro (1986).

> except for a few very obvious things (such as that the heavens are several and visible, that the stars are luminous, that the sun is most luminous, and so on), and except for some generalities (which can be known with probability, precisely in the manner that Aristotle advances them). Indeed, in regard to the more difficult subjects, whoever has a better intellect can describe the heavens as he wishes, and no one can show him evidence to the contrary. Thus, at a serious conference of scholars I have heard a gentleman who, as a joke, started to defend the view that the heavens are made of milk; and, because of his subtle intellect, he did it very well and even answered very strong objections without falling into any major absurdity or contradiction. Further, in the domain of law and human actions (which derive from causes which are finite, connected with us, and dependent on us), it happens that, despite the most eloquent speaker, once our emotions have calmed down, not only do we arrive at the truth, but anyone can make a resolute determination of it. And who in everyday life does not know, more or less, how to distinguish truth from falsehood, once he has heard the reasons of both sides of a civil controversy? On the other hand, who among the innumerable series of intelligent men has been able to determine anything for certain about the recondite aspects of the heavens? And if that had been done, why would there be so many controversies? Now, I do not deny that there is a necessary truth in such controversies, but I say there is no man who knows it; nor is it enough that it should be knowable, for even the supreme God is supremely knowable, but almost unknown by us. Our poor mind is more blind for the understanding of the nature of things than are the eyes of a bat for seeing the rays of the sun. But finally, if the truth is necessary and demonstrably so, as you say, produce the evidence, give the reasons and causes, abandon persuasion in the style of the rhetoricians, and no one will contradict you.[37]

In fairness to Galileo, I would only add that in selection 3, immediately after the passage to which Rocco objects, Galileo does indeed "produce the evidence, give the reasons and causes, [and] abandon persuasion in the style of the rhetoricians." He gives a conclusive demonstration that the sunspots are on the solar surface based on the fact that they appear to be wider and to go faster near the center of the solar disk than near the edges. This is no proof that the earth moves, but it does destroy the earth-heaven dichotomy.

We can also examine how the message in that initial Galilean passage relates to others. For example, since the passage contrasts the rhetoric being rejected with necessary demonstration, it can be taken as part of the book's rhetoric of strict demonstration, previously mentioned. If carried out systematically, this type of comparative analysis would bring us close to identifying one element of the actual content of the work, in regard to the methodology of scientific inquiry. The result would be, as we have seen, that the *Dialogue* advocates primarily the importance of critical reasoning and a judicious use of mathematics rather than demonstrative proof and the mathematicist conflation of the world of numbers with the world of nature.

The anti-rhetorical passage initially mentioned is not the only example of an attempt to project an anti-rhetorical image. There are two others. One of them occurs at the beginning of the book,[38] in a passage omitted from our selections but already mentioned for another purpose earlier (appendix 2.5).

The *Dialogue* begins by discussing whether and why the universe is perfect and three-dimensional and by criticizing Aristotle's relevant views. Aristotle is

37. Favaro (7:628–29).
38. Favaro (7:33–38); Galileo (1967, 9–14).

said to argue that the universe is perfect because it has three spatial dimensions and three is a perfect number due to these reasons: three is the number of parts which every complete thing has (namely, beginning, middle, and end); three is the number used in making sacrifices to the gods; and three is the minimum number of things required before the word "all" can be used to refer to them collectively. Salviati reacts to this by saying:

> To be frank, in all these considerations I do not feel compelled to grant anything other than that whatever has a beginning, a middle, and an end may and should be called perfect; but I feel no inclination to grant that, because beginning, middle, and end are three in number, therefore the number three is a perfect number and has the faculty of conferring perfection to whatever has it. For example, I do not understand how or believe that in regard to number of legs three is more perfect than four or two; nor is it the case that the number four is an imperfection for the elements, and that it would be a greater perfection if they were three in number. Therefore, it would have been better to leave these subtleties to the rhetoricians and prove his conclusions with a necessary demonstration, for this is the appropriate thing to do in the demonstrative sciences.[39]

Here Galileo seems to view rhetoricians as traffickers in silly arguments like Aristotle's about the perfection of the number three, and as having no serious business to transact in science, or at least in "demonstrative" science.

Notice, however, that he expresses this anti-rhetorical sentiment at the end of a brilliant piece of rhetoric on his part, namely, a convincing refutation of Aristotle's argument. Galileo is therefore practicing rhetoric in the sense of persuasive argumentation; his key point is that Aristotle's argument is unpersuasive and dependent on a silly assumption.

It could be said therefore that Galileo does not practice what he preaches, or, more precisely, that he practices the opposite of what he preaches. We would then face the problem of whether we should derive a lesson from his words or his deeds. Because his rhetorical practice is so prevalent and his anti-rhetorical pronouncements so few, the choice would be easy.

While this would solve the relevant difficulties, there may be a better way of handling the situation. Galileo's anti-rhetorical pronouncement is inconsistent with his rhetorical practice only if that pronouncement is unduly universalized into the rule that persuasive argumentation is *never* appropriate in science. But perhaps he is saying merely that persuasive argumentation is better avoided or is of secondary importance; this is consistent with it being *occasionally* appropriate. One of these occasions is when another scientist attempts some persuasive argumentation that turns out to be incorrect for some reason. Then, like Galileo replying to Aristotle's argument about the number three, we may have to practice rhetoric to deal with rhetoric.

This account is perhaps the most rhetorically favorable one for this passage. But from an evaluative standpoint we may want to argue that we are still not giving persuasive argumentation its due. For there is much truth in computer scientist Joseph Weizenbaum's claim that "scientific demonstrations, even mathematical proofs, are fundamentally acts of persuasion."[40]

39. Favaro (7:35); cf. Galileo (1967, 11).
40. Weizenbaum (1976, 15).

Be that as it may, the lesson emerging from this passage is that rhetoric in the sense of persuasive argumentation has *some* role to play in science. One of these roles is to answer rhetoric with rhetoric. But this is not the only role. A long time ago Thomas Kuhn argued that persuasion is an essential activity when a scientist is faced with paradigm choice, that is, with having to choose between one worldview and a radically different one.[41] Of course, this was exactly Galileo's predicament. So we find his whole book to be essentially a long and complex persuasive argument designed to justify the superiority of the geokinetic view to the geostatic one.

3.4. A similar tension between anti-rhetorical words and rhetorical deeds exists within the only other anti-rhetorical passage in the *Dialogue*. That passage is even more valuable as an illustration of eloquent expression. The context is as follows.

As we have seen, the Copernican Revolution involved not only the issues of the motion and location of the earth and sun, but also the rejection of the doctrine that there is an essential difference between heavenly and terrestrial bodies, the so-called earth-heaven dichotomy. According to this doctrine, heavenly and terrestrial bodies had many different physical properties: heavenly bodies were made of a weightless substance called aether, whereas terrestrial bodies were made of the elements earth, water, air, and fire; heavenly bodies were luminous, but terrestrial ones were not; unlike terrestrial bodies, heavenly bodies did not undergo any chemical or physical changes (except regular circular motion). Moreover, and this is the important point in the present context, heavenly bodies had a purity, perfection, and nobility which terrestrial bodies did not possess.

One version of the anti-Copernican objection from the earth-heaven dichotomy was that the earth cannot revolve around the sun because if it did it would do so in the third planetary orbit (between the smaller orbit of Venus and the larger one of Mars); if the latter were true, the earth would be a heavenly body, and so there would be no separation between those bodies which are unchangeable, pure, and perfect, and those which are the opposite.

As we have seen, Galileo had many criticisms of this argument. One was that it was groundless because the doctrine of heavenly unchangeability is false, as the evidence of sunspots shows. This was a type of scientific criticism. But the existence and interpretation of this evidence were controversial, and so argumentation was needed to decide such issues as the epistemological legitimacy and empirical reliability of the telescope (from which that evidence stemmed) and the location of the spots. Whenever long and complex argumentation is needed, questions of persuasiveness become important. That brings us to the realm of rhetoric (in one sense of the term).

Another Galilean criticism was that this anti-Copernican argument begs the question.[42] For it bases the conclusion that the earth's planetary revolution is impossible on the premise that the earth and the heavenly bodies are essentially dif-

41. Kuhn (1970, 152). Works which are less explicitly rhetorically oriented, but which have definite implications in this direction are Brown (1977; 1988), Margolis (1987), and Shapere (1984).

42. Favaro (7:63–65); Galileo (1967, 38–41).

ferent. If we examine how the anti-Copernicans justified this dichotomy, we see that their primary supporting argument, and the one to which others are ultimately reduced, was based on the difference of natural motion between terrestrial and heavenly bodies; according to Aristotelian physics, the natural motion of terrestrial bodies is straight up and down (away from and toward the center of the universe), whereas the natural motion of heavenly bodies is circular (around the center of the universe). But this law of natural motion is another way of stating the conclusion that is at issue.

This second criticism is a good example of critical reasoning; the critique is logical insofar as it examines inferential relationships among propositions. But there is an irreducibly rhetorical dimension here because the difficulty with begging the question is not that the conclusion does not follow from the premises, but that it follows too well, for it *is* one of the premises. Thus, the failure of begging the question is in the realm of persuasion: we obviously cannot be persuaded to accept a claim by means of an argument which requires that we accept that same claim. This is why, in my classification of types of criticism above (appendix 1.3), I categorized question-begging as a case of "persuasive disconnection."

Besides this scientific-rhetorical and this logical-rhetorical criticism, Galileo advances the following more strictly rhetorical objection:

> Further, of the emptiness of such rhetorical conclusions, we have spoken many times. Is there anything more foolish than saying that the Earth and terrestrial elements are relegated and separated from the heavenly spheres, but confined inside the lunar orb? Is not the lunar orb a heavenly sphere and, as they themselves agree, located in the middle of all the others? What a way to separate the pure from the impure and the sick from the healthy—to give those who are infected room at the heart of the city! And I thought that the pesthouse should be located as far away as possible! Copernicus admires the arrangement of the parts of the universe because God placed the great lamp, which was to give the most light everywhere in his temple, at its center and not on one side. . . . But, please, let us not confuse these rhetorical flowers with solid demonstrations, and let us leave them to the orators, or rather to the poets, who with their pleasantries know how to praise highly things that are very vile and even pernicious.[43]

This passage begins and ends with the type of anti-rhetorical expressions that I call the rhetoric of anti-rhetoric. But for the most part it elaborates a type of rhetorical criticism of the anti-Copernican position. It makes fun of the idea of

43. Favaro (7:292–93); cf. Galileo (1967, 268–69). It is interesting and ironical that Galileo himself had the reputation of being very talented at this sort of thing. For example, during the trial of 1633, the Tuscan ambassador to the Holy See reported an interesting conversation with the pope's nephew Cardinal Francesco Barberini in which the latter had said he was very warmly disposed toward Galileo (which is true) but nevertheless found the *Dialogue* objectionable because of its "reporting much more validly what favors the side of the earth's motion than what can be adduced for the other side. I [the ambassador] said that perhaps the nature of the situation indicated this, and therefore he [Galileo] was not to blame; but his Eminence answered that I was aware that *he knew how to express exquisitely and how to justify wonderfully whatever he wanted*" (Finocchiaro 1989, 246, italics added).

separating the pure from the impure by placing the impure in the middle, surrounded by the pure. I find the criticism extremely effective, and I would categorize it as primarily eloquent expression.

Furthermore, I see nothing methodologically wrong with employing this type of rhetoric the way Galileo does here; that is, with using it in addition to empirical considerations, critical reasoning, and persuasive argumentation. Of course, it may be employed improperly if, for example, it were used as a substitute for persuasive argumentation, critical reasoning, and empirical observation; but then the problem would be not with eloquent expression per se, but with the impropriety stemming from elsewhere. Other techniques, methods, and procedures are also liable to potential misuse or abuse.

Consider another example of eloquent expression. In Galileo's time, the phenomenon of the tides continued to puzzle natural philosophers, and no adequate explanation was available. In selection 15, he updates and elaborates a theory that explains them in terms of the earth's motion, thus also providing what he felt to be one of his best arguments for Copernicanism. The passage begins by briefly criticizing alternative explanations. One of these is that lunar heat increases the temperature of sea water and causes it to expand and thus to rise. One of his criticisms is the empirical one of inviting us to test the temperature of water at high and low tides, to see that there is no difference. Then he adds, referring to proponents of this theory, to "tell them to start a fire under a boiler full of water and keep their right hand in it until the water rises by a single inch due to the heat, and then to take it out and write about the swelling of the sea" (p. [446]).

The role of eloquent expression is not solely or primarily critical or destructive. It is even more useful for constructive purposes. One of the best examples of constructive eloquence is the statement of the basic argument in favor of the heliocentrism of planetary motion (selection 11]).[44]

As a final example of eloquence, it is instructive to examine a passage whose purpose is different from those considered so far. At one point in the discussion of heliocentrism Galileo hurls an insult to some opponents by calling them "men whose definition contains only the genus but lacks the difference" (selection 11, [355]). Here Galileo is calling some of his opponents simply animals, or perhaps irrational animals (see note to the text). Though clever, this is little more than name calling.

Obviously this sort of rhetoric has no place in scientific inquiry, although in the heat of a dispute individual scientists may be unable to resist it. However, it is equally obvious that such rhetoric makes the *Dialogue* a more interesting and readable book. It gives the book aesthetic value and makes it a work of art capable of being appreciated as literature. The difference between this clever insult

44. See also the statement of the explanation of the seasons, not included here (Favaro 7:416–23; Galileo 1967, 389–97); the statement of the explanation of the monthly and annual periods of the tides, also omitted (Favaro 7:470–86; Galileo 1967, 444–61); and the statement of the pro-Copernican argument from analogy based on the luminosity of bodies, found in Galileo's "Reply to Ingoli" (Favaro 5:559–61; Finocchiaro 1989, 196–97). But in the nature of the case, these cannot be summarized without destroying the eloquence of the original or recreating another original instance of eloquent expression.

and the previous examples of eloquent expression lies in their purpose; whereas the image of a pesthouse in midtown and the invitation to burn one's hand were aimed at the criticism of ideas, to call some opponents animals is to insult people. The difference might be used as a basis for a different categorization of such rhetoric. But we need not pursue any more extensively the systematization of such activities.

On the other hand, this introduction to the rhetorical reading of the *Dialogue* (and of scientific works in general) would be incomplete without discussing an important issue suggested by our discussion, namely, the relationship of rhetoric to critical reasoning and to methodological reflection. This issue involves primarily the questions of how rhetorical image projection is related to methodological reflection, and how critical reasoning is related to persuasive argumentation and eloquent expression.

3.5. Consider Galileo's assertion that "in the natural sciences the art of oratory is ineffective";[45] this may be rephrased as the claim that rhetoric is ineffective in natural science and implies that one should not use rhetoric in science. Earlier I regarded this as part of his attempt to project an anti-rhetorical image, of his rhetoric of anti-rhetoric, so to speak. And consider an assertion occurring in one of the above mentioned passages that contain his rhetoric of strict demonstration: "that it is impossible for them to happen otherwise . . . such is the character or mark of true natural phenomena" (selection 15, [450]); this means that the mark of true natural phenomena is their necessary truth, which implies that the aim of physical science is to provide strict demonstrations of natural phenomena. I have already suggested that these two assertions are sides of the same coin because he seems to regard rhetoric and strict demonstration as opposites. The new point to stress now is that these assertions are instances of methodological reflection, for they formulate principles about the nature of scientific knowledge which we may or may not follow in the search for truth.

In this regard, they are like many other methodological remarks in the *Dialogue*. For example, as we have seen, the book contains the assertion that there is a one-to-one correspondence between physical truth and good reasoning on the one hand and physical falsity and bad reasoning on the other (selection 7, [156]). We have also seen the remark that there is nothing "more shameful in a public discussion dealing with demonstrable conclusions than to see someone slyly appear with a textual passage (often written for some different purpose) and use it to shut the mouth of an opponent" (selection 5, [138–39]). There is also his claim that it is fitting that sensory experience should have priority over intellectual theorizing (selection 3, [75]). And there is the claim that "to want to treat physical questions without geometry is to attempt to do the impossible" (selection 9, [229]).

We have also seen that such assertions can be interpreted so as to attribute to Galileo a methodology he does not espouse. The first can be taken as an expression of naive rationalism unappreciative of the importance of critical reasoning. The second can be interpreted to imply a complete rejection of all authority. The third can be construed as a commitment to simpleminded empiricism. The fourth can be regarded as a version of extreme mathematicism.

45. Favaro (7:78); cf. Galileo (1967, 54).

Moreover, I argued that such attributions misinterpret Galileo's methodology. The simpleminded correspondence between physical truth and human reasoning is asserted by Simplicio, is later appropriately qualified by Sagredo, and is contradicted by Galileo's whole procedure. The complete rejection of authority involves taking the quoted remark out of context, thus divorcing it from other remarks about authority and other things Galileo does. The simpleminded empiricism reflects only one side of the methodological situation and unduly neglects his inclinations toward intellectual theorizing. The mathematicist ideal is an exaggeration of the quoted sentence and cannot do justice to the qualitative and nonmathematical side of his procedures.

Whether my own interpretation and my criticism of alternative construals are correct or not, it is important to realize that understanding the methodological aspect of the *Dialogue* is difficult and generates controversy. Earlier, when I discussed the book's methodological reflections, I rejected the alternative interpretations as being one-sided, exaggerated out of proportion, taken out of context, or failing to correspond to Galilean practice as a whole. I did not talk of rhetoric then, as I did later for Galileo's rhetoric of anti-rhetoric and his rhetoric of strict demonstration. Why not? Is there a difference between the earlier cases and these?

To answer this question, note that we could apply to the earlier cases the notion of rhetoric as the projection of impressions. We could say that there is also in the book a rhetoric of naive rationalism (simple correspondence between physical truth and human reasoning), a rhetoric of total rejection of authority, a rhetoric of simple empiricism, and a rhetoric of mathematicism. Using such rhetorical terminology, the key point of my earlier analysis was that none of these rhetorics correspond to reality, namely, the reality of Galileo's methodological practice and the totality of his methodological reflections. For, besides a rhetoric of naive rationalism, there is also a rhetoric and a practice of judicious rational-mindedness and critical reasoning; besides a rhetoric of total antiauthoritarianism, there is a rhetoric and a practice of using authorities as sources of arguments and methodological principles; besides a rhetoric of empiricism, there is a rhetoric of apriorism and a practice of judiciously combining empirical and intellectual procedures; and besides a rhetoric of mathematicism, there is a rhetoric of the pitfalls of applying mathematical truths to physical reality and a practice of a limited use of the mathematical approach.

Similarly, I criticized the rhetoric of anti-rhetoric as internally incoherent and as inconsistent with Galileo's overall practice. I criticized the rhetoric of strict demonstration as also contradicted by his practice; and it can also be criticized as inconsistent with his rhetoric of epistemological modesty (selection 4) and his rhetoric of probabilism (selections 1 and 6).

Thus, there is no intrinsic difference between the cases of methodological reflections examined earlier (appendix 2) and the rhetoric of anti-rhetoric and of strict demonstration examined later. The main difference lies in my evaluation of the corresponding methodological principles and of the corresponding methodological interpretations of Galileo. In my earlier discussion, I focused on passages and issues from which we can derive useful lessons, and in the process I criticized various one-sided readings that were based on corresponding rhetoric on his part. In the present discussion, I have been focusing on examples of methodological reflections that do not lend themselves to deriving useful lessons, in part because the corresponding Galilean pronouncements largely contradict his own practice and so should be regarded as mere rhetoric (in the sense of attempts at public relations).

In fact, there is an overlap between methodological reflection and rhetoric in the sense of image projection. Some rhetoric (when the subject matter is methodological procedures) is simply part of methodological reflection. We could expand the definition of methodological reflection to include the communication of methodological principles, in addition to their formulation, analysis, evaluation, and application. The communication of methodological principles (namely, the projection of methodological images) is not always effective. In some cases, for some reason, the methodological assertions fail to get really projected. The reasons may involve considerations of context, overall balance of textual evidence, correspondence between words and deeds, one-sidedness, and exaggeration. When rhetoric is ineffective, one may speak of "mere" rhetoric. To use such a locution is to express a negative evaluation of the attempt at communication being considered. However, not all rhetoric is mere rhetoric.

3.6. Rhetoric in the sense of image projection raises, as we have just seen, the question of its connection with methodological reflection, given that the projection of methodological images is an area of overlap. Analogously, rhetoric in the sense of persuasive argumentation raises the question of its relationship to critical reasoning since it is even more obvious that argumentation provides an analogous area of overlap.

Recall that reasoning is the process of interrelating thoughts in such a way that some depend on or follow from others; that argument is the special case of reasoning when conclusions are supported by reasons; and that critical reasoning is the special case of reasoning when arguments are analyzed or evaluated. Now, persuasive argumentation may be defined as the special case of argument having the property of persuasiveness or persuasive force.

Persuasiveness is, first, a positive evaluative property, as distinct from a structural or analytical property. That is, to say that an argument is persuasive is to appraise it as good or valuable in some respect; it is not to analyze it into parts and determine how they interrelate so as to provide an understanding of the argument. Second, persuasiveness is a matter of degree rather than an all-or-none affair; it is continuous or gradual, not discrete. That is, arguments are capable of being more or less persuasive.

The persuasive force of an argument refers to the extent to which the audience comes to believe its conclusion on the basis of its reasons. The extent of such audience acceptance depends on at least two factors: the number of persons who did not previously accept the conclusion but do so now as a result of the argument, and the intensification or increase in strength of their belief in the conclusion. That is, an argument is persuasive if and to the extent it increases the acceptance of the conclusion by an audience, namely, the extent to which the argument causes the conclusion to become accepted more widely, more strongly, or both.

As just defined, persuasiveness refers to psychological and sociological phenomena about people's beliefs, and in that sense to a factual and empirical situation. But these facts are mental entities that are not open to direct inspection. Therefore, we must resort to indirect methods to discover the extent to which belief in the conclusion is more widely or intensely held. One of these indirect methods involves exploring the extent to which the audience *should* come to believe the conclusion as a result of the argument. The assumption here is that people are rational in the sense that they actually believe what they should. This assumption would be questionable if it were construed as a universal generalization to the effect that every person always does this on all occasions. But it is plausible if taken to mean that most people most of the time do this; that they normally or typically do this.

Thus, it is preferable to expand the definition of persuasive force so that the persuasiveness of an argument is the extent to which the audience comes or should come to believe the conclusion on the basis of its reasons. More generally, we may say that persuasiveness is the extent to which the audience comes, should come, would come, or is likely to come to believe it.

Because of this broader definition and because people's mental activities are not directly observable, the assessment of persuasive force consists to some extent of argument and counterargument about the argument being considered. How large this extent is depends on one's orientation. Scholars in communication studies would tend to focus on the empirical approach, whereas philosophers would tend to focus on conceptual considerations. Such conceptual considerations need not be purely logical or formal or apriorist; to require them to be so would be an attempt to reduce persuasive force to logical force or formal validity. On the other hand, one key point about persuasiveness is that the increased acceptance of the conclusion may be due to logical factors, but also to psychological, aesthetic, or other nonlogical factors. One of these nonlogical factors may be eloquent expression.

Let us apply these ideas to the relevant rhetorical passages from the *Dialogue* that were introduced earlier. The passage on the perfection of the number three was discussed earlier as an illustration of both the rhetoric of anti-rhetoric and persuasive argumentation. The persuasive argumentation consisted of the following criticism: "I do not understand how or believe that in regard to number of legs three is more perfect than four or two; nor is it the case that the number four is an imperfection for the elements, and that it would be a greater perfection if they were three in number."[46] The argument may be reconstructed as follows: it is wrong to think that "the number three is a perfect number and has the faculty of conferring perfection to whatever has it"[47] because having three legs does not make animals more perfect than having two or four legs; similarly, the existence of four elements (in the terrestrial region) according to the geostatic view has never been thought by anyone to be a sign of imperfection for that view. The persuasive force is due to the fact that Galileo refutes a generalization about the number three by citing two counterinstances where the number three does not confer perfection. This refutation is a matter of logic, and this logical force generates the argument's persuasiveness.

46. Favaro (7:35); cf. Galileo (1967, 11).
47. Ibid.

Another rhetorical example was Galileo's comparison of a geocentric universe to a pesthouse in midtown. This passage was given as an example of both rhetoric of anti-rhetoric and eloquent expression. The eloquence lies in the clever image of a pesthouse in midtown and the striking character of its comparison to the geostatic universe where the impure earth is in the middle of the pure heavenly bodies. The eloquence of these images increases one's tendency to think of the geostatic system as an inappropriate arrangement. To see the persuasive force due to eloquence, we must reconstruct the passage as the following argument from analogy: the geostatic universe is unlikely to exist because it is like having a pesthouse in midtown, and such city planning is inappropriate. This argument seems to have a persuasive force beyond any which it might derive from the analogy logically and formally considered; the extra persuasiveness is due to the cleverness and eloquence of the analogy.

The third example of eloquent expression given above was the criticism of the heat theory of tides. Its proponents were invited to test it by burning their hand in a kettle of boiling water while waiting for the level of the water to rise in a tide-like fashion due to the heat. We have already seen that there is a purely empirical component to Galileo's criticism, which he expressed in a neutral manner, by referring to the temperature of water during high and low tides. But the clever invitation has a life of its own and adds to the criticism. That is, the passage can be reconstructed as the following argument: it is false to account for the tides by saying that they are caused by heat because, first, seawater at high tide is no warmer than at low tides, and second, the expansion of water due to heat is insufficient to cause the level of seawater to rise as much as it does during high tides; this can be seen by performing the kettle experiment. The persuasive force of the last subargument is largely due to the eloquence of the suggestion.

These examples show that rhetoric in the sense of eloquent expression is related to rhetoric in the sense of persuasive argumentation. They also show that persuasive argumentation is simply a special case of argument and is in this sense related to critical reasoning.

3.7. To summarize, rhetoric is relevant to science in at least three ways, depending on three different senses of "rhetoric." One sense involves the projection of images, the production of impressions, or the communication of messages by verbal means. This rhetoric is ubiquitous in science because when scientists communicate with peers or with the public, they constantly engage in it in regard to both substantive topics of research and the significance of their accomplishments. For example, there is generally a rhetoric of anti-rhetoric in science, and we find it also in Galileo's *Dialogue*; further, in this work, we also find a rhetoric of indeterminacy, a rhetoric of religious apologetics, and a rhetoric of strict demonstration. From this viewpoint, the rhetorical analysis of science becomes an important part of the understanding of science because it attempts to distinguish (and interrelate) appearance and reality within science. One important question I have not examined is why the rhetoric of anti-rhetoric is so common in science, although I have suggested that it may be an expression of the Aristotelian (and positivistic) ideal of demonstration, or a rhetorical expression of misgivings vis-à-vis eloquent expression or at least the misuse and abuse of it; but the question needs further investigation.

Another meaning of "rhetoric" is the art of persuasive argumentation. In this sense, rhetoric is an essential component of scientific inquiry, and there is a growing consensus among scientists, historians, and philosophers that this is so. This type of rhetoric is especially important when scientists disagree over fundamental issues and when they must choose between two or more general theories. One general example of such rhetoric I mentioned is the whole Copernican argument in the *Dialogue*, although there was no space for the details; a more specific example I analyzed involved Galileo's criticism of Aristotle's idea that three is a perfect number.

A third sense of "rhetoric" is the one pertaining to eloquent expression. This has a small but real role in science, as long as it is not regarded as a substitute for other primary activities such as persuasive argumentation, experimentation, observation, critical reasoning, methodological reflection, judgment, and mathematical analysis. To be sure, it is dispensable, and normally the eloquence of an argument or an experimental report does not outweigh its substantive deficiencies; but other things being equal, eloquence adds to persuasiveness. The *Dialogue* is a masterpiece of eloquent expression in both the critical and constructive mode, but only three examples were reported here: his remarks that the geostatic universe was like having the pesthouse in the middle of town; his suggestion that proponents of the heat explanation of tides burn their hands in boiling water in order to test it; and his calling some anti-Copernicans animals, by the clever description that they are men whose definition includes only the genus but lacks the species.

I have not denied, but rather implicitly hinted at, other senses of rhetoric that also deserve study. In fact, the last example of name-calling and insult could perhaps be categorized differently. But the more important point here is that these other rhetorical aspects of the *Dialogue* lend the book an aesthetic and literary dimension, worthy of appreciation for its own sake.

The kinds of rhetoric discussed here are related to the two other main dimensions of the *Dialogue* examined earlier. The projection of images is a special case of methodological reflection when the images in question pertain to methodological issues. But like all rhetoric of public relations, the projection of methodological images can be effective or ineffective. Often it is the ineffective projection of methodological images that is explicitly labeled rhetoric; whereas when the communication of methodological images is effective, one frequently speaks of methodological reflection pure and simple, or to be more precise, sound methodological reflection.

Finally, persuasive argumentation is a special case of reasoning when arguments have persuasiveness or persuasive force. An argument has persuasive force when it makes its conclusion more widely or more intensely accepted. Far from being opposed to logical force, persuasive force may depend on it. But it may also depend on extra logical factors, an important one of which is eloquent expression. Thus, eloquent expression is indirectly related to critical reasoning by being a factor in the persuasive force of an argument.

# Glossary

**Academician.** A term referring to Galileo, used by him in some of his books written in dialogue form. It is meant to remind readers that he was a member of the Lincean Academy (or Academy of the Lynx-eyed). This was the first modern scientific academy, having been founded in 1603 by Prince Federico Cesi (1585–1630), although it fell apart soon after his death. Galileo was made a member in 1611, became a friend of Cesi, and received support from the Academy for the publication of many of his works.

**action and reaction.** See *Newton's laws of motion*.

**affirming the consequent.** In logic, a formally invalid type of argument in which a conclusion is derived from two premises; one premise is a conditional proposition; the other premise affirms the then-clause (called consequent) of this conditional; and the conclusion affirms the if-clause (called antecedent) of the conditional. That is, an argument of the form: "if *p* then *q*; *q*; therefore, *p*."

**annual orbit.** In the geostatic worldview, this was the circle along which the sun moved in an eastward path among the stars and which took one year to complete. In the Copernican system, the annual orbit is simply the earth's orbit around the sun, along which the earth moves in the same (eastward) direction. Another term meaning the same thing and used by both worldviews is *ecliptic*.

**anthropocentrism.** The view that mankind is the center of everything that exists, where the notion of center may have a physical, biological, methodological, spiritual, or other meaning. Geocentrism may thus be viewed as essentially a special case of anthropocentrism, namely, as a type of physical anthropocentrism. But the abandonment of geocentrism does not necessarily imply the abandonment of other forms; for example, one can be a Copernican and yet hold that the human species has a central and special place in the animal kingdom and the biological world; and even if one accepts both Copernicanism and biological evolution, there remain other more philosophical ways of giving mankind the central place in the order of things.

**apriorism.** The methodological doctrine that genuine knowledge of the world may be acquired by intellectual theorizing, pure thinking, or speculative reflection, independently of observation, experiment, or sensory experience; also the attitude which in the acquisition of knowledge tends to give priority to intellectual theorizing over sensory experience, or overemphasize the former vis-à-vis the latter. It should be contrasted with empiricism. Apriorism is sometimes equated with rationalism, but this equation should be avoided because rationalism also pertains to reasoning, rational-mindedness, and rationality; hence, this equation would hold only for one of the many meanings of rationalism.

**arc.** A part of the circumference of a circle (or more generally, of any curved line); the subdivision of the circumference into 360 parts is the basis for the system of angular measurement defining a degree as the size of an angle formed at the center of a circle by the lines drawn from an arc equal to the 360th part of the circumference.

**Archimedes of Syracuse** (287–212 B.C.). Greek mathematician and physicist, and a younger contemporary of Aristarchus; most famous for such things as the proof of theorems about the volume and surface of spheres and cylinders, a method for approximating the value of *pi*, the law of the lever, and the principle of hydrostatics that bears his name. This principle states that a body immersed in a fluid is buoyed up by a force equal to the weight of the displaced fluid; thus, an object weighs less when immersed in water than it does in air. The law of the lever states that for a rigid bar free to rotate around a fixed point (the fulcrum), equilibrium is reached when the two forces on either side of the fulcrum have magnitudes inversely proportional to their respective distances from it.

**argument.** A piece of reasoning trying to prove something (called a conclusion) on the basis of reasons (also called premises). An argument must be expressed by means of at least two statements, the conclusion and one reason; a single statement by itself cannot express an argument. But normally an argument is expressed by a structured series of interconnected statements. In an argument, the conclusion may be said to be based on or supported by or justified by or implied by or proved by or derived from or inferred from or drawn from the reasons (or premises).

**Aristarchus of Samos** (c. 310–250 B.C.). Greek astronomer who elaborated the theory that the earth moves around the sun.

**Aristotle** (384–322 B.C.). Greek thinker who made significant contributions to many subjects: he is one of the greatest philosophers who ever lived; in biology and the science of logic, his key ideas were not superseded until the nineteenth century; his writings also contain the earliest systematization of the geostatic worldview.

***The Assayer.*** A book published by Galileo in 1623 in which he discussed primarily the nature of comets, but also a wide range of scientific and methodological topics. He defended an erroneous view according to which comets have a terrestrial origin, but in the process he managed to formulate several important insights in methodology, logic, and rhetoric. The book was part of a polemic with a Jesuit mathematician and astronomer named Orazio Grassi (1590–1654) that generated several other works. This controversy also poisoned Galileo's relationship with the Jesuits and was a factor (among many) that led to his trial and condemnation in 1633.

**astronomy.** The branch of science that studies the nature, origin, positions, motions, and configurations of the heavenly bodies. It is the most ancient of the sciences, having already reached in antiquity a high degree of precision and systematicity.

**begging the question.** An error in an argument consisting of assuming what the argument is trying to prove. Normally in an argument it is the conclusion which is in question, in the sense that it is the claim that needs to be supported. To beg the question means that, instead of providing support for the conclusion, one implicitly assumes it among the premises that are taken for granted.

**Bellarmine, Robert** (1542–1621). Jesuit theologian, perhaps the most influential Catholic churchman of his time, and now a saint. Besides being a cardinal, he also served as a professor at the Collegio Romano (the Jesuit university in Rome), an archbishop, the pope's theologian, and a consultant to the Inquisition.

**Brahe, Tycho.** See *Tycho Brahe*.

**Bruno, Giordano** (1548–1600). Italian philosopher and martyr for freedom of thought. When put on trial by the Inquisition, he refused to recant and so was executed by being burned alive at the stake. Little is known about his final trial; but his heresies included belief in a plurality of worlds. His works contain a confusing mixture of scientific and unscientific elements, and of philosophical and mystical speculations; he was inclined to believe in Copernicanism and in the infinity of the universe. He was a Dominican, but left the order in 1576 and fled Rome to escape the Inquisition; for the next fifteen years he wandered throughout Europe, lecturing at many universities and publishing many books; finally, in 1591 he returned to Italy, where he was soon arrested and tried by the Inquisition in Venice and then transferred to Rome and kept in prison there for another eight years until his final trial.

**centrifugal force.** In modern physics, the apparent force experienced by a rotating or revolving body tending to make it move away from the center; the force is apparent and not real in the sense that it results from the body's inertial tendency to move uniformly in a straight line. The term means literally "center-fleeing force." In such a situation, the force that is real and not apparent is the one which causes the body to curve away from a rectilinear path and toward the center; this is called *centripetal* force, which means literally "center-seeking force." For example, on a rotating earth, a body lying on its surface and rotating with it experiences an upward pull away from the earth's center; this decreases the body's weight (slightly) but is not caused by a real force pulling the body away from the earth's center; rather, the only real force is the earth's gravitational attraction, which causes the body's weight; in other words, due to the earth's rotation, terrestrial bodies weigh slightly less than they would on a motionless earth. Another example is provided by the earth's orbital revolution around the sun; here, the centripetal force is the sun's gravitational attraction. In a stable rotating or revolving system, the centrifugal and the centripetal forces exactly counterbalance each other, and so one can speak interchangeably of either one, remembering that they are equal in magnitude but opposite in direction. One of the strongest mechanical objections against the earth's rotation involved the problem of centrifugal force, and Galileo discussed it in terms of the "extruding power of whirling"; the criticism of this objection required the development of the laws of circular motion and centrifugal force; Galileo's criticism is groping toward these laws, and Huygens soon thereafter worked them out in a rigorous and systematic manner. In circular motion, the centrifugal force is directly proportional to the square of the linear velocity and inversely proportional to the radius ($F = KV^2/R$); alternatively, since the linear velocity equals the product of the angular speed and the radius, the force is directly proportional to the square of the angular speed and to the radius ($F = KW^2R$).

**centripetal force**. See *centrifugal force*.

**charity**. The principle of charity states that in criticizing an opposing argument or view, one should first formulate a charitable interpretation and then make the latter the target of one's criticism; a charitable interpretation is one that portrays the original argument or view in a reasonably favorable light, as possessing some strength that must be taken seriously, and as free of insignificant or trivial errors which do not affect the main issue. Two good examples of Galileo's use of this principle occur at the beginning of his critique of the extruding-power objection (selection 9) and of his critique of Aristotle's geocentrism (selection 11).

**Chiaramonti, Scipione** (1565–1652). Professor of philosophy at the University of Pisa from 1627 to 1636. He is mentioned favorably in Galileo's *Assayer* (1623) in regard to the nature of comets; but he is frequently and sharply criticized in the *Dialogue* for the anti-Copernican views advanced in his *Anti-Tycho* (1621) and in his 1628 book on the interpretation of novas.

**comet**. A large heavenly body appearing as a luminous mass to which is attached a long tail, and visible for only brief periods ranging from a few days to several months. Though comets had been observed since antiquity, in Galileo's time their nature and origin remained controversial; the main issue was whether they were heavenly bodies or atmospheric phenomena; in *The Assayer* Galileo elaborated a theory according to which comets originated in the terrestrial atmosphere but moved into interplanetary space. Nowadays, comets are known to be bodies of great volume but very small mass, to consist mostly of ice, and to follow definite (elliptical or parabolic) orbits around the sun; furthermore, the periodic recurrence of some of them can be predicted with great accuracy; but many more details remain controversial or unknown.

**conceptual**. Pertaining to intellectual theorizing, thinking, and speculation, as distinct from observation, experiment, and sensory experience; here, there is no connotation that intellectual theorizing is privileged in a one-sided fashion or carried to an extreme, as done by apriorism. That is, conceptual is the counterpart of empirical, whereas apriorism is the opposite of empiricism.

**conclusion**. Assertion that is part of an argument and is supported by reasons or derived from premises.

**conservation of motion**. In Galileo's work, the principle of the conservation of motion states that once a body has acquired a certain degree of speed in an horizontal direction, it conserves that speed unless external disturbances interfere with it. This principle is an approximation to both the law of the conservation of momentum and the law of inertia in modern physics.

**"Considerations of the Copernican Opinion"**. A set of notes and drafts written by Galileo in 1615 when he was under investigation by the Inquisition for his Copernican views. In them Galileo sketches some clarifications of the problem of the instrumentalist versus the realist interpretation of Copernicanism as well as some answers to the theological objections against Copernicanism and to Cardinal Bellarmine's objections.

**contradiction**. In logic, the relationship between two propositions such that they cannot both be true and cannot both be false, but rather one must be true and the other false. When two propositions contradict each other, they are called contradictories of one another. For example, the proposition that the earth stands still contradicts the proposition that the earth moves.

**contraposition**. In logic, the relationship between two conditional propositions when their if-clauses and then-clauses are interchanged and negated; each proposition is then called the contrapositive of the other. That is, "if not-*q* then not-*p*" is the contrapositive of "if *p* then *q*." Contraposition is a formally valid procedure, and a conditional proposition is logically equivalent to its contrapositive.

**contrariety**. In logic, the relationship between two propositions such that they cannot both be true but can both be false; the two propositions are called contraries of each other; for example, the propositions that the earth is at the center of the universe and that the sun is at the same center are contraries. In Aristotelian natural philosophy, contrariety referred to the opposition between such pairs as hot and cold, moist and dry, up and down, and light and heavy; such contrariety was considered to be the source of all qualitative change.

**conversion**. In logic, the relationship between two conditional propositions when their if-clauses and then-clauses are interchanged; each proposition is called the converse of the other. That is, "if *q* then *p*" is the converse of "if *p* then *q*." Conversion is formally invalid, and a conditional proposition does not logically entail its converse.

**Copernicanism**. The worldview of Copernicus and his followers. Its essential point is that the earth is a sphere which rotates daily on its own axis and revolves yearly around the sun, and thus it is neither standing still nor located at the center of the universe.

**Copernicus, Nicolaus** (1473–1543). Polish astronomer, author of *On the Revolutions of the Heavenly Spheres*, whose original title was *De revolutionibus orbium celestium* (Nuremberg, 1543). This was the first work to elaborate the technical details of the theory that the earth spins on its axis and moves around the sun, an idea originally conceived and discussed by the ancient Greeks.

**Corfu**. A small island off the northwestern coast of Greece lying about halfway between Venice at the northern end of the Adriatic Sea and Syria at the eastern end of the Mediterranean.

**cosmology**. The study of the structure, origin, and most general features of the universe as a whole, including the place of the earth and of mankind in it.

**critical reasoning**. A mental skill consisting of reasoning aimed at the analysis, evaluation, or self-reflective formulation of arguments.

**critical thinking**. Thinking consisting of critical reasoning and/or methodological reflection.

**cubit**. An ancient unit of distance corresponding to the length of a forearm, and thus approximately one and one-half to two feet. This is the term used to translate Galileo's term *braccio*.

**deductively valid**. A deductively valid argument is one such that if its premises are true then it is impossible for its conclusion to be false. The term is essentially equivalent to *formally valid*.

**degree** (of arc). A unit of angular distance and measurement defined as the 360th part of a circle, whose complete circumference is thus said to be divided into 360 degrees; a degree is in turn subdivided into 60 minutes, each minute into 60 seconds, and each second into 60 thirds.

**demonstration**. In the strictest sense, a demonstration is a rigorous proof or perfectly valid argument, such as found in mathematics. By extension, a demonstration also means an argument that is conclusive insofar as it establishes its conclusion beyond any reasonable doubt. More generally, a demonstration may also mean an argument that *claims* to be especially strong, even if it does not actually have this property.

**demonstrativism**. The methodological doctrine that a scientific theory must be provided with a demonstration in the strict sense of the word—namely, a perfectly conclusive and valid proof—in order for it to be scientific. This contrasts with fallibilism but should not be confused with realism; that is, demonstrativism is incompatible with fallibilism, but in principle compatible with both instrumentalism and realism.

**denying the consequent**. In logic, a formally valid type of argument in which a conclusion is derived from two premises; one premise is a conditional proposition; the other premise denies the then-clause (called consequent) of this conditional; and the conclusion denies the if-clause (called antecedent) of the conditional. That is, an argument of the form: "if *p* then *q*; not-*q*; therefore, not-*p*."

**Descartes, René** (1596–1650). French thinker whose contributions include the invention of analytic geometry, the clarification of the laws of motion (such as inertia), and the formulation of some metaphysical and epistemological problems widely discussed even today (such as the relationship between mind and body).

**"Discourse on the Tides"**. An essay written and privately circulated by Galileo in 1616 at the request of a cardinal who wanted to be informed of his tidal argument in favor of Copernicanism; this was done during the first phase of the Galileo Affair, when Galileo was under investigation by the

Inquisition, but just before Church authorities concluded the investigation by giving him a private warning (through Cardinal Bellarmine) and by issuing an anti-Copernican decree (through the Congregation of the Index). The bulk of the essay is repeated almost verbatim in the "Fourth Day" of the *Dialogue*.

**dogmatism**. See *skepticism*.

**elemental**. Pertaining to the four terrestrial elements (earth, water, air, and fire).

**ellipse**. A closed plane curve such that the sum of the distances from any point on it to two fixed points (called foci) is constant. The shape is that generated by the intersection of a cone and a plane inclined to but not touching its base. As Kepler discovered, the path of a planet revolving around the sun is an ellipse with the sun as one of the foci. But Galileo paid no attention to this finding.

**empirical**. Pertaining to observation, experiment, or sensory experience, without the implication that there is or should be an exclusive or excessive emphasis on these activities vis-à-vis thinking, theorizing, reflection, and speculation. See also *empiricism*.

**empiricism**. The methodological doctrine that all knowledge of the world is or should be based exclusively on observation, experiment, and sensory experience; also the attitude which, in the quest for knowledge, gives priority to these activities over thinking or theorizing or overstresses these activities vis-à-vis thinking or theorizing. It should be contrasted with apriorism. Empiricism should not be equated with being empirical since it is best to construe *empirical* as pertaining to observation, experiment, and the senses in general, without any implication of exclusivity or excessive emphasis; one may use the term *empiricist* for this purpose, and thus equate empiricism with being empiricist, and consequently also distinguish empirical and empiricist.

**epistemological modesty**. An attitude which on the one hand recognizes the value of the human mind and senses in providing true and useful knowledge about the world, but which on the other also recognizes that such knowledge as we may acquire is at best limited, incomplete, susceptible to revision, and infinitesimally small compared to the knowledge we lack. Such an attitude represents a balanced mean between skepticism and dogmatism, and it is strongly advocated by Galileo in the *Dialogue*.

**epistemological realism.** See *realism*.

**epistemological reflection**. See *methodological reflection*.

**epistemology**. Systematic epistemology is the branch of philosophy that studies the nature, origin, scope, reliability, and limitations of knowledge; for example, the relationship between knowledge and other concepts such as truth, belief, certainty, reason, and evidence, and the methods and procedures effective for acquiring knowledge. It is also called theory of knowledge. See also *methodology*; *methodological reflection*.

**equinox**. Either of the two times of the year when night and day are equally long (twelve hours) everywhere on earth, and the sun's apparent path among the stars makes it cross the celestial equator. In the Northern Hemisphere, the *vernal* equinox occurs about March 21 when the sun crosses it in a northward direction, while the *autumnal* equinox occurs about September 23 when the sun crosses it in a southward direction. These terms are also used to refer to the two points on the celestial sphere where the ecliptic (namely, the apparent path of the sun) intersects the celestial equator. See also *precession of the equinoxes*.

**equivocation**. Flaw in reasoning stemming from two different meanings for a word or sentence; the two meanings are exchanged in such a way as to suggest that a particular conclusion is more justified than it really is.

**fallacy**. Flawed reasoning or an argument having a serious flaw. The flaws can take various forms, such as the premises not supporting the conclusion, the conclusion being assumed among the premises, the premises contradicting each other or the conclusion, or one or more of the premises being false or unjustified in some obvious or peculiar manner.

**fallibilism**. The methodological doctrine that a scientific theory is always subject to revision, improvement, or falsification in the light of further evidence, and thus can never be regarded as being absolutely, categorically, or unconditionally correct. So defined, fallibilism is compatible with both instrumentalism and realism, though not with demonstrativism.

**first law of motion**. See *Newton's laws of motion*.

**Florence**. A city in central Italy where Galileo lived for the second half of his life (1610–1642), while he held the position of Philosopher and Chief Mathematician to the Grand Duke of Tuscany. Florence was then the capital of a small independent state ruled by the House of Medici and including Pisa and the surrounding region. The *Dialogue* was first published there in 1632.

**force**. In modern physics, a force is defined by means of Newton's second law of motion, and so it is a cause of *changes* of speed or direction of motion. In Aristotelian physics, a force is a cause of motion and can be internal or external; internal forces cause natural motions, external forces cause violent motions. In Galileo's work, force has a less clear and less precise meaning that overlaps with both the Aristotelian and the Newtonian concepts, as well as with the concept of energy; although he was groping toward the Newtonian concept, he did not really possess it; Galileo's notion is also interwoven in confusing ways with his talk of "power" and "moment."

**geocentric**. Pertaining to the earth being at the center of the universe. The geocentric worldview claims that the earth is located at the center of the universe. This term corresponds in large measure to the term *geostatic*, which refers to the earth being motionless; thus, the worldview criticized by Galileo is one that is both geocentric and geostatic. But the meaning of the two terms is different, and it is possible to devise a system that is geocentric but not geostatic or one that is geostatic but not geocentric. For example, some astronomers placed the earth at the center of the universe but made it rotate daily on its axis, though the planets and the sun revolved around the earth. Similarly, the term *geocentric* is often contrasted with *heliocentric* (which refers to the sun being at the center) because in Galileo's discussion the only two alternatives are whether to place the earth or the sun at the center; but *geocentric* and *heliocentric* are not exactly contradictory since it is possible that neither the earth nor the sun should be at the exact center (which happens to be the case in physical reality).

**geokinetic**. Pertaining to the earth's motion or claiming that the earth moves. The geokinetic worldview claims that the earth rotates daily on its axis from west to east and revolves yearly around the sun in the same direction. This term is contrasted with the term *geostatic* and may be taken to correspond to *Copernican*.

**geostatic**. Pertaining to the earth standing still or claiming that the earth stands still. The geostatic worldview claims that the earth is motionless at or near the center of the universe and that all heavenly bodies revolve around it. This is contrasted with the term *geokinetic* and may be taken to correspond to *Ptolemaic*.

**gravitation**. In modern physics, a concept defined by Newton's law of universal gravitation, which states the following: any two bodies or particles of matter attract each other with a force which is directly proportional to the product of their masses and inversely proportional to the square of their mutual distance, and which acts along the straight line between their centers. Together with Newton's laws of motion, this law provides the essential theoretical proof of Copernicanism and the essential mechanical explanation of the tides (as due to the moon's gravitational attraction).

**gravity**. A term used interchangeably with *weight* and *heaviness* in the *Dialogue*. In the Aristotelian worldview, gravity is the property of the elements earth and water whereby they tend to move toward the center of the universe; it manifests itself either as weight or free fall; and it is contrasted with a property called *levity* which is attributed to the elements air and fire, which consists of the tendency to move away from the center of the universe, and which manifests itself as buoyancy or spontaneous upward motion; bodies with gravity are called *heavy bodies*, and those with levity are called *light bodies*; it follows that light bodies go up because of their intrinsic property of levity, and not because they weigh less than the surrounding medium; in short, bodies with levity are thought to have no weight. Galileo abandoned the dichotomy between gravity and levity and held that all bodies have weight, thus explaining buoyancy and spontaneous upward motion in terms of the relative weight or specific gravity of the bodies involved; for him gravity was a property belonging to all bodies in the universe (heavenly as well as terrestrial), but consisted of the tendency to go toward the center of the whole of which one was a part, so that a rock on the moon would tend to move toward the center of the moon. Thus, for both Aristotle and Galileo gravity could be labeled a universal property, but in different senses; for Aristotle it was universal in the sense that it was defined in terms of the center of the universe, a unique point yielding an absolute frame of reference; for Galileo it was universal in the sense that it characterized all material bodies in the universe; but even Galileo did not conceive of gravity as universal in the sense of Newton's *gravitation*, namely in terms of mutual attraction among all bodies in the universe, and thus as acting between the earth and the moon.

**great circle**. A circle on a sphere whose center coincides with the center of the sphere. For example, on the earth and on the stellar sphere, the equator and the meridians are great circles.

**heavenly unchangeability**. The Aristotelian thesis that the heavenly bodies undergo no qualitative changes (but only move in regular circular motion).

**Huygens, Christiaan** (1629–1695). Dutch physicist, astronomer, mathematician, and inventor who is one of the founders of modern science. He elaborated the wave-like nature of light (a point which eluded even Newton), gave a correct explanation of Saturn's rings (which had baffled Galileo), improved the telescope, invented the pendulum clock, and was one of the pioneers of the mathematical theory of probability. His chief work (*Horologium oscillatorium*, Paris, 1673) studies the pendulum clock and in the process gives a systematic and rigorous elaboration of the laws of circular motion and centrifugal force in their essentially modern form.

**impetus**. In Aristotelian physics, the impetus of a projectile was the power to move which had been transferred to it by the projector and which would be gradually lost. Galileo uses the term to refer to the power that a body has due either to the quantity of motion it embodies or the tendency it has to move in particular ways. Thus, the Galilean meaning is inexact and corresponds partly to the Aristotelian meaning and partly to what modern physics would call either momentum, kinetic energy, or even potential energy.

**inch**. A term used in this book to translate Galileo's term *dito*, which literally means finger; the *dito* was an ancient inexact unit of length corresponding to the breadth of a finger.

**independent-mindedness**. The willingness and ability to think for oneself and have a judicious attitude toward authorities, thus avoiding the extremes of a slavish, blind, and total acceptance on the one hand and of an uncritical and total disregard on the other. Galileo had this sort of independence of mind toward Aristotle, the Bible, and the Catholic Church.

**inertia**. See *Newton's laws of motion*.

**inferior planets**. Planets having orbits smaller than the annual orbit, namely Mercury and Venus.

**Ingoli, Francesco** (1578–1649). See "*Reply to Ingoli*."

**Inquisition**. This is the common name for the Congregation of the Holy Office, which is the department of the Catholic Church whose purpose is to defend and uphold faith and morals; it was officially instituted in 1542 by Pope Paul III, and one of its duties was to take over the suppression of heresies and heretics begun by the Medieval Inquisition; by the time Galileo got into religious trouble, the notion of heresy had been given a legal definition and inquisitorial procedures had been codified.

**instrumentalism**. The methodological doctrine that a scientific theory is an instrument, tool, or device for making mathematical calculations or observational predictions; that it may be more or less useful or convenient; but that it is not a description of physical reality capable of being true or false. Instrumentalism is contrasted with realism, but is consistent with fallibilism and demonstrativism. Many instrumentalists do not apply their doctrine to all scientific theories but only to a particular one; this happens when they are convinced a particular theory is descriptively false but do not want to deny it has value as an instrument of calculation and prediction. For example, many anti-Copernicans adopted an instrumentalist interpretation of Copernicanism but not of the Ptolemaic view; this was the type of instrumentalism held by Cardinal Bellarmine and Pope Urban VIII.

**intermediate proposition.** A proposition that functions simultaneously as a premise in one subargument and as a conclusion in another subargument, in the context of a bigger more complex argument.

**judiciousness.** The willingness and ability to be impartial, balanced, and moderate; that is, to avoid one-sidedness (by properly taking into account all distinct aspects of an issue) and to avoid extremism (by properly taking into account the two opposite sides of any one aspect). This does not mean that anything goes, that one indiscriminately accepts all points of view as equally good, or that one does not accept any one point of view as better; nor does it mean that one mechanically splits the differences separating the several aspects and the opposite sides; instead, the view which one accepts must be "properly" balanced and moderate. Galileo's attitude toward Copernicanism is an example of judiciousness.

**Jupiter.** A planet whose orbit is bigger than the annual orbit and that takes about twelve years to complete it. In the geostatic system, it is the sixth planet from the earth; in the geokinetic system, it is the fifth planet from the sun. This difference in numerical sequence is due to the fact that the moon is counted as the first planet in the geostatic system, whereas it is not counted as a planet but as a satellite of the earth in the geokinetic system. With the telescope, Galileo discovered that Jupiter has four satellites revolving around it at different distances and with different periods; he named them *Medicean stars* in honor of the House of Medici, rulers of his native Tuscany.

**Kepler, Johannes** (1571–1630). German scientist who is one of the founders of modern science. He held the position of Mathematician to the Holy Roman Emperor and elaborated Copernicanism in new and important ways. The most significant elaboration involved the formulation of three laws of planetary motion, which were based on a painstaking analysis of Tycho Brahe's novel observational data. Kepler was one of the first supporters of Galileo's telescopic discoveries, although the latter never did reciprocate in kind; Galileo did not even pay attention to Kepler's laws, in part because they were buried in books filled with many other mystical and metaphysical speculations that Galileo found objectionable and that turned out to have no intrinsic value.

**Kepler's laws of planetary motion.** In modern astronomy, the three laws first discovered by Kepler. The first law states that the planets move around the sun in elliptical orbits with the sun located at one of the foci of these ellipses (implying that the Copernican system in the versions elaborated by Copernicus himself and by Galileo is only approximately but not literally true because they only used circles). The second law states that the line connecting a planet to the sun sweeps over equal areas of the ellipse in equal times (implying that the orbital speed of a given planet is not constant but increases in a precise way when its distance to the sun decreases). The third law states a mathematical relationship between the average distance of a planet to the sun and the period of time required by the same planet to complete its revolution around the sun: for any two planets, the squares of their periods are proportional to the cubes of their average distances (implying that the period itself varies not simply in accordance with the average distance but in accordance with the 3/2 power of that distance).

**kinetic energy**. In modern physics, a form of energy stemming from the motion of a body; it is defined as one-half the product of the mass and the square of the speed ($mv^2/2$). In Galileo's time this concept was being developed, and kinetic energy was often confused with other physical quantities, especially momentum; momentum is now defined as the product of mass and velocity ($mv$). The problem at that time was how to best measure the "quantity of motion," whether in terms of the quantity $mv$ or $mv^2$; this was eventually solved by saying that both are legitimate measures and which one is used depends on the context and purpose at hand. Note that the velocity mentioned in the modern definition of momentum is a so-called "vector" quantity, namely, a quantity consisting of a number and a direction, so that momentum is an indication of direction as well as of amount of motion; whereas kinetic energy is a so-called "scalar" quantity because in the quantity $mv^2/2$ the speed has just a numerical magnitude but no spatial direction.

**law of squares**. A law about falling bodies that is usually regarded as one of Galileo's major scientific contributions; it states that in free fall the distance fallen is proportional to the square of the time elapsed ($d = kt^2$). This is also equivalent to saying that the instantaneous velocity is proportional to the time elapsed; or alternatively, that there is acceleration in free fall and the acceleration is constant. This law was not only a key element of Galileo's science of falling bodies, but later led Newton to say that the constant acceleration was evidence of a constant force and that the latter was the earth's gravitational attraction toward the falling body.

**law of universal gravitation**. See *gravitation*; *Newton*.

**"Letter to the Grand Duchess Christina"**. A long letter written and privately circulated by Galileo in 1615, addressed to Christina of Lorraine, the widow of the previous grand duke of Tuscany and the mother of the then-ruling grand duke. Galileo was then under investigation by the Inquisition as a result of formal complaints against him, and he hoped that the letter would answer those complaints. It is Galileo's most considered and eloquent discussion of the relationship between science and religion, and takes the form of a criticism of the biblical objection to Copernicanism.

**Lincean Academy**. See *Academician*.

**Locher, Ioannes G.** A student of Scheiner and author in 1614 of a book critical of Copernicanism. This book is extensively and severely criticized by Galileo in the *Dialogue*.

**logic**. The branch of scholarship that studies the nature of reasoning, its relationship to other things, and what distinguishes correct from incorrect reasoning. This includes principles for the analysis and the evaluation of such things as arguments, demonstrations, proofs, fallacies, paralogisms, and sophisms.

**maieutics**. See *Socratic method*.

**Mars**. A planet whose orbit is bigger than the annual orbit and that takes about two years to complete it. In the geostatic system, it is the fifth planet from the earth; in the geokinetic system, it is the fourth planet from the sun. This difference in numerical sequence is due to the fact that the moon is the first planet in the geostatic system but is a satellite of the earth in the geokinetic system. The word Mars is also the name of the god of war in the religious mythology of ancient Rome.

**mathematicism**. A methodological procedure or doctrine that involves an excessive or one-sided emphasis on the use or power of mathematics in natural science.

**Medicean stars**. A term used by Galileo to refer to Jupiter's satellites, which he discovered. He named the new bodies in honor of the House of Medici, which ruled Florence and the Grand Duchy of Tuscany.

**Mediterranean Sea**. A large sea between southern Europe, north Africa, and the Middle East. It surrounds the Italian peninsula on three sides, and its only natural connection with other bodies of water is the Strait of Gibraltar, joining it to the north Atlantic Ocean.

**Mercury**. A planet that revolves in its orbit in such a way that it always appears close to the sun (closer than the planet Venus which behaves in the same way). In the Copernican system, Mercury is the first planet from the sun and completes its orbit in about three months. In the geostatic system, opinions differed about whether it was the second, third, or fourth planet from the earth; it was most commonly regarded as the second (between the moon and Venus).

**meridian**. A circle on the surface of a sphere passing through both poles and cutting the equator at right angles. The sphere in question could be either the earth or the stellar sphere, each being the projection of the other; for any point on the earth there is such a meridian, and when crossing it a heavenly body reaches the highest apparent elevation in its course.

**metaphysics**. A branch of philosophy that studies the most general features of reality and the most general concepts involving being and nothingness, appearance and reality, and existence and nonexistence. For example, it studies such issues as whether or not God exists; whether human beings have immortal souls; whether every event has a cause; and whether the good, the true, and the beautiful are ultimately distinct or identical.

**methodological reflection**. Thinking aimed at the formulation, analysis, evaluation, or application of principles about the proper procedure to follow in the search for truth and the quest for knowledge; such thinking occurs while one is practically involved in investigating a concrete scholarly problem. For most purposes, the term is interchangeable with epistemological reflection.

**methodology**. Systematic methodology is the branch of philosophy that studies the proper methods, procedures, and principles for acquiring knowledge and discovering the truth. For example, it studies what is the role of observation, thinking, authority, and mathematics in the search for truth; whether observation is more important than thinking; whether all knowledge of the physical world must be expressed in mathematical terms; whether the Bible is a scientific authority; and whether the simplicity of a theory is an indication of its truth because nature operates by the simplest possible means. For most purposes, the terms systematic methodology and systematic epistemology may be used interchangeably.

**middle term.** In an argument with two premises and one conclusion, the middle term is the term which is common to both premises but does not appear in the conclusion; for example, in an argument of the form "all A are B; all C are A; so, all C are B," the middle term is *A*. By extension, the middle term also refers to the clause which is common to both premises but does not appear in the conclusion; for example, in an argument of the form "if P then not-Q; Q; so, not-P," the middle term is *Q*.

**minute** (of arc). Unit of angular distance and measurement defined as the 60th part of a degree, which in turn is the 360th part of a circle; thus, a circle is divided into 360 degrees, and a degree into 60 minutes.

**moment.** Aside from the obvious connotation of an instant of time, this word is used by Galileo with several other meanings. One is an approximation to the *momentum* of modern physics. Another is synonymous with the terms *magnitude* or *intensity* or *degree*, as in the phrase "the moment of the speed a body possesses."

**momentum.** In modern physics, momentum is defined as the product of a body's mass and velocity, taking velocity as a vector quantity (which has both a numerical magnitude and a spatial direction); the law of conservation of momentum states that in a closed system the total amount of momentum neither increases nor decreases but remains constant. Galileo's counterpart of this law is his principle of conservation of motion; but the correspondence is inexact because he had no conception that momentum is a vector and because he did not clearly distinguish between momentum and kinetic energy.

**natural motion.** In Aristotelian natural philosophy, natural motion is the motion that a body has by nature; that is, motion that the body has *because* of its nature; namely, motion caused by the moving body's inherent nature; or again, motion caused by a force internal or inherent to the moving body. Thus, the natural motion of a terrestrial body is the motion it spontaneously tends to undergo in order to reach its natural place of rest, if it is not already there; for example, the natural motion of the elements earth and water is straight toward the center of the universe, and the natural motion of the elements air and fire is straight away from the center of the universe. Natural motion is contrasted with violent motion. Galileo partly accepted and partly modified this notion. He continued to speak of spontaneous (or internally caused) motion as one kind of natural motion, but dissociated it from the doctrine of natural places; so, for him the oscillation of a pendulum on the earth or the free fall of a rock on the moon would be as natural as the free fall of a rock on the earth. He sometimes added another meaning to the concept of natural motion, namely motion that can last forever. And he contrasted natural motion with violent, but also spoke of a third kind which is neither natural nor violent and which he labeled *neutral*; an example of the latter would be horizontal motion on a frictionless surface.

**natural philosophy.** The branch of learning that studies the most general features of the physical world. Until Galileo's time, such learning was regarded as a branch of philosophy because philosophy was equated with learning in general. Since that time, the study of physical phenomena has gradually separated from philosophy and has become subdivided into many branches such as physics, astronomy, chemistry, biology, and geology.

**natural sciences**. Disciplines such as astronomy, biology, and physics, which study different kinds of natural phenomena; they are contrasted with the mathematical sciences (arithmetic, geometry), the social and behavioral sciences (psychology, sociology, anthropology), and the humanities (philosophy, history, literature).

**new star**. See *nova*.

**Newton, Isaac** (1642–1727). British scientist who made many important contributions to several fields, and is thus not only one of the founders of modern science but also one of the greatest scientists of all time. For example, he was one of the inventors of a new branch of mathematics (calculus); he discovered the spectral composition of white light; and he invented a new type of telescope. His greatest contribution was the systematization of the basic laws of motion and their rigorous application to the motion of terrestrial and heavenly bodies; this led him to the discovery of the law of universal gravitation and to the effective resolution of the Copernican controversy; he accomplished this in a monumental work first published in 1687 and entitled *Mathematical Principles of Natural Philosophy*.

**Newton's law of gravitation**. See *Newton*; *gravitation*.

**Newton's laws of motion**. In modern physics, the three laws of motion that were systematically and rigorously elaborated by Newton. The first law asserts that every body persists in its state of rest or of uniform motion in a straight line unless compelled by an external force to change that state; this is also called the law of inertia because inertia means the inherent tendency of a body to remain in its natural state, where natural state is defined indifferently as either rest or constant rectilinear motion. The second law states that a force causes a change in velocity in the direction of the straight line along which it acts, and the change in velocity per unit time is proportional to the force; this is also called the law of force because it may be seen as a physical definition of force, saying that a force is measured by the rate of change in velocity it causes (namely, the rate of change in speed or direction). The third law says that for every action there is an equal and opposite reaction (as when the firing of a gunshot makes the gun recoil); this is also called the law of action and reaction.

**nova**. A phenomenon seen by the naked eye as the sudden appearance of a new star or the sudden tremendous increase in brilliance of an old star, followed by its gradual fading away; the initial increase occurs in a few days and ranges from thousands to millions of times brighter, whereas the eventual decrease takes a few months or years. In 1572 and 1604 two especially spectacular such phenomena were seen and generated the controversy of whether they were heavenly phenomena or occurrences in the earth's atmosphere; Galileo, together with many other astronomers, regarded them as heavenly phenomena, thus undermining heavenly unchangeability, the earth-heaven dichotomy, and the geostatic worldview; the Aristotelians tried to explain away the evidence in order to place them in the space below the moon, and thus preserve those same claims. These phenomena are today explained as cases of explosions of stars, which cause a tremendous increase in the light and energy emitted; they are classified into novas and supernovas depending on the magnitude of the increase. In Latin, the term *nova* literally means "new" and in this context is applied to *star*; thus, in the *Dialogue* Galileo often refers to them simply as "new stars."

**open-mindedness.** The willingness and ability to know, understand, and learn from the arguments, evidence, and reasons against one's own views. Galileo's attitude toward Copernicanism is an example of open-mindedness, insofar as he knows and understands the anti-Copernican arguments, even though he thinks they are wrong and is inclined to accept Copernicanism.

**oracle.** In classical Greece and Rome, a pronouncement believed to originate from the gods and to contain some hidden meaning which required much interpreting and deciphering to ascertain.

**orb.** A term that is partly synonymous with the term *orbit*, namely the path followed by one heavenly body around another; the term *orb* also refers to the region of the heavens where a given orbital path is located; and the term *orb* also refers to the spherical layer encompassing an orbital path; orbs in this last sense were sometimes taken to have physical reality and be composed of the element aether. Thus, to each planet there corresponded an orb that was said to belong to it or to be occupied by it. For example, the "lunar orb" could refer either to the path of the moon around the earth, to the region of space surrounding the earth at a distance equal to that of the moon, or to the spherical layer of aether in which the moon was embedded and whose rotation made it revolve around the earth.

**orbit.** The path followed by a heavenly body as it moves among the other bodies, usually around some particular body or point that is regarded as the center or focus of the orbit.

**Padua.** A city in northern Italy where Galileo lived for eighteen years (1592–1610) while he was professor of mathematics at its university. The city was then part of the Republic of Venice and its university was a state institution supported by public funds. As befitted the wealth, power, and illustrious history of the Venetian republic, the University of Padua was then one of the greatest universities in the world, attracting students, teachers, and scholars from throughout Europe.

**palm.** A term used in this book to translate literally Galileo's word *palmo*; this was an ancient inexact unit of length corresponding to either the width of the palm of a hand, or the length of a hand, or the distance from the tip of the thumb to the tip of the little finger when extended.

**parabola.** In mathematics, a plane curve defined as the set of all points equidistant from a fixed straight line (called directrix) and a fixed point (called focus). The shape generated is that of the intersection of a cone and a plane parallel to its side. In *Two New Sciences* Galileo showed that the path of a projectile is a parabola. Another approximate example is the shape of the cable in a suspension bridge.

**parallax.** The change in the apparent position of an observed object due to a change in the actual position of the observer. The amount of the apparent movement of the object is a function of both its distance from the observer and the distance separating the two observations; such apparent positions are usually measured in angles (relative to a given frame of reference), so that triangulation and trigonometry enable us to compute the distance of an object by observing it from two different places separated by a known distance; for example, the peak of a mountain appears located at different angles when viewed from two different places in the valley below, and its distance can be calculated; similarly, by measuring the apparent position of the sun or moon from two different locations on the earth's surface, ancient astronomers had given reasonable estimates of the sun's and moon's distance from the earth.

**parallel.** This term has two relevant meanings, as an adjective and as a noun. Two lines lying in the same plane are said to be parallel to each other when they never meet regardless of how far they are extended; similarly, two planes are said to be parallel to each other when they never meet regardless of how far they are extended. Used as a noun, a parallel is a circle on the surface of a sphere (such as the earth or the stellar sphere) which is parallel to the equator; these parallels become smaller and smaller as one moves on the sphere's surface from the equator to the poles.

**paralogism.** An argument that is illogical for some reason; a fallacy whose error consists in the serious violation of some logical principle.

**Peripatetic.** A Greek word whose literal meaning refers to a person who walks around; a nickname given to Aristotelians at the time of Galileo. They acquired this nickname because in the school founded by Aristotle, the teachers had the habit of walking around while lecturing and discussing.

**phases.** The phases of a nearby heavenly body (like the moon and Venus) are the periodic changes in its apparent shape from round disk, to semicircle, to crescent, and back to semicircle and round disk. They are caused by changes in the relative position between the sun, the earth, and the other body: a crescent is seen when the body is in the region between the earth and the sun; a semicircle is seen when the line connecting the three bodies forms an angle close to a right angle; and a full disk is seen when the body's entire surface illuminated by the sun can be seen from the earth, either because the earth is between the sun and the body (as in the case of the moon) or when the sun is between the earth and the body (as in the case of Venus).

**physics.** The most fundamental of the natural sciences, which studies such phenomena as motion, heat, light, matter, and energy.

**Pisa.** A city in central Italy where Galileo was born and at whose university he studied and then held his first position teaching mathematics (1589–1592). Pisa was then part of the Grand Duchy of Tuscany, whose capital was Florence. The Leaning Tower was already hundreds of years old, and it is alleged that it provided Galileo and his opponents with a handy site to experiment with falling bodies by dropping them from its top; but the reality, significance, and conclusiveness of such experiments are questionable.

**planet.** In the geostatic worldview, a heavenly body that appears to move both around the earth and in relation to other heavenly bodies; that is, a heavenly body which simultaneously performs two motions around the earth, the diurnal motion from east to west every day, and another revolution from west to east in a definite period of time which varies from one planet to another. There were seven planets, and their arrangement in the order of increasing orbit and period was as follows: moon, one month; Mercury, Venus, and sun, one year; Mars, two years; Jupiter, twelve years; and Saturn, twenty-nine years. In the Copernican view, a planet is a heavenly body which revolves around the sun, again in a definite period of time which varies from one planet to another. In Galileo's time there were six known primary planets whose arrangement in the order of increasing orbit and period was: Mercury, three months; Venus, six months; earth, one year; Mars, two years; Jupiter, twelve years; and Saturn, twenty-nine years. There is also one secondary planet (the moon), which is secondary because it revolves around the sun due to the fact that it revolves once a month around the earth while the latter revolves once a year around the sun. The Copernican controversy involved essentially the question of whether the earth or the sun is a planet, for this is equivalent to whether the earth revolves around the sun or vice versa. In Galileo's time, the planets Uranus, Neptune, and Pluto had not yet been discovered.

**Plato** (c. 427–347 B.C.). Greek thinker who dealt mostly with metaphysics, epistemology, ethics, and political theory. His writings laid the foundations for Western philosophy; he was a pupil of Socrates and the teacher of Aristotle, and wrote his works in dialogue form, portraying Socrates as one of the speakers.

**precession of the equinoxes.** Precession means a motion (like that of a spinning top) such that the axis around which a body rotates is not fixed but itself rotates around another axis, thus describing the shape of a cone whose vertex is the intersection of the two axes; the equinoxes here should be conceived not as times of the year but as points on the celestial sphere, namely the points where the ecliptic and the celestial equator intersect; the precession of the equinoxes is the phenomenon that the equinoxes appear to move slowly westward along the ecliptic, with a period now known to be about 26,000 years. The phenomenon was first noted by Hipparchus around 120 B.C., and its period was first estimated to be about 36,000 years; with the benefit of more observational data, Copernicus gave the correct figure of 26,000 years; thus, the vernal equinox took place in the zodiacal constellation of Aries at the time of Hipparchus, but it now occurs in Pisces, which is the next constellation to the west (having moved westward about 1/12 of a circle in about 2,000 years). The phenomenon was first explained by Newton in terms of the gravitational attraction of the sun and moon on the earth's equatorial bulge, which causes the axis of terrestrial rotation to precess.

**premise.** In an argument, a statement on which the conclusion is based; when the reasoning is correct, the premises are said to imply or support the conclusion. The terms premise and reason (in one of its meanings) are used interchangeably.

**Prime Mobile**. A term meant to convey the idea of the "first body in motion." In Aristotelian natural philosophy, the Prime Mobile was a sphere lying outside the stellar sphere and was acted upon by the First Unmoved Mover; by rotating daily, the sphere of the Prime Mobile carried along all the other spherical layers inside it, namely all the other heavenly bodies (except the earth). The Prime Mobile was needed by those Aristotelians for whom the stellar sphere could not be a source of the diurnal motion; in fact, there was evidence that the stellar sphere had another slower movement (the precession of the equinoxes), and the idea was to have a distinct sphere for each distinct movement.

**proposition**. A thought which is expressed in a complete sentence or in a subordinate clause, which is part of an argument or could be made part of an argument, and which is capable of being accepted or rejected. It is synonymous with the terms statement, assertion, thesis, and claim.

**Ptolemy, Claudius** (second century A.D.). Greek mathematician, astronomer, geographer, and author of the *Almagest*; this work contains a complete, detailed, mathematical, and systematic exposition of the geostatic worldview and is the classical synthesis of ancient astronomy.

**Pythagoras** (c. 580–c. 500 B.C.). Greek thinker famous for having founded a school whose followers included the earliest known proponents of a theory of the moving earth. He is also renowned for the discovery of the geometrical theorem (named after him) that in a right triangle the square of the hypotenuse (the side opposite the right angle) equals the sum of the squares of the other two sides.

**question begging**. See *begging the question*.

**rationalism**. A term sometimes used to refer to the methodological principle advocating an excessive emphasis on the use and the power of pure reason and intellectual theorizing in the search for truth. When so defined, rationalism is a synonym for apriorism and the opposite of empiricism, and it should not be confused with reasoning as such or with rational-mindedness.

**rational-mindedness**. The willingness and ability to accept the views supported by the best arguments and strongest evidence.

**realism**. The epistemological doctrine that a scientific theory refers to or describes real entities and processes in the physical world in a manner which is more or less accurate, so that the theory is capable of being more or less true or false. Realism does not deny that a scientific theory is also a more or less convenient instrument of calculation and prediction; but it does deny that it is *merely* an instrument; since the latter is what instrumentalism claims, realism and instrumentalism are incompatible. Realism only attributes a descriptive or referential character to a scientific theory, and does not specify the type of description or reference involved. Thus, a realist need not, but may, also claim that a scientific theory must be true and proved to be true in order to be scientific; this is a distinct methodological doctrine which I call demonstrativism. On the other hand, a realist may, but need not, also claim that a scientific theory can never be absolutely true, but only approximately true, more or less probable, more or less close to the truth, and always falsifiable or revisable in the light of further evidence; this is a distinct methodological doctrine called fallibilism. That is, realism may be combined with either demonstrativism or fallibilism (though not with both since the latter two are mutually incompatible); we thus get two doctrines that may be called, respectively, demonstrativist realism and fallibilist realism. Galileo was clearly a realist in his interpretation of Copernicanism, and the *Dialogue* suggests that he was also a fallibilist.

**reason**. One meaning of this term is equivalent to *premise*. In another one of its meaning, reason is the human ability to engage in *reasoning*. A third meaning, avoided in this book, is that reason is the aspect of the mind that engages in *apriorism*.

**reasoning**. A form of thinking consisting of giving reasons for conclusions, reaching conclusions on the basis of reasons, or drawing consequences from premises; that is, reasoning is the interrelating of our thoughts in such a way as to make some thoughts dependent on others, and this interdependence can take the form of some thoughts being based on others or some thoughts following from others.

**reductio ad absurdum**. Type of argument in which the conclusion is proved by assuming its falsity and showing that this assumption implies an absurdity; this absurdity can be an explicit contradiction, namely that a particular proposition is both true and false; or it can be the truth of the conclusion to be proved, which is absurd from the viewpoint of the assumption made; or it can be a proposition that is false in some obvious sense.

**relativity of motion**. In modern physics, the principle of the relativity of motion states that motion is relative in the sense that uniform motion can be detected only in relation to something which does not share it, so that there is no way of determining whether a system is in uniform motion by observations within the system. This principle mentions only uniform motion, and so accelerated motion is not subject to this restriction; further, the principle needs to be generalized and revised in various ways, along the lines developed by Einstein.

**"Reply to Ingoli"**. An essay written and privately circulated by Galileo in 1624 when he began working actively on a book that later became the *Dialogue*. It is a critical examination of the anti-Copernican arguments in Ingoli's 1616 essay "On the Location and Rest of the Earth, Against the Copernican System"; it was, for the most part, incorporated into the *Dialogue*. It discusses both annual and diurnal motion, and both physical and astronomical arguments (but no methodological or theological arguments). Francesco Ingoli (1578–1649) was a clergyman who by 1624 had become secretary of the newly created Congregation for the Propagation of the Faith.

**revolution**. A term used to refer to a type of motion consisting of a body moving in an orbit around another. In the geostatic system, all heavenly bodies perform revolutions around the earth; in the Copernican system, only the moon performs a monthly revolution around the earth, while the earth and all the other planets perform revolutions around the sun. In a revolution, one body is said to *revolve* around another. This ought to be distinguished from the term *rotation*, which refers to the motion of a body around itself. Though distinct, revolution and rotation are not incompatible; for example, in the

Copernican system the earth simultaneously rotates daily around its own axis and revolves yearly in a heliocentric orbit.

**rhetoric**. The skill consisting of creating and using verbal persuasion, expression, and communication; or the discipline that studies forms and principles of verbal persuasion, expression, and communication.

**Rome**. A city in central Italy, seat of the ancient Roman Republic and Roman Empire and of the Roman Catholic Church; until 1870 capital of the Papal States, a political entity ruled by the pope; since 1870 capital of a united Italy. As a Catholic, Galileo had a special relationship to Rome, but as a native and resident of Tuscany, he was something of a foreigner in Rome (because of the political subdivision of Italy into many independent states).

**rotation**. A useful term to refer specifically to that type of motion consisting of a body spinning around itself, namely around an imaginary line going through the body; this imaginary line is called the *axis* of rotation. In the Copernican system, the earth rotates daily around a line going through its center and through the north and south poles. This is to be distinguished from the term *revolution*, which refers to the motion of a body around another. Though distinct, the two motions are not incompatible; for example, in the Copernican system the earth simultaneously rotates daily around its own axis and revolves yearly in an orbit around the sun.

**Saturn**. A planet whose orbit is bigger than the annual orbit and that takes about twenty-nine years to complete it. In the geostatic system, Saturn is the seventh planet from the earth; in the Copernican system, it is the sixth planet from the sun. This difference in numerical sequence is due to the fact that the moon is a planet only in the geostatic system, whereas it is a satellite of the earth in the Copernican system. In Galileo's time, Saturn was regarded as the outermost planet in both systems, because the planets Neptune, Uranus, and Pluto were not yet discovered. As for the case of all other planets, this word is also the name of one of the ancient Roman gods; Saturn was the father of Jupiter.

**Scheiner, Christopher** (1573–1650). A Jesuit professor of mathematics who engaged in a dispute with Galileo over priority in the discovery of sunspots and over their interpretation. One of his students was Ioannes G. Locher, the author in 1614 of the anti-Copernican "booklet" widely and severely criticized in the *Dialogue*.

**second** (of arc). A unit of angular distance and measurement defined as the 60th part of a minute of arc, which in turn is the 60th part of a degree, which is the 360th part of a circle; thus, a circle is subdivided into 360 degrees, a degree into 60 minutes, and a minute into 60 seconds.

**second law of motion**. See *Newton's laws of motion*.

***Sidereal Messenger***. A short book Galileo published in 1610 announcing his first telescopic discoveries: lunar mountains, Jupiter's satellites, and the stellar composition of nebulas and of the Milky Way.

**simplicity**. The methodological principle of simplicity holds that a simpler theory is better than or preferable to a less simple one. Some versions of this are unobjectionable. But there is controversy about the relative importance of simplicity vis-à-vis other criteria and about the exact meaning of simplicity.

For example, does simplicity require only the smallest possible number of assumptions and entities? The idea of simplicity is itself far from simple.

**skepticism**. The methodological view that knowledge is impossible and unattainable (though it may perhaps be sought after). It is opposed to dogmatism, which is the uncritical acceptance of an authority or source of knowledge, presupposing the idea that the acquisition of knowledge is easy and unproblematic. It is possible to reject both skepticism and dogmatism and emphasize that, although the path to knowledge is tortuous and difficult, fallible knowledge is attainable and that not all knowledge claims are equally worthless or equally legitimate.

**Socrates** (469–399 B.C.). Greek thinker who is one of the greatest philosophers, teachers, and moralists of all times, and is known to us only from the accounts of others (like his pupil Plato) because he did not write down his ideas. He was especially interested in questions of good and evil and the meaning of life, and it is thus ironic that his fellow citizens in Athens tried and executed him for atheism and corrupting the youth. Two basic principles he held and lived by are that the unexamined life is not worth living and that true wisdom requires an awareness of the limitations of one's knowledge. These principles were the result of a lifelong search to clarify the meaning of the Delphic Oracle which had declared that no one in the world was wiser than Socrates.

**Socratic method**. A method of teaching and of justification (invented and practiced by Socrates) in which the teacher or proponent engages in a dialogue with the pupil or opponent, focuses on asking questions rather than giving answers, stresses negative criticism of the other party's answers (to awaken his curiosity), and leads him gradually to arrive at the truth or work out the correct answer himself (as if he were discovering it himself). The last feature is especially important and is called "maieutics" (meaning "midwifery"), for in such a process the teacher is an intellectual midwife who helps to give birth to ideas, but the other person is the one who must undergo the labor of bringing them forth.

**solipsism**. This term has several distinct but related meanings. Taken metaphysically, it is the thesis that the self and only the self exists. Taken methodologically, it is the thesis that the self can be aware and have knowledge of only its own experiences, namely that the individual self is sufficient unto itself for the acquisition of knowledge and so does not need to rely on the efforts and authority of others. Methodological solipsism is thus independent-mindedness carried to the absurd extreme of completely disregarding all authorities, so as to reinvent the wheel by oneself. True independent-mindedness avoids going that far, just as it avoids going to the opposite extreme of uncritically accepting authority.

**solstice**. Either of the two times of the year when the apparent position of the sun is farthest from the celestial equator, so that for about one day the sun does not seem to move either northward or southward. In the Northern Hemisphere, this occurs about June 21 (called the *summer solstice*) and about December 22 (called the *winter solstice*). The term is also used to refer to the corresponding points in the apparent orbit of the sun on the celestial sphere.

**subargument**. An argument that is part of another bigger and more complex argument.

**sunspot**. A phenomenon discovered by Galileo and others consisting of dark patches visible in the solar disk, which appear and disappear seemingly at random, but which seem to move with some regularity while they last. This discovery raised two main scientific issues. The first was whether sunspots are real or (for example) optical illusions of the telescope. The second was the interpretation of sunspots: Scheiner wanted to interpret them as swarms of small, previously undetected planets that obscure various parts of the solar disc when a sufficiently large number of them get in the line of sight of a terrestrial observer; this (incorrect) interpretation enabled Scheiner to retain heavenly unchangeability, and with it the earth-heaven dichotomy and the geostatic worldview; Galileo argued in support of the (correct) interpretation that the spots are physical processes originating in the body of the sun and taking place on its surface; this makes them part of the sun, thus undermining heavenly unchangeability, the earth-heaven dichotomy, and the anti-Copernican objection based thereupon; and from their motion he arrived at the further discovery that the sun is not motionless but rotates on its axis once a month.

***Sunspot Letters***. A book published by Galileo in 1613, stemming from a series of letters he exchanged with Jesuit astronomer Christopher Scheiner by way of a German intermediary named Mark Welser. Galileo's dispute with Scheiner involved two aspects: the question of priority in the discovery of sunspots and the issue of how to interpret these phenomena. The controversy did not stop in 1613, and the *Dialogue* is one of several works where it reappears.

**superior planets**. Those planets that have orbits larger than the ecliptic or annual orbit, such as Mars, Jupiter, and Saturn. They are contrasted with the *inferior planets*, namely Mercury and Venus, whose orbits are smaller than the ecliptic or annual orbit.

**supernova**. See *nova*.

**superposition of motion**. A basic principle of modern physics, specifying how to combine two motions to yield a resultant one, or conversely how to analyze a given motion into its component parts. The key idea is that when a body is moving simultaneously in two different directions, one draws a parallelogram whose sides lie along these directions and whose lengths equal respectively the magnitudes of the two speeds along these directions; then the body's actual trajectory is along the diagonal of this parallelogram, and its speed is measured by the length of this diagonal. Galileo had a clear intuition of this principle and used it widely, although at the time it was controversial because it conflicted with Aristotelian physics. For example, on a ship moving forward on a calm sea, if one drops a rock from the top of its mast, the rock's motion of fall will combine with the original forward horizontal motion in such a way as to make it strike the deck at the foot of the mast, just as would happen on a motionless ship; more exactly, as Galileo demonstrated in his *Two New Sciences*, the motion of a projectile follows a parabolic path because a parabola results from the superposition of the projectile's constant horizontal motion and its accelerated downward fall (where the acceleration is such that the distance fallen increases as the square of the time elapsed, in accordance with the law of squares).

**syllogism**. As a technical term, a syllogism is an argument with two premises, one conclusion, and three terms, consisting entirely of statements of the following four types: all A are B, no A are B, some A are B, and some A are not B. Examples of syllogisms are: "all A are B, all B are C, so all A are C" and "no A are B, some C are A, so some C are not B." More generally, the term syllogism is also used to refer to any explicitly stated argument that claims to be especially attentive to logical rules.

**tangent**. In geometry, a straight line touching the circumference of a circle in just one point and at a right angle to the radius to that point.

**teleology**. The belief in or the study of the existence of purpose or design in nature. For example, a teleological view of the world may involve the claim that everything was created or designed by God for some purpose, such as the good of mankind. The following two Aristotelian arguments provide good examples of teleological thinking: (1) the heavens are unchangeable because heavenly changes would be of no benefit to mankind, and hence they would be useless (namely, they would be without purpose); (2) the fixed stars cannot be at such a distance as 10,000 times farther than the sun because in that case there would be a lot of empty space between Saturn (the outermost planet) and the stars, and such space would be useless.

**telescope**. An optical instrument consisting of an arrangement of lenses, mirrors, or both which magnifies the image of distant objects so that they appear larger or nearer, thus rendering our vision more powerful and enabling us to see things which cannot be seen by the naked eye. Invented in Holland in 1608 for unclear purposes, the following year it was significantly improved by Galileo and turned into an effective scientific instrument for acquiring new knowledge about the world. With it he soon discovered the moon's mountains, Jupiter's satellites, the stellar composition of nebulas, the phases of Venus, and sunspots; this new evidence led him and others to a favorable reevaluation of Copernicanism. At first the new instrument was controversial and raised questions about whether it was legitimate to use, whether it was reliable, how and why it worked, how it could be replicated, how one could learn to use it, and even what it should be called.

**theology**. The scholarly study of the nature of God and such related topics as religion, the Bible, churches, revelation, the supernatural, and religious experience. Some related issues in the Copernican controversy were whether theological arguments have any role in the scientific study of the physical world, and whether theological or scientific arguments have priority in cases where there is a conflict between the two.

**third law of motion**. See *Newton's laws of motion*.

**tides**. A phenomenon observable at most seashores and consisting of the periodic alternate motion of sea water; in a harbor the sea level can be seen to rise and fall a number of times every day, whereas on a beach one notices primarily a sequence of water flowing inland and ebbing back to sea; the most common interval between high and low tide is six hours, which is to say that there are two high tides and two low tides per day. This phenomenon had long puzzled natural philosophers, and in Galileo's time there was still no satisfactory explanation of how and why the tides occur. He elaborated and added a new twist to an idea first advanced by others; he claimed that the tides are caused by the earth's motion, so that their occurrence provides indirect physical evidence for Copernicanism; this causal explanation is erroneous, although his supporting argument is not worthless, and it is not clear where his reasoning goes wrong. Nowadays, following Newton, physicists explain the tides in terms of the gravitational force exerted on the oceans by the moon (and sun), which attract different parts of seawater with unequal force; but this general theory is incapable of predicting the precise details, which must be calculated from a mathematical analysis of empirical records.

**Tuscany**. A region in central Italy with Florence as capital and extending to Pisa and other nearby cities. In Galileo's time it was an independent state called a grand duchy because it was ruled by a grand duke. Galileo regarded it as his homeland, even in comparison with other parts of Italy, having been born in Pisa into a family which had moved there from Florence.

***Two New Sciences.*** A book Galileo published in Holland in 1638 that formulates some fundamental principles of the science of motion and the science of the strength of materials; it contains no discussion of Copernicanism or astronomy. It is his most important book from a purely scientific viewpoint; the part on the laws of motion is primarily an exposition of research carried out as a university professor before 1609, when his telescopic discoveries thrust him into the Copernican controversy.

**Tycho Brahe** (1546–1601). Danish astronomer, best known as an incomparable observer and collector of data and as the originator of the so-called Tychonic system, according to which the earth is motionless at the center of the universe, but the planets revolve around the sun and are carried by it around the earth. Kepler worked with him and inherited his data.

**Urban VIII** (1568–1644). Pope from 1623 to 1644, whose personal name was Maffeo Barberini. At first he was a supporter and admirer of Galileo; for example, in 1616 he was instrumental in preventing stronger measures against Copernicanism, and in 1620 he dedicated a poem to Galileo; in 1623 the latter, in turn, dedicated *The Assayer* to him. But Urban's attitude soured after the *Dialogue* was published, partly for nonintellectual reasons involving the politics of the Thirty Years War (1618–1648) between Catholics and Protestants; another reason involved the interpretation of Copernicanism, in regard to which Urban was inclined toward skepticism and instrumentalism.

**Venice**. A seaport in northern Italy which in Galileo's time was the capital of an independent republic comprising many coastal areas of northeastern Italy and the eastern Mediterranean. As professor of mathematics at the University of Padua, he was a resident of this republic, a state employee, and a frequent visitor of the capital city for eighteen years (1592–1610). The *Dialogue* is written as a discussion taking place in Venice.

**Venus**. A planet which revolves in its orbit in such a way that it always appears close to the sun (but not as close as Mercury, which behaves similarly). In the Copernican system, Venus is the second planet from the sun and completes its orbit in about six months. In the geostatic system, opinions differed about whether it was the second, third, or fourth planet from the earth; but it was most commonly regarded as the third (between Mercury and the sun). With the telescope, Galileo discovered the phases of Venus, which are changes in its apparent shape similar to those which the moon exhibits each month; this proved conclusively that Venus revolves around the sun as Copernicanism claims, although the rest of the Copernican system required other evidence for a conclusive demonstration. The word Venus also refers to the goddess of beauty and love in ancient Roman religion.

**violent motion**. In Aristotelian natural philosophy, violent motion is motion that occurs because of the influence of some external force; examples are the motion of a cart pulled by a horse, the motion of a rowboat pushed by rowing, and the lifting of a weight with a pulley; violent motion is contrasted with *natural motion*. Galileo partly accepted and partly modified this doctrine of violent motion. But even to the extent that there was agreement about the general concept, there often was disagreement about particular cases; for example, one aspect of the Copernican controversy was whether the earth's daily axial rotation and its annual orbital revolution are instances of natural motion or violent motion.

# Bibliography


Aiton, E. J. 1954. Galileo's Theory of the Tides. *Annals of Science* 10:44–57.

———. 1963. On Galileo and the Earth-Moon System. *Isis* 54:265–66.

———. 1965. Galileo and the Theory of the Tides. *Isis* 56:56–61.

Angell, R. B. 1964. *Reasoning and Logic*. New York: Appleton.

Ariew, R. 1987. The Phases of Venus before 1610. *Studies in History and Philosophy of Science* 18:81–92.

Barenghi, G. 1638. *Considerazioni sopra il Dialogo*. Pisa.

Barone, F. 1972. Galileo e la logica. In *Saggi su Galileo Galilei*, ed. C. Maccagni, 52–70. Florence: Barbèra.

Battersby, M. E. 1989. Critical Thinking as Applied Epistemology. *Informal Logic* 11:91–100.

Biagioli, M. 1993. *Galileo Courtier*. Chicago: University of Chicago Press.

Blackwell, R. J. 1991. *Galileo, Bellarmine, and the Bible*. Notre Dame: University of Notre Dame Press.

———, ed. and trans. 1994. *A Defense of Galileo, the Mathematician from Florence by Thomas Campanella*. Notre Dame: University of Notre Dame Press.

Brahe, Tycho. 1596. *Epistolae astronomicae*. Uraniborg.

———. 1602. *Astronomiae instauratae Progymnasmata*. Uraniborg.

Brown, H. I. 1976. Galileo, the Elements, and the Tides. *Studies in History and Philosophy of Science* 7:337–51.

———. 1977. *Perception, Theory and Commitment*. Chicago: Precedent Publishing Co.

———. 1985. Galileo on the Telescope and the Eye. *Journal of the History of Ideas* 46:487–501.

———. 1987. *Observation and Objectivity*. New York: Oxford University Press.

———. 1988. *Rationality*. London: Routledge.

Burstyn, H. L. 1962. Galileo's Attempt to Prove that the Earth Moves. *Isis* 53:161–85.

———. 1963. Galileo and the Earth-Moon System. *Isis* 54:400–401.

——. 1965. Galileo and the Theory of the Tides. *Isis* 56:61–63.
Burtt, E. A. 1954. *Metaphysical Foundations of Modern Physical Science*. Garden City, New York: Doubleday.
Butterfield, H. 1949. *The Origins of Modern Science, 1300–1800*. London: Bell.
Butts, R. E., and J. C. Pitt, eds. 1978. *New Perspectives on Galileo*. Dordrecht: Reidel.
Chalmers, A. 1985. Galileo's Telescopic Observations of Venus and Mars. *British Journal for the Philosophy of Science* 36:175–83.
——. 1993. Galilean Relativity and Galileo's Relativity. In *Correspondence, Invariance and Heuristics*, ed. S. French and H. Kamminga, 189–205. Dordrecht: Kluwer.
Chalmers, A., and R. Nicholas. 1983. Galileo and the Dissipative Effects of a Rotating Earth. *Studies in History and Philosophy of Science* 14:315–40.
Chiaramonti, S. 1621. *Anti-Tycho*. Venice.
——. 1628. *De tribus novis stellis quae annis 1572, 1600, 1604 comparuere* . . . . Cesena.
——. 1633. *Difesa . . . al suo Antiticone e Libro delle tre nuove Stelle*. Florence: Landini.
Clavelin, M. 1964. Galilée et le refus de l'equivalence des hypotheses. *Revue d'Histoire des Sciences* 17:305–30.
——. 1974. *The Natural Philosophy of Galileo*. Trans. A.J. Pomerans. Cambridge: M. I. T. Press.
Coffa, J. A. 1968. Galileo's Concept of Inertia. *Physis* 10:261–81.
Cohen, H. F. 1994. *The Scientific Revolution*. Chicago: University of Chicago Press.
Cohen, I. B. 1960. *The Birth of a New Physics*. Garden City, New York: Doubleday.
——. 1967. Newton's Attribution of the First Two Laws of Motion to Galileo. In *Symposium Internazionale di Storia, Metodologia, Logica e Filosofia della Scienza*, xxv–xliv. Florence: Gruppo Italiano di Storia delle Scienze.
Cohen, L. J. 1981. Can Human Irrationality Be Experimentally Demonstrated? *The Behavioral and Brain Sciences* 4:317–31, 359–67.
——. 1989. *An Introduction to the Philosophy of Induction and Probability*. Oxford: Clarendon Press.
Colombe, L. delle. 1611. *Contro il Moto della Terra*. In Favaro, 1890–1909, 3:251–90.
Copernicus, N. 1992. *On the Revolutions*. Trans. and ed. E. Rosen. Baltimore: Johns Hopkins University Press.
Coyne, G. V., M. Heller, and J. Zycinski, eds. 1985. *The Galileo Affair*. Vatican City: Specola Vaticana.
Crombie, A. C. 1967. The Mechanistic Hypothesis and the Scientific Study of Vision. In *Historical Aspects of Microscopy*, ed. S. Bradbury and G. L'E. Turner, 3–112. Cambridge, England: Heffer.
Drake, S., ed. and trans. 1957. *Discoveries and Opinions of Galileo*. Garden City, New York: Doubleday.
——, ed. 1967. *Dialogue Concerning the Two Chief World Systems*. 2nd revised edition. Berkeley: University of California Press.

——. 1970. *Galileo Studies*. Ann Arbor: University of Michigan Press.
——, ed. and trans. 1976. *Galileo against the Philosophers*. Los Angeles: Zeitlin & Ver Brugge.
——. 1978. *Galileo at Work*. Chicago: University of Chicago Press.
——. 1979. History of Science and Tide Theories. *Physis* 21:61–69.
——. 1980. *Galileo*. New York: Hill & Wang.
——. 1981. *Cause, Experiment and Science*. Chicago: University of Chicago Press.
——. 1982. Foreword and Additional Notes to Sexl and Von Meyenn (1982, XIII–XXVI and 575–88).
——. 1983. *Telescopes, Tides, and Tactics*. Chicago: University of Chicago Press.
——. 1984. Galileo, Kepler and the Phases of Venus. *Journal of the History of Astronomy* 15:198–208.
——. 1986a. Galileo and the Projection Argument. *Annals of Science* 43:77–79.
——. 1986b. Reexamining Galileo's *Dialogue*. In Wallace (1986, 155–75).
——. 1990. *Galileo: Pioneer Scientist*. Toronto: University of Toronto Press.
Drake, S., and C. D. O'Malley, eds. and trans. 1960. *The Controversy on the Comets of 1618*. Philadelphia: University of Pennsylvania Press.
Dreyer, J. L. E. 1953. *A History of Astronomy from Thales to Kepler*. Revised by W. H. Stahl. New York: Dover.
Duhem, P. 1954. *The Aim and Structure of Physical Theory*. Trans. P. P. Wiener. Princeton: Princeton University Press.
——. 1969. *To Save the Phenomena*. Trans. E. Doland and C. Maschler. Chicago: University of Chicago Press.
Einstein, A. 1934. On the Method of Theoretical Physics. In *The World As I See It*. New York: Covici Friede Publishers.
Favaro, A. 1883. *Galileo e lo Studio di Padova*. Florence.
——, ed. 1890–1909. *Le opere di Galileo Galilei*. 20 vols. National Edition. Florence: Barbèra.
Feldhay, R. 1995. *Galileo and the Church*. Cambridge: Cambridge University Press.
Feyerabend, P. K. 1964. Realism and Instrumentalism. In *The Critical Approach to Science and Philosophy*, ed. M. Bunge, 280–308. New York: Free Press.
——. 1975. *Against Method*. Atlantic Highlands, New Jersey: Humanities Press.
Finocchiaro, M. A. 1973. *History of Science as Explanation*. Detroit: Wayne State University Press.
——. 1974. The Concept of *Ad Hominem* Argument in Galileo and Locke. *The Philosophical Forum* 5:394–404.
——. 1975. Dialectical Aspects of the Copernican Revolution. In Westman (1975a, 204–12).
——. 1977a. Logic and Rhetoric in Lavoisier's Sealed Note. *Philosophy and Rhetoric* 10:111–22.
——. 1977b. Review of *Galileo against the Philosophers*, by S. Drake. *Isis* 68:645–47.
——. 1980. *Galileo and the Art of Reasoning: Rhetorical Foundations of Logic and Scientific Method*. Boston: Reidel.
——. 1985. Wisan on Galileo and the Art of Reasoning. *Annals of Science* 42:613–16.

——. 1986. The Methodological Background to Galileo's Trial. In Wallace (1986, 241–72).
——. 1988a. Empiricism, Judgment, and Argument. *Argumentation* 2:313–35.
——. 1988b. Galileo's Copernicanism and the Acceptability of Guiding Assumptions. In *Scrutinizing Science*, ed. A. Donovan et al., 49–67. Dordrecht: Kluwer.
——. 1988c. *Gramsci and the History of Dialectical Thought*. Cambridge: Cambridge University Press.
——, ed. and trans. 1989. *The Galileo Affair: A Documentary History*. Berkeley: University of California Press.
——. 1990. Varieties of Rhetoric in Science. *History of the Human Sciences* 3:177–93.
——. 1992a. Asymmetries in Argumentation and Evaluation. In *Argumentation Illuminated*, ed. F. H. van Eemeren et al., 62–72. Amsterdam: Stichting International Centre for the Study of Argumentation.
——. 1992b. To Save the Appearances: Duhem on Galileo. *Revue internationale de philosophie* 46:291–310.
——. 1994. Two Empirical Approaches to the Study of Reasoning. *Informal Logic* 16:1–21.
Fiorentino, P. 1868. *Pietro Pompanazzi*. Florence.
Franklin, A. 1976. *The Principle of Inertia in the Middle Ages*. Boulder: Colorado Associated University Press.
——. 1986. *The Neglect of Experiment*. Cambridge: Cambridge University Press.
——. 1990. *Experiment, Right or Wrong*. Cambridge: Cambridge University Press.
——. Forthcoming. Mechanics, Aristotelian. In *Encyclopedia of Philosophy*, ed. E. Craig. London: Routledge.
Freeman, J. B. 1988. *Thinking Logically*. Englewood Cliffs, New Jersey: Prentice Hall.
——. 1991. *Dialectics and the Macrostructure of Arguments*. New York: Foris Publications.
Galilei, G. 1632. *Dialogo di Galileo Galilei Linceo . . . .* Florence: Landini.
——. 1890–1909. *Le Opere di Galileo Galilei*. 20 vols. National Edition by A. Favaro. Florence: Barbèra.
——. 1967. *Dialogue Concerning the Two Chief World Systems*. Trans. and ed. S. Drake. 2nd revised edition. Berkeley: University of California Press.
——. 1974. *Two New Sciences*. Trans. and ed. S. Drake. Madison: University of Wisconsin Press.
Galileo. See "Galilei, G."
Galluzzi, P., ed. 1984. *Novità celesti e crisi del sapere*. Florence: Giunti Barbèra.
Gardner, M. R. 1983. Realism and Instrumentalism in Pre-Newtonian Astronomy. In *Testing Scientific Theories*, ed. J. Earman, 201–66. Minneapolis: University of Minnesota Press.
Gaukroger, S. 1978. *Explanatory Structures*. Atlantic Highlands, New Jersey: Humanities Press.
Geymonat, L. 1965. *Galileo Galilei: A Biography and Inquiry into his Philosophy of Science*. Trans. S. Drake. New York: McGraw-Hill.

Gingerich, O. 1992. *The Great Copernicus Chase and Other Adventures in Astronomical History*. Cambridge, Mass.: Sky Publishing Corporation.

———. 1993. *The Eye of Heaven*. New York: American Institute of Physics.

Goosens, W. K. 1980. Galileo's Response to the Tower Argument. *Studies in History and Philosophy of Science* 11:215–27.

Govier, T. 1985. *A Practical Study of Argument*. Belmont, Calif.: Wadsworth.

Gross, A. G. 1990. *The Rhetoric of Science*. Cambridge: Harvard University Press.

Hacking, I. 1983. *Representing and Intervening*. Cambridge: Cambridge University Press.

Hall, A. R. 1954. *The Scientific Revolution, 1500–1800*. London: Longmans, Green, & Co.

Harris, W. H., and J. S. Levey, eds. 1975. *The New Columbia Encyclopedia*. New York: Columbia University Press.

Hatfield, G. 1990. Metaphysics and the New Science. In Lindberg and Westman (1990, 93–166).

Herivel, J. 1965. *The Background to Newton's Principia*. Oxford: Clarendon.

Hill, D. K. 1984. The Projection Argument in Galileo and Copernicus. *Annals of Science* 41:109–33.

———. 1986. Galileo's Work on 116v. *Isis* 77:283–91.

———. 1988. Dissecting Trajectories. *Isis* 79:646–68.

Jardine, N. 1976. Galileo's Road to Truth and Demonstrative Regress. *Studies in History and Philosophy of Science* 7:277–318.

———. 1984. *The Birth of History and Philosophy of Science*. Cambridge: Cambridge University Press.

John Paul II. 1979. Deep Harmony which Unites the Truths of Science with the Truths of Faith. In *Galileo Galilei*, ed. P. Poupard, 195–200. Trans. I. Campbell. Pittsburgh: Duquesne University Press, 1987.

Johnson, R. H., and J. A. Blair. 1977. *Logical Self Defense*. Toronto: McGraw-Hill Ryerson.

King, H. C. 1955. *The History of the Telescope*. London: C. Griffin.

Kitchener, K. S. 1983a. Cognition, Metacognition, and Epistemic Cognition. *Human Development* 26:222–32.

———. 1983b. Educational Goals and Reflective Thinking. *Educational Forum* 48(1):75–95.

Kitchener, K. S., and K. W. Fisher. 1990. A Skill Approach to the Development of Reflective Thinking. In *Developmental Perspectives on Teaching and Learning Thinking Skills*, ed. D. Kuhn, 48–62. Basel, Switzerland: Karger.

Kitcher, P. 1983. *The Nature of Mathematical Knowledge*. New York: Oxford University Press.

Koestler, A. 1959. *The Sleepwalkers*. New York: Macmillan.

Koyré, A. 1965. *Newtonian Studies*. London: Chapman & Hall.

———. 1966. *Etudes galiléennes*. Paris: Hermann.

———. 1978. *Galileo Studies*. Trans. J. Mepham. Hassocks, England: The Harvester Press.

Kuhn, T S. 1957. *The Copernican Revolution*. Cambridge: Harvard University Press.

———. 1970. *The Structure of Scientific Revolutions*. 2nd edition. Chicago: University of Chicago Press.

Lakatos, I. 1978. *Mathematics, Science, and Epistemology*. Ed. J. Worrall and G. Currie. Cambridge: Cambridge University Press.

Lakatos, I., and E. Zahar. 1975. Why Did Copernicus' Research Program Supersede Ptolemy's? In Westman (1975a, 354–83).

Langford, J. J. 1966. *Galileo, Science and the Church*. Ann Arbor: University of Michigan Press.

Laudan, L. 1977. *Progress and Its Problems*. Berkeley: University of California Press.

Levere, T. H., and W. R. Shea, eds. 1990. *Nature, Experiment, and the Sciences*. Dordrecht: Kluwer.

Lindberg, D. C. 1992. *The Beginnings of Western Science*. Chicago: University of Chicago Press.

Lindberg, D. C., and R. S. Westman, eds. 1990. *Reappraisals of the Scientific Revolution*. Cambridge: Cambridge University Press.

Locher, I. G. 1614. *Disquisitiones mathematicae de controversiis ac novitatibus astronomicis*. Ingolstadt.

Lupoli, A. 1986. Il *Dialogo* e la filosofia implicita di Galilei. *Rivista di storia della filosofia* 41:75–89.

Machamer, P. K. 1973. Feyerabend and Galileo. *Studies in History and Philosophy of Science* 4:1–46.

———. 1978. Galileo and the Causes. In Butts and Pitt (1978, 161–80).

MacLachlan, J. 1973. The Test of an 'Imaginary' Experiment of Galileo's. *Isis* 64:374–79.

———. 1977. Mersenne's Solution for Galileo's Problem of the Rotating Earth. *Historia Mathematica* 4:173–82.

———. 1990. Drake against the Philosophers. In Levere and Shea (1990, 123–44).

Margolis, H. 1987. *Patterns, Thinking, and Cognition*. Chicago: University of Chicago Press.

———. 1991. Tycho's System and Galileo's *Dialogue*. *Studies in History and Philosophy of Science* 22:259–75.

———. 1993. *Paradigms and Barriers*. Chicago: University of Chicago Press.

McGuiness, C. 1990. Talking about Thinking. In *Lines of Thinking*. Vol. 2, *Skills, Emotion, Creative Processes, Individual Differences and Teaching Thinking*, ed. K. J. Kilhooly et al., 301–12. New York: John Wiley,

McMullin, E., ed. 1967. *Galileo: Man of Science*. New York: Basic Books.

———. 1976. The Fertility of Theory and the Unit of Appraisal in Science. In *Essays in Memory of Imre Lakatos*, ed. R. S. Cohen et al., 395–432. Dordrecht: Reidel.

———. 1978. The Conception of Science in Galileo's Work. In Butts and Pitt (1978, 209–57).

———. 1990. Conceptions of Science in the Scientific Revolution. In Lindberg and Westman (1990, 27–92).

Meichenbaum, D. 1986. Metacognitive Methods of Instruction. In *Facilitating Cognitive Development*, ed. M. Schwebel and C. A. Maher, 23–32. New York: Haworth.

Mertz, D. W. 1980. On Galileo's Method of Causal Proportionality. *Studies in History and Philosophy of Science* 11:229–42.

Millman, A. B. 1976. The Plausibility of Research Programs. In *PSA 1976: Proceedings of the 1976 Biennial Meeting of the Philosophy of Science Association*, ed. F. Suppe and P. D. Asquith, 1:140–48. East Lansing, Mich.: Philosophy of Science Association.

Morpurgo-Tagliabue, G. 1981. *I processi di Galileo e l'epistemologia*. Rome: Armando.

Moss, J. D. 1993. *Novelties in the Heavens*. Chicago: University of Chicago Press.

Naylor, R. H. 1974a. Galileo and the Problem of Free Fall. *British Journal for the History of Science* 7:105–34.

———. 1974b. Galileo's Simple Pendulum. *Physis* 16:23–46.

———. 1976. Galileo: Real Experiment and Didactic Demonstration. *Isis* 67:398–419.

———. 1990. Galileo's Method of Analysis and Synthesis. *Isis* 81:695–707.

Newton, I. 1934. *Mathematical Principles of Natural Philosophy*. Motte's translation revised by F. Cajori. Berkeley: University of California Press.

Nickles, T. 1980. Introductory Essay. In *Scientific Discovery, Logic, and Rationality*, ed. T. Nickles, 1–59. Boston: Reidel.

Oregius, A. 1629. *De Deo uno*. Rome.

Pagano, S. M., ed. 1984. *I documenti del processo di Galileo Galilei*. Vatican City: Pontificia Academia Scientiarum.

Pagnini, P., ed. 1964. *Galileo Galilei: Opere*. 5 vols. Florence: Salani.

Paris, S. G., and P. Winograd. 1990. How Metacognition Can Promote Academic Learning and Instruction. In *Dimensions of Thinking and Cognitive Instruction*, ed. B. F. Jones and L. Idol, 15–51. Hillsdale, New Jersey: Lawrence Erlbaum.

Pera, M. 1994. *The Discourses of Science*. Chicago: University of Chicago Press.

Perelman, C., and L. Olbrechts-Tyteca. 1969. *The New Rhetoric*. Trans. J. Wilkinson and P. Weaver. Notre Dame: University of Notre Dame Press.

Perkins, D. N. 1989. Reasoning as It Is and as It Could Be. In *Thinking Across Cultures*, ed. D. N. Topping et al., 175–94. Hillsdale, New Jersey: Lawrence Erlbaum.

Perkins, D. N., R. Allen, and J. Hafner. 1983. Difficulties in Everyday Reasoning. In *Thinking: The Expanding Frontier*, ed. W. Maxwell, 177–89. Philadelphia: The Franklin Institute Press.

Pitt, J. C. 1992. *Galileo, Human Knowledge, and the Book of Nature*. Boston: Kluwer.

Popper, K. R. 1959. *The Logic of Scientific Discovery*. New York: Harper.

Prelli, L. J. 1989. *A Rhetoric of Science*. Columbia: University of South Carolina Press.

Ptolemy, C. 1984. *Ptolemy's Almagest*. Ed. and trans. G. J. Toomer. New York: Springer.

Redondi, P. 1987. *Galileo Heretic*. Trans. R. Rosenthal. Princeton: Princeton University Press.

Reston, J., Jr. 1994. *Galileo: A Life*. New York: HarperCollins.

Ronan, C. A. 1974. *Galileo*. New York: G. P. Putnam's Sons.

Ronchi, V. 1958. *Il cannochiale di Galilei e la scienza del seicento*. 2nd edition. Turin: Einaudi.

Rosen, E. 1947. *The Naming of the Telescope*. New York: H. Schuman.
——, ed. 1959. *Three Copernican Treatises*. 2nd edition. New York: Dover.
——, ed. and trans. 1992. *Nicholas Copernicus: On the Revolutions*. Baltimore: Johns Hopkins.
Rosenkrantz, R. D. 1977. *Inference, Method and Decision*. Dordrecht: Reidel.
Salusbury, T. 1661. *Mathematical Collections and Translations*. Tome 1, pt. 1. London.
Santillana, G. de, ed. 1953. *Dialogue on the Great World Systems*. Rev. edition of T. Salusbury's translation. Chicago: University of Chicago Press.
——. 1955a. *The Crime of Galileo*. Chicago: University of Chicago Press.
——, ed. 1955b. *Dialogue on the Great World Systems*. Abridged text edition. Chicago: University of Chicago Press.
Schilpp, P. A., ed. 1951. *Albert Einstein, Philosopher-Scientist*. Evanston, Ill.: Library of Living Philosophers.
Scriven, M., and A. Fisher. Forthcoming. *Critical Thinking*. Newbury Park, Calif.: Sage.
Seeger, R. J. 1966. *Galileo Galilei, His Life and His Works*. Oxford: Pergamon Press.
Segre, M. 1980. The Role of Experiment in Galileo's Physics. *Archive for History of Exact Science* 23:227–52.
——. 1991. *In the Wake of Galileo*. New Brunswick, New Jersey: Rutgers University Press.
Settle, T. B. 1961. An Experiment in the History of Science. *Science* 133:19–23.
Sexl, R., and K. von Meyenn, eds. 1982. *Dialog über die beiden hauptsachlichsten Weltsysteme, das ptolemaische und das kopernikanische*. Trans. E. Strauss (1891). Stuttgart: B. G. Teubner.
Shapere, D. 1974. *Galileo: A Philosophical Study*. Chicago: University of Chicago Press.
——. 1984. *Reason and the Search for Knowledge*. Dordrecht: Kluwer.
Shea, W. R. 1972. *Galileo's Intellectual Revolution*. New York: Science History Publications.
——. 1991. *The Magic of Numbers and Motion*. Canton, Mass.: Science History Publications.
Skyrms, B. 1975. *Choice and Chance*. 2nd edition. Belmont, Calif.: Dickenson.
Sosio, L., ed. 1970. *Dialogo sopra i due massimi sistemi*. Turin: Einaudi.
Stich, S. P. 1985. Could Man Be an Irrational Animal? *Synthese* 64:115–35.
Strauss, E., ed. and trans. 1891. *Dialog über die beiden hauptsachlichsten Weltsysteme*. Leipsig: Teubner.
Thomas, S. N. 1986. *Practical Reasoning in Natural Language*. 3rd edition. Englewood Cliffs, New Jersey: Prentice Hall.
Thomason, N. 1992. Could Lakatos, Even with Zahar's Criterion for Novel Fact, Evaluate the Copernican Research Programme? *British Journal for the Philosophy of Science* 43:161–200.
——. 1994a. The Power of ARCHED Hypotheses. *British Journal for the Philosophy of Science* 45:255–64.
——. 1994b. Sherlock Holmes, Galileo, and the Missing History of Science. In *PSA: Proceedings of the 1994 Biennial Meeting of the Philosophy of Science Associa-*

*tion*, ed. D. Hull et al., 1:323–33. East Lansing, Mich.: Philosophy of Science Association.
——. Forthcoming. 1543—The Year That Copernicus Didn't . . . . In *1543 and All That*, ed. A. Conones and G. Freeland. Dordrecht: Kluwer.
Toulmin, S., and J. Goodfield. 1961. *The Fabric of the Heavens*. New York: Harper.
Treddenick, H., ed. 1969. *The Last Days of Socrates*. New York: Penguin.
Van Helden, A. 1984. Galileo and the Telescope. In Galluzzi (1984, 149–58).
——, ed. and trans. 1989. *Sidereus Nuncius, or the Sidereal Messenger*. Chicago: University of Chicago Press.
——. 1994. Telescopes and Authority from Galileo to Cassini. *Osiris*, second series, 9:7–29.
Vickers, B. 1983. Epideictic Rhetoric in Galileo's *Dialogo*. *Annali dell'Istituto e Museo di Storia della Scienza di Firenze* 8:69–102.
Wallace, W. A. 1974. Three Classics of Science. In *The Great Ideas Today 1974*. Chicago: Encyclopedia Britannica.
——, ed. and trans. 1977. *Galileo's Early Notebooks*. Notre Dame: University of Notre Dame Press.
——. 1981a. *Prelude to Galileo*. Dordrecht: Reidel.
——. 1981b. Aristotle and Galileo: The Use of *Hupothesis* (*Suppositio*) in Scientific Reasoning. In *Studies in Aristotle*, ed. D. J. O'Meara, 47–77. Washington: Catholic University of America Press.
——. 1983. The Problem of Causality in Galileo's Science. *Review of Metaphysics* 36:607–32.
——. 1984. *Galileo and His Sources*. Princeton: Princeton University Press.
——, ed. 1986. *Reinterpreting Galileo*. Washington: Catholic University of America Press.
——. 1992a. *Galileo's Logical Treatises*. Dordrecht: Kluwer.
——. 1992b. *Galileo's Logic of Discovery and Proof*. Dordrecht: Kluwer.
Webbe, Joseph. [1635 ?]. English translation of Galileo's *Dialogue*. British Library manuscript Harleian MS 6320.
Weizenbaum, J. 1976. *Computer Power and Human Reason*. San Francisco: Freeman.
Westman, R. S. 1972. Kepler's Theory of Hypothesis and the Realist Dilemma. *Studies in History and Philosophy of Science* 3:233–64.
——, ed. 1975a. *The Copernican Achievement*. Berkeley: University of California Press.
——. 1975b. The Melanchthon Circle, Rheticus, and the Wittenberg Interpretation of the Copernican Theory. *Isis* 66:165–93.
——. 1984. The Reception of Galileo's *Dialogue*. In Galluzzi (1984, 329–71).
——. 1994. Two Cultures or One? A Second Look at Kuhn's *The Copernican Revolution*. *Isis* 85:79–115.
Wisan, W. L. 1978. Galileo's Scientific Method: A Reexamination. In Butts and Pitt (1978, 1–57).
——. 1984a. On Argument *Ex Suppositione Falsa*. *Studies in History and Philosophy of Science* 15:227–36.
——. 1984b. On the Art of Reasoning. *Annals of Science* 41:483–87.

# Index

Compositor: Impressions, Inc.
Text: 10/13 Galliard
Display: Galliard
Printer and binder: Edwards Brothers, Inc.